Feynman Lectures on Computation

The last lecture course that Nobel Prize winner Richard P. Feynman gave to students at Caltech from 1983 to 1986 was not on physics but on computer science. The first edition of the *Feynman Lectures on Computation*, published in 1996, provided an overview of standard and not-so-standard topics in computer science given in Feynman's inimitable style. Although now over 20 years old, most of the material is still relevant and interesting, and Feynman's unique philosophy of learning and discovery shines through.

For this new edition, Tony Hey has updated the lectures with an invited chapter from Professor John Preskill on "Quantum Computing 40 Years Later". This contribution captures the progress made toward building a quantum computer since Feynman's original suggestions in 1981. The last 25 years have also seen the "Moore's law" roadmap for the IT industry coming to an end. To reflect this transition, John Shalf, Senior Scientist at Lawrence Berkeley National Laboratory, has contributed a chapter on "The Future of Computing beyond Moore's Law". The final update for this edition is an attempt to capture Feynman's interest in artificial intelligence and artificial neural networks. Eric Mjolsness, now a Professor of Computer Science at the University of California Irvine, was a Teaching Assistant for Feynman's original lecture course and his research interests are now the application of artificial intelligence and machine learning for multi-scale science. He has contributed a chapter called "Feynman on Artificial Intelligence and Machine Learning" that captures the early discussions with Feynman and also looks toward future developments.

This exciting and important work provides key reading for students and scholars in the fields of computer science and computational physics.

T0143288

Richard P. Feynman. Photograph courtesy of Michelle Feynman and Carl R. Feynman

Feynman Lectures on Computation

Anniversary Edition

Richard P. Feynman

Edited by

Tony Hey

With a Foreword by Bill Gates

CRC Press
Taylor & Francis Group
Boca Raton London New York

CRC Press is an imprint of the
Taylor & Francis Group, an **informa** business

Anniversary edition published 2023
by CRC Press
6000 Broken Sound Parkway NW, Suite 300, Boca Raton, FL 33487-2742

and by CRC Press
4 Park Square, Milton Park, Abingdon, Oxon, OX14 4RN

CRC Press is an imprint of Taylor & Francis Group, LLC

© 2023 Carl. R. Feynman and Michelle Feynman

Editor's Preface, Afterword, and Reminiscences © 2023 Tony Hey

Chapter 7 and Reminiscences © 2023 John Preskill

Chapter 9 © 2023 John Shalf

Chapter 10 © 2023 Eric Mjolsness

Reminiscences © 2023 Michael Douglas

First published 1996 by Westview Press

First edition published by CRC Press 2018

Library of Congress Cataloging-in-Publication Data

Names: Feynman, Richard P. (Richard Phillips), 1918-1988, author. | Hey, Anthony J. G., editor.
Title: Feynman lectures on computation / Richard P. Feynman ; edited by Tony Hey.
Description: Anniversary edition. | Boca Raton : CRC Press, 2023. | Series: Frontiers in physics | Includes Feynman's lectures with some new supplementary essays. | Includes bibliographical references and index.
Identifiers: LCCN 2022046479 (print) | LCCN 2022046480 (ebook) | ISBN 9781032415888 (hardback) | ISBN 9780367857332 (paperback) | ISBN 9781003358817 (ebook)
Subjects: LCSH: Electronic data processing. | Computer science. | Feynman, Richard P. (Richard Phillips), 1918-1988.
Classification: LCC QA76 .F45 2023 (print) | LCC QA76 (ebook) | DDC 004--dc23/eng/20221122
LC record available at https://lccn.loc.gov/2022046479
LC ebook record available at https://lccn.loc.gov/2022046480

ISBN: 978-1-032-41588-8 (hbk)
ISBN: 978-0-367-85733-2 (pbk)
ISBN: 978-1-003-35881-7 (ebk)

DOI: 10.1201/9781003358817

Typeset in Palatino
by Deanta Global Publishing Services, Chennai, India

Contents

Reminiscences

Afterword

Foreword by Bill Gates

I was working hard at Microsoft when Richard Feynman gave the lectures in this book. But it wasn't his thoughts about computing that first got my attention – it was his Messenger Lectures on physics.

In the mid 1980s, I was planning a trip with a friend and decided to add some learning to our relaxation. I found one of the Messenger Lectures in a local university's film collection and checked it out. I loved it so much that I ended up watching it twice and immediately went looking for more! Years later I bought the rights to those lectures and worked with Microsoft to get them posted online for free.

Feynman had an amazing knack for making physics clear and fun at the same time. He took obvious delight in knowledge – during the lectures, you could see his face light up when he got excited about something. And he made the ideas so clear that anyone could understand them. I didn't read his lectures on computation until some time later, but they have all the same joyful qualities.

Given how quickly the computing industry has developed, you may be wondering how a series of thirty-year-old lectures could still be relevant today. After all, as I write this foreword, anyone with an Internet connection has access to an artificial intelligence agent – what Feynman called "advanced applications" – that can hold its own in a conversation and will soon be able to do even more.

There are several reasons. For one thing, some of the original chapters are as useful today as they were three decades ago. The first three, for example, are as clear and concise an explanation of how computers work as I have seen. Anyone who wants to do research, write code, or just understand the workings of their laptop machine should read them.

In addition, my friend and former colleague Tony Hey has done an admirable job of updating this edition with new material on neural networks, quantum computing, and the end of Moore's Law. It's a testament to Feynman's thinking that these subjects fit so naturally into this book. He not only foresaw developments like robotics, computer vision, and speech that were still decades away; he also created an intellectual framework that was powerful and flexible enough to accommodate them.

Finally, all of us can learn from the high standards that Feynman held himself to. He insisted that he didn't really understand something until he had figured it out for himself, checked his work in multiple ways, and explained it in simple terms to someone else. And he hated the idea that he might be fooling himself – a quality that I've been trying to emulate for years, in my business career and now in my philanthropy. Even if you don't care about computing, you should care about testing your own assumptions, and no one questioned his own assumptions more rigorously than Richard Feynman.

Editor's Preface

The original printing of the *Feynman Lectures on Computation* was in 1996. Although the published lectures are now nearly 25 years old, much of the material is Feynman's treatment of many standard topics in computer science, such as computability and information theory, as well as some not-so-standard topics, such as reversible computing and quantum computers. In May 2019, I was pleasantly surprised to be contacted by a commissioning editor, Carolina Antunes, from Taylor and Francis about the possibility of publishing a new edition of the lectures to mark their 25th anniversary. Due to the Covid-19 pandemic, publication of the new edition was delayed.

Forty years ago, in his talk at a conference at MIT in May 1981, Feynman raised the possibility of building a new type of computer made up of intrinsically quantum mechanical elements that could be used to simulate large quantum systems that could not be simulated on a classical computer. Remarkably, Feynman stated explicitly that such a computer was "not a Turing machine, but a machine of a different kind". In the *Lectures on Computation*, I included his later, more detailed, analysis of how one might design such a quantum computer. With the present excitement about progress in quantum computing, there has been a continuing and growing interest in Feynman's lectures.

In the 25 years since the original publication of the lectures, what else has changed? After a brief introduction to logic gates and computers, Feynman examines fundamental limits to computation – from mathematics, from information theory, from thermodynamics, from quantum mechanics, and from semiconductor technology. Two major changes stand out. First, the progress made toward actually building quantum computers capable of performing useful calculations. Second, Moore's Law, as embodied in the IT industry's technological roadmap that has historically charted the course for the industry to make digital systems faster, smaller, and cheaper roughly every two years, is coming to an end. It therefore seems a good idea to include new material that reflects the significant progress and change in these two areas.

I am therefore delighted to include two new chapters written by expert guest authors on these advances. John Preskill, the Richard P. Feynman Professor of Theoretical Physics at Caltech, has contributed a new chapter on "Quantum Computing 40 Years Later", which details the progress made since Feynman's original 1981 talk at MIT. John Shalf, Senior Scientist at the Lawrence Berkeley National Laboratory, has contributed a chapter on "The Future of Computing beyond Moore's Law" that looks at the different opportunities and strategies to continue computing performance improvements given the breakdown of the historical technology drivers.

I have also looked through the transcripts of the audio recordings and my original notes on the lectures to see if there was any significant additional content that could be included. Figure 0.1 shows Feynman's original

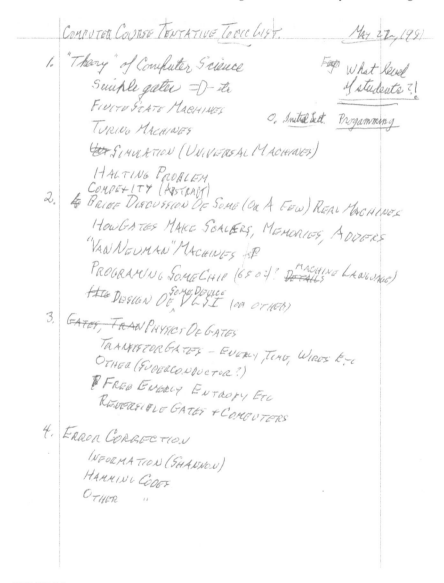

FIGURE 0.1
Feynman's 1981 draft course outline. Images are reproduced by kind permission of the Caltech archive.

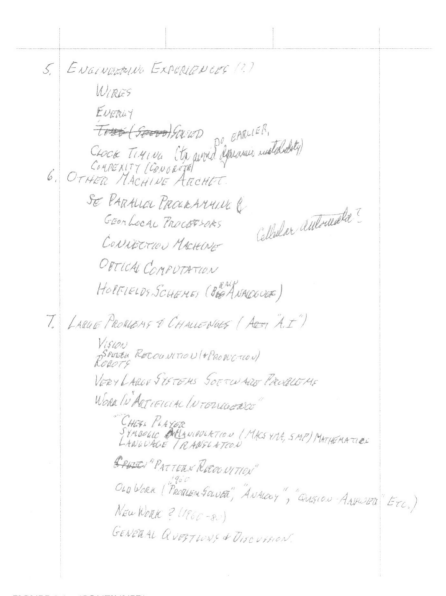

FIGURE 0.1 (CONTINUED)
Feynman's 1981 draft course outline. Images are reproduced by kind permission of the Caltech archive.

1981 notes for the topics he wanted to cover in the computation course. Over the next few years, he made several versions of such topic lists and he also made week-by-week schedules of which lectures he would give, as well as those that would be given by guest lecturers such as Gerry Sussman and Sandy Frey. See Figure 0.2 for the allocations for the last

version of the course in 1985/1986. Looking at these lecture allocations, I
think that the published "Lectures on Computation" give a pretty accurate account of the major topics that Feynman himself lectured on in the
course. It is intriguing to note that in his 1985/1986 schedule, Feynman
devotes only one lecture to the topic of "Quantum Mechanical Computer".

```
OUTLINE:CS/PH 187                           (Revised - Oct. 4, 1985)


INTRODUCTION                                Feynman: Oct. 2

THE VON NEUMANN COMPUTER

Fundamental Operation                       Feynman: Oct. 4
     File Clerk Model
     Instruction Fetch
     Registers and Instruction Set

Gates and Combinatorial Logic               Feynman: Oct. 7
     Simple ALU Parts, ROM's, PLA's

Registers and Sequential Machines           Feynman: Oct. 9-11
     Flip-flops, RAM's, Counters, Shift Registers
     Clocking, Timing, and Self-timing

Computing a Simple Function                 Frey:    Oct. 14-18
     GCD Machine (Breadboard of Chips and LED's)
     Registers and Data Paths
     State Machine Controller

Specifying More Complex Functions           Frey:    Oct. 21-25
     Data Hierarchy;
        Addressing, Indirection, and Typing
     Software Hierarchy;
        Processes, Macros, and Sub-routines
        Operating Systems

Modern Languages (Kajiya-3 hrs.)            Kajiya:  Oct. 28-30-
                                                     Nov. 1


LOGICAL LIMITATIONS OF COMPUTING ENGINES

Theory of Computation - What Can Be Computed?   Feynman:  Nov. 4-11
     Finite-State Machines                                  '6-8'
        Transducers, Recognizers
        Regular Expressions
     Universal Machines - Infinite Memory
        Turing Machines, Post Machines, RAM Machines
        Interpretation is the Major Idea!
     Uncomputable Functions
        Counting Arguments
        Halting Theorem
     Logical Classes of Machines
        Grammars

Physical Basis of Computing Engines         Feynman:  Nov. 13-25
     Free Energy                                      15-17-20-24'
     Bennett's Ideas
     Maxwell's Daemon
     Information as a Source of Free Energy
     Reversible Computation
        Fredkin Gates
```

FIGURE 0.2
Feynman's 1985 course outline and lecture allocations. Images are reproduced by kind permission of the Caltech archive.

```
              Billiard-Ball Computer
        Free Energy vs. Speed

Information and Communications
        More Information and Entropy       Feynman:   Nov. 27-29
        Noise and Communications
            Shannon's Theorem
            Data Compression              Feynman:   Dec. 2
            Hamming Codes
            More Error Correcting Codes    Frey:      Dec. 4

Algorithms and Complexity
        Time and Space                     Frey:      Dec. 6
        Some Examples
        NP-Completeness                    Frey:      Dec. 9

Digital Signalling                         Frey:      Dec. 11-13
        Sampling Theorem
        Spectral Analysis


PHYSICAL ASPECTS OF COMPUTATION

Semiconductor Physics                      Feynman:   Jan. 8
        PN, NPN

CMOS Inverter                              Feynman:   Jan. 10
        Timing
        Energetics                         Feynman:   Jan. 13
        Pass Transistors
        Hot Clocking                       Feynman:   Jan. 15
        Use of Inductance

Integrated Circuit Technology (VLSI example)
        Planar Process Fabrication Technology   Feynman:   Jan. 17
            Stick figures
            Layout and design rules
            Examples - Registers, PLA,...
        2-Phase Clocks and Synchronous Design   Feynman:   Jan. 20

Practical Limitations                      Feynman:   Jan. 22
        Wires - Rence's rule
        Cray Design Ideas (other than algorithmic)   Feynman:   Jan. 24
            Measured terminated transmission lines
            Getting the heat out
            Power supplies

Other Fundamental Technologies
        Bipolar                            Feynman:   Jan. 27
        Gallium Arsinide
        Three-dimensional VLSI             Feynman:   Jan. 29
        Superconductors - Josephson Junction
        Optical Computing                  Feynman:   Jan. 31
            Two-dimensional parallel                  -Feb. 3
                Linear
                Non-linear amplifiers
                Memory storage
            Optical fibers
            Electro-optical transducers
        Quantum Mechanical Computer        Feynman:   Feb. 5
```

FIGURE 0.2   (CONTINUED)

Feynman's 1985 course outline and lecture allocations. Images are reproduced by kind permission of the Caltech archive.

ORGANIZATIONAL ASPECTS OF COMPUTING: Architectures of Computing Engines

```
The Von Neumann Architecture                          Frey:    Feb. 7
        The memory interface bottleneck                        -10-12
        Cache memory
        RISC machines
        Pipelines and look-ahead
        I/O, interrupts, and instruction sequencing
        Wide horizontal microcode
                Specialized processing units
                        SCHEME chip
                        Floating Point Processing Element

Concurrent Processor Architectures

        Microscopic parallelism
                Local communication only
                        Systolic arrays of Kung and Leiserson  Douglas: Feb. 14
                        Cellular automata                      Feynman: Feb. 17
                Connection Machine                             Feynman: Feb. 19-21
                        Memory
                        Processors
                        Routing communications
                Hopfield type machines                         Feynman: Feb. 24

        Process concurrency
                Special purpose hardware                       Douglas: Feb. 26
                        Orrery example
                        QCD Machines
                Cosmic cube                                    Frey:    Feb. 28
                        Fine grain                                      -Mar. 3
                        Large grain
                Ultra and the IBM RP3                          Frey:    Mar. 5
                Application/process decomposition              Frey:    Mar. 7

        Macroscopic parallelism                                Frey:    Mar. 10
                Networks of large computers

        Communication problems and limits                     Frey:    Mar. 12-14
                Global communication schemes
                        Batcher sorting nets
                        Shuffles
                        Hypercubes
                Assynchronous arbiters
                Networks
                General properties of array machines
                        Error detection and correction


MORE DIFFICULT COMPUTER APPLICATIONS


Computer Intelligence
        Example: Chess Playing Machines

Robotics
        Closed-loop, Real-time Control Systems
```

FIGURE 0.2   (CONTINUED)
Feynman's 1985 course outline and lecture allocations. Images are reproduced by kind permission of the Caltech archive.

```
Vision
     Creating Scenes
     Recognition
     Stereopsis
     Examples:
          Fingerprint Identification
          Grabbing Errant Satellites for Repair
          Selecting Parts from Bins

Speech
     Creation
     Recognition

Natural Language Understanding
     Syntax
     Semantics
     Translation

Searching Algorithms
     Monte Carlo
     Spin-Glass Annealing
     Alpha-Beta (Example, Chess playing machines)

Pattern Directed Search
     Representation of Knowledge
          Semantic networks
          Frames
          Mathematical logic as a representational system
               Planner, Prolog
          Analogical Representation
               Example, Blocks world
          Heuristic Search Rules
          Examples; Eurisco, Wilkins chess player

Useful Applications
     Sumbolic Manipulation (MACSYMA, SMP)
     Computer Aided Design and Manufacture (CAD-CAM)
     Expert Systems
          Example, MYCIN, why not so useful
     Chip and Board Layout Machines
     Spelling Correctors, Indexers, Translators

Summary and Conclusions
     What the big problems are today
     How we might go about solving them
```

FIGURE 0.2 (CONTINUED)
Feynman's 1985 course outline and lecture allocations. Images are reproduced by kind permission of the Caltech archive.

The new chapter by John Preskill fills in the gaps between Feynman's proposal in 1981, his lecture course in the early 1980s, and where we are now.

Feynman had worked as a consultant for Thinking Machines Corporation in the summer of 1983. Danny Hillis's ambitious goal for the company was to turn his PhD thesis ideas on massively parallel computer architectures into a successful commercial product called the "Connection Machine". The early 1980s was a time of great experimentation in parallel computer architectures and Figure 0.2 shows a section headed "Concurrent Processor Architectures". Besides including lectures from Feynman on the Connection Machine, there were lectures on Caltech's own Cosmic Cube distributed memory parallel computer as well as on IBM's RP3 project, an experimental shared memory parallel computer. Nearly 40 years later, it is clear that Caltech's David beat IBM's Goliath in the sense that the Cosmic Cube's style of distributed memory, message passing, parallel architecture now forms the basis for almost all present-day supercomputers. There seems little point in including details of the many bold, but ultimately commercially unsuccessful, explorations of parallel computer architecture. Instead, the new chapter by John Shalf gives us an overview of possible post-Moore's law computer architectures and technologies. Figure 0.3 shows Jack Dongarra's famous collection of failed parallel computer company mugs.

In his original preface to the lectures, Feynman talks briefly about the field of "artificial intelligence" or AI – which he preferred to call "advanced applications". Figure 0.2 shows a list of "More difficult computer applications" – ranging from computer vision and robotics to speech and natural language understanding. In the original preparation of these lectures, I benefited greatly from previous work on some of the course material done by one of his teaching assistants for the course, Eric Mjolsness. Eric is now a Professor in Computer Science and Mathematics at the University of California in Irvine and his research interests are

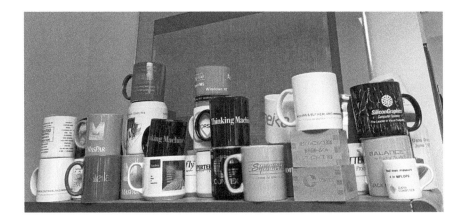

FIGURE 0.3
Jack Dongarra's collection of failed computer company mugs. Reproduced here by kind permission of Jack Dongarra.

neural networks, machine learning, and symbolic AI, with applications to multi-scale scientific problems. Eric was a PhD student researching neural networks with John Hopfield at the time of the lectures and was an observer of debates between Feynman, advocating a neural network approach to AI, and Gerry Sussman, defending the more established practice of symbolic AI.

I suggested that the two major changes in the last 25 years were in quantum computing and the approaching end of Moore's law. In fact, I think there is a third major change that should be included. This is what Terry Sejnowski, one of the early pioneers of neural networks, calls "The Deep Learning Revolution" [1]. This revolution was triggered by Fei-Fei Li's ImageNet computer vision competition, and by AlexNet, the winning "deep neural network" entry from Geoffrey Hinton's team from Toronto, in 2012. I am very pleased that Eric agreed to write a chapter about "Feynman on Artificial Intelligence and Machine Learning" in which he would include some of his first-hand experiences and discussions with Feynman on his lectures.

Finally, I have included two new short reminiscences of life with Feynman at Caltech. One is from John Preskill with some of his memories of Feynman as a colleague at Caltech. Curiously, in his article for this new edition of the lectures, John reveals his regret that he never took the opportunity to discuss quantum computation with Feynman. The second is from Michael Douglas, who worked with Gerry Sussman on his Digital Orrery project and was a student on Feynman's original course on the "Potentialities and Limitations of Computing Machines". Michael was a teaching assistant on later versions of the course and is now a distinguished practitioner of string theory in the Simons Center for Geometry and Physics at Stony Brook University.

Acknowledgements

I would like to include belated acknowledgements to two individuals who helped correct a number of errors and misprints in the 1996 edition of the Lectures. The first is to Professor Yasuo Hara, then at Heisei Teikyo University in Japan, who translated Feynman's Lectures for a Japanese language edition. The second is to Joel Chavas, then a neurobiology student at the Max Planck Institute for Biophysical Chemistry in Goettingen, who corrected the UTM specification in Chapter 3.

Dedication

As a conclusion to the Editor's Preface for this new edition of Feynman's *Lectures on Computation*, I would like to dedicate this edition to the memory of Helen Tuck. Helen started as the secretary for the two Caltech Nobel Prize winners, Richard Feynman and Murray Gell-Mann, in 1981. She took over this interesting but intimidating job after the retirement of their previous secretary, the legendary Julie Curcio. I was on sabbatical leave at Caltech in 1981 and was fortunate to get to know Helen quite well as she settled into her new role. Her role in the computing book project was that she knew that Feynman really wanted his lectures written up and published. However, several people had tried to do this and then given up on the task. She recommended me as a possible editor in November 1987. When I visited Caltech for the Hypercube computing conference in January 1988, I met with Feynman and we agreed that I would edit his lecture notes for publication. I knew Feynman was ill with cancer but I was still shocked when he died so soon after our agreement. It was Helen who made sure that I received all the relevant material for Feynman's computing course – mainly the course notes and audio tapes. Although it took me a long time to sort out and edit all this into a form suitable for publication, I was determined to justify Helen's belief that I would finish the task. That Feynman's lectures on computing finally appeared in book form is certainly largely due to Helen's loyalty and persistence.

Tony Hey
Southampton

.

Reference

1. Terrence J. Sejnowski, *The Deep Learning Revolution*, The MIT Press, Cambridge, MA and London, 2018.

Feynman's Preface

When I produced the *Lectures on Physics*, some 30 years ago now, I saw them as an aid to students who were intending to go into physics. I also lamented the difficulties of cramming several hundred years' worth of science into just three volumes. With these *Lectures on Computation*, matters are somewhat easier, but only just. Firstly, the lectures are not aimed solely at students in computer science, which liberates me from the shackles of exam syllabuses and allows me to cover areas of the subject for no more reason than that they are interesting. Secondly, computer science is not as old as physics; it lags by a couple of hundred years. However, this does not mean that there is significantly less on the computer scientist's plate than on the physicist's: younger it may be, but it has had a far more intense upbringing! So there is still plenty for us to cover.

Computer science also differs from physics in that it is not actually a science. It does not study natural objects. Neither is it, as you might think, mathematics; although it does use mathematical reasoning pretty extensively. Rather, computer science is like engineering – it is all about getting something to do something, rather than just dealing with abstractions as in pre-Smith geology.* Today in computer science we also need to "go down into the mines" – later we can generalize. It does no harm to look at details first.

But this is not to say that computer science is all practical, down to earth bridge-building. Far from it. Computer science touches on a variety of deep issues. It has illuminated the nature of language, which we thought we understood: early attempts at machine translation failed because the old-fashioned notions about grammar failed to capture all the essentials of language. It naturally encourages us to ask questions about the limits of computability, about what we can and cannot know about the world around us. Computer science people spend a lot of their time talking about whether or not man is merely a machine, whether his brain is just a powerful computer that might one day be copied; and the field of "artificial intelligence" – I prefer the term "advanced applications" – might have a lot to say about the nature of 'real' intelligence and mind. Of course, we might get useful ideas from studying how the brain works, but we must

---

* William Smith was the father of modern geology; in his work as a canal and mining engineer he observed the systematic layering of the rocks and recognized the significance of fossils as a means of determining the age of the strata in which they occur. This led to his formulation of the superposition principle in which rocks are successively laid down upon older layers. Prior to Smith's great contribution, geology was more akin to armchair philosophy than an empirical science. [Editor]

remember that automobiles do not have legs like cheetahs nor do airplanes flap their wings! We do not need to study the neurologic minutiae of living things to produce useful technologies; but even wrong theories may help in designing machines. Anyway, you can see that computer science has more than just technical interest.

These lectures are about what we can and can't do with machines today, and why. I have attempted to deliver them in a spirit that should be recommended to all students embarking on the writing of their PhD theses: imagine that you are explaining your ideas to your former smart, but ignorant, self, at the beginning of your studies! In very broad outline, after a brief introduction to some of the fundamental ideas, the next five chapters explore the limitations of computers – from logic gates to quantum mechanics! The second part consists of lectures by invited experts on what I've called advanced applications – vision, robots, expert systems, chess machines, and so on.[*]

---

[*] A companion volume to these lectures called "Feynman and Computation" was published in 1999. This contains articles from many of the same experts who contributed to Feynman's course as well reprints of Feynman's classic articles "There's Plenty of Room at the Bottom" and "Simulating Physics with Computers". [Editor]

Author and Editor Biographies

The late **Richard P. Feynman** was Richard Chace Tolman Professor of Theoretical Physics at the California Institute of Technology. He was awarded the Nobel Prize in 1965 for his work on the development of quantum electrodynamics and made many other fundamental contributions to physics. What is less well-known is his contribution to computer science with his ideas about quantum computing. He was one of the most famous and beloved figures of the 20th century, both in physics and in the public arena.

Tony Hey is Chief Data Scientist at the UK's Rutherford Appleton Laboratory at Harwell. After an academic career including Dean of Engineering at the University of Southampton in the UK, he became Director of the UK's pioneering eScience initiative. After ten years as a Vice President in Microsoft Research in Redmond in the United States, he returned to the UK and now leads a group applying Deep Learning neural networks to the analysis of experimental scientific data. He is also co-author of *The Computing Universe: A Journey through a Revolution*, a popular introduction to the development of computer science.

Contributors

John Preskill
Richard P. Feynman Professor of
  Theoretical Physics
California Institute of Technology
Pasadena, California, USA

John Shalf
Senior Scientist
Lawrence Berkeley National
  Laboratory
Berkeley, California, USA

Eric Mjolsness
Professor of Computer Science and
  Mathematics
University of California, Irvine
Irvine, California, USA

Michael Douglas
Professor in the Simons Center for
  Geometry and Physics
Stony Brook University
Stony Brook, New York, USA

1

Introduction to Computers

CONTENTS

Computers can do lots of things. They can add millions of numbers in the twinkling of an eye. They can outwit chess grandmasters. They can guide weapons to their targets. They can book you onto a plane between a guitar-strumming nun and a non-smoking physics professor. Some can even play the bongos. That's quite a variety! So, if we're going to talk about computers, we'd better decide right now which of them we're going to look at and how.

In fact, we're not going to spend much of our time looking at individual machines. The reason for this is that once you get down to the guts of computers you find that, like people, they tend to be more or less alike. They can differ in their functions, and in the nature of their inputs and outputs – one can produce music, another a picture, while one can be set running from a keyboard, another by the torque from the wheels of an automobile – but at heart, they are very similar. We will hence dwell only on their innards. Furthermore, we will not assume anything about their specific input/output (I/O) structure, about how information gets into and out of the machine; all we care about is that, however the input gets in, it is in digital form, and whatever happens to the output, the last the innards see of it, it's digital too; by digital, I mean binary numbers: ls and 0s.

What does the inside of a computer look like? Crudely, it will be built out of a set of simple, basic elements. These elements are nothing special – they could be control valves, for example, or beads on an abacus wire – and there are many possible choices for the basic set. All that matters is that they can be used to build everything we want. How are they arranged? Again, there will be many possible choices; the relevant structure is likely to be determined by considerations such as speed, energy dissipation, aesthetics, and what have you. Viewed this way, the variety in computers is a bit like the variety in houses: a Beverly Hills condo might seem entirely different from a garage in Yonkers, but both are built from the same things – bricks, mortar, wood, sweat – only the condo has more

DOI: 10.1201/9781003358817-1

of them, and arranged differently according to the needs of the owners. At heart, they are very similar.

Let us get a little abstract for the moment and ask: *how* do you connect up *which* set of elements to do the *most* things? It's a deep question. The answer again is that, up to a point, it doesn't matter. Once you have a computer that can do a few things – strictly speaking, one that has a certain "sufficient set" of basic procedures – it can do basically anything any other computer can do. This, loosely, is the basis of the great principle of "Universality". Whoa! You cry. My pocket calculator can't simulate the Great Red Spot on Jupiter like a bank of Cray supercomputers! Well, yes it can: it would need rewiring, and we would need to soup up its memory, and it would be damned slow, but if it had long enough it could reproduce anything the Crays do. Generally, suppose we have two computers **A** and **B**, and we know all about **A** – the way it works, its "state transition rules" and whatnot. Assume that machine **B** is capable of merely *describing* the state of **A**. We can then use **B** to simulate the running of **A** by describing its successive transitions; **B** will, in other words, be mimicking **A**. It could take an eternity to do this if **B** is very crude and **A** very sophisticated, but **B** will be able to do whatever **A** can, eventually. We will prove this later in the course by designing such a **B** computer, known as a Turing machine.

Let us look at universality another way. Language provides a useful source of analogy. Let me ask you this: which is the *best* language for describing something? Say, a four-wheeled gas-driven vehicle. Of course, most languages, at least in the West, have a simple word for this; we have "automobile", the English say "car", the French "voiture", and so on. However, there will be some languages that have not evolved a word for "automobile", and speakers of such tongues would have to invent some, possibly long and complex, descriptions for what they see, in terms of their basic linguistic elements. Yet none of these descriptions is inherently "better" than any of the others: they all do their job and will only differ in efficiency. We needn't introduce democracy just at the level of words. We can go down to the level of alphabets. What, for example, is the best alphabet for English? That is, why stick with our usual 26 letters? Everything we can do with these, we can do with three symbols – the Morse code, dot, dash, and space; or two – a Baconian cipher, with *A* through *Z* represented by five-digit binary numbers. So we see that we can choose our basic set of elements with a lot of freedom, and all this choice really affects is the efficiency of our language, and hence the sizes of our books: there is no "best" language or alphabet – each is logically universal, and each can model any other. Going back to computing, universality in fact states that the set of complex tasks that can be performed using a "sufficient" set of basic procedures is independent of the specific, detailed structure of the basic set.

For today's computers to perform a complex task, we need a precise and complete description of how to do that task in terms of a sequence of simple basic procedures – the "software" – and we need a machine to carry out these procedures in a specifiable order – this is the "hardware". This instructing has to be exact and unambiguous. In life, of course, we never tell each other *exactly* what we want to say; we never need to, as context, body language, familiarity with the speaker, and so on, enable us to "fill in the gaps" and resolve any ambiguities in what is said. Computers, however, can't yet "catch on" to what is being said, the way a person does. They need to be told in excruciating detail exactly what to do. Perhaps one day we will have machines that can cope with approximate task descriptions, but in the meantime, we have to be very prissy about how we tell computers to do things.

Let us examine how we might build complex instructions from a set of rudimentary elements. Obviously, if an instruction set *B* (say) is very simple, then a complex process is going to take an awful lot of description, and the resulting "programs" will be very long and complicated. We may, for instance, want our computer to carry out all manner of numerical calculations, but find ourselves with a set *B* that doesn't include multiplication as a distinct operation. If we tell our machine to multiply 3 by 35, it says "what?" But suppose *B* does have addition; if you think about it, you'll see that we can get it to multiply by adding lots of times – in this case, add 35 to itself twice. However, it will clearly clarify the writing of *B* programs if we augment the set *B* with a separate "multiply" instruction, *defined* by the chunk of basic *B* instructions that go to make up multiplication. Then when we want to multiply two numbers, we say "computer, 3 times 35", and it now recognizes the word "times" – it is just a lot of adding, which it goes off and does. The machine breaks these compound instructions down into their basic components, saving us from getting bogged down in low-level concepts all the time. Complex procedures are thus built up stage by stage. A very similar process takes place in everyday life; one replaces with one word a set of ideas and the connections between them. In referring to these ideas and their interconnections, we can then use just a single word and avoid having to go back and work through all the lower-level concepts. Computers are such complicated objects that simplifying ideas like this are usually necessary, and good design is essential if you want to avoid getting completely lost in details.

We shall begin by constructing a set of primitive procedures and examine how to perform operations such as adding two numbers or transferring two numbers from one memory store to another. We will then go up a level, to the next order of complexity, and use these instructions to produce operations like multiply and so on. We shall not go very far in this hierarchy. If you want to see how far you can go, the article on Operating Systems by P.J. Denning and R.L. Brown (*Scientific American*, September

1984, pp. 96–104) identifies 13 levels! This goes from level 1, that of electronic circuitry – registers, gates, buses – to number 13, the Operating System Shell, which manipulates the user programming environment. By a hierarchical compounding of instructions, basic transfers of ls and 0s on level one are transformed, by the time we get to 13, into commands to land aircraft in a simulation or check whether a 40-digit number is prime. We will jump into this hierarchy at a fairly low level, but one from which we can go up or down.

Also, our discussion will be restricted to computers with the so-called "Von Neumann architecture". Don't be put off by the word "architecture"; it's just a big word for how we arrange things, only we're arranging electronic components rather than bricks and columns. Von Neumann was a famous mathematician who, besides making important contributions to the foundations of quantum mechanics, also was the first to set out clearly the basic principles of modern computers. We will also have occasion to examine the behavior of several computers working on the same problem, and when we do, we will restrict ourselves to computers that work in sequence, rather than in parallel; that is, ones that take turns to solve parts of a problem rather than work simultaneously. All we would lose by the omission of "parallel processing" is speed, nothing fundamental.

We talked earlier about computer science not being a real science. Now we have to disown the word "computer" too! You see, "computer" makes us think of arithmetic – add, subtract, multiply, and so on – and it's easy to assume that this is all a computer does. In fact, conventional computers typically have one place where they do their basic math, and the rest of the machine is for the computer's main task, which is shuffling bits of paper around – only in this case, the paper notes are digital electrical signals. In many ways, a computer is reminiscent of a bureaucracy of file clerks, dashing back and forth to their filing cabinets, taking files out and putting them back, scribbling on bits of paper, passing notes to one another, and so on; and this metaphor, of a clerk shuffling paper around in an office, will be a good place to start to get some of the basic ideas of computer structure across. We will go into this in some detail, and the impatient among you might think too much detail, but it is a perfect model for communicating the essentials of what a computer does and is hence worth spending some time on.

1.1 The File Clerk Model

Let's suppose we have a big company, employing a lot of salesmen. An awful lot of information about these salesmen is stored in a big filing system somewhere, and this is all administered by a clerk. We begin with the idea that the clerk knows how to get the information out of the filing system. The data is stored on cards, and each card has the name of the salesman, his location, the number and type of sales he has made, his salary, and so on.

```
┌──────────────────────────────────────┐
│                                      │
│   Salesman:  ───────────             │
│   Sales:     ───────────             │
│   Salary:    ───────────             │
│   Location:  ───────────             │
│                 .                    │
│                 .                    │
│                 .                    │
│                                      │
└──────────────────────────────────────┘
```

Now suppose we are after the answer to a specific question: "What are the total sales in California?" Pretty dull and simple, and that's why I chose it: you must start with simple questions in order to understand difficult ones later. So how does our file clerk find the total sales in California? Here's one way he could do it:

> Take out a card
> If the "location" says *California*, then
> Add the number under "sales" to a running count called
>   "total"
> Put "sales" card back
> Take next card and repeat.

Obviously, you have to keep this up until you've gone through all the cards. Now let's suppose we've been unfortunate enough to hire particularly stupid clerks, who can read, but for whom the above instructions assume too much: say, they don't know how to keep a running count. We need to help them a little bit more. Let us invent a "total" card for our clerk to use. He will use this to keep a running total in the following way:

> Take out next "sales" card
> If *California*, then
> Take out "total" card
> Add sales number to number on card
> Put "total" card back
> Put "sales" card back
> Take out next "sales" card and repeat.

This is a very mechanical rendering of how a crude computer could solve this adding problem. Obviously, the data would not be stored on cards, and the machine wouldn't have to "take out a card" – it would read the stored information from a register. It could also write from a register to a "card" without physically putting something back.

Now we're going to stretch our clerk! Let's assume that each salesman receives not only a basic salary from the company, but also gets a little commission from sales. To find out how much, we multiply his sales by the appropriate percentage. We want our clerk to allow for this. Now he is cheap and fast, but unfortunately too dumb to multiply.* If we tell him to multiply 5 by 7, he says "what?" So we have to teach him to multiply. To do this, we will exploit the fact that there is one thing he does well: he can get cards very, very quickly.

We'll work in base two. As you all probably know, the rules for binary arithmetic are easier than those for base ten; the multiplication table is so small it will easily fit on one card. We will assume that even our clerk can remember these; all he needs are "shift" and "carry" operations, as the following example makes clear.

```
In decimal:      22 x 5 = 110

In binary:       10110            In decimal:  22
                  101                            5
                 10110
                 10110   (shift twice)
                _____                        ____
                1101110                        110
```

So as long as our clerk can shift and carry, he can, in effect, multiply. He does it very stupidly, but he also does it very quickly, and that's the point of all this: the inside of a computer is as dumb as hell but it goes like mad! It can perform very many millions of simple operations a second and is just like a very fast dumb file clerk. It is only because it is able to do things so fast that we do not notice that it is doing things very stupidly. (Interestingly enough, neurons in the brain characteristically take milliseconds to perform elementary operations, which leaves us with the puzzle of why is the brain so smart. Computers may be able to leave brains standing when it comes to multiplication, but they have trouble with things even small children find simple, like recognizing people or manipulating objects.)

To go further, we need to specify more precisely our basic set of operations. One of the most elementary is the business of transferring

---

* As an aside, although our dense file clerk is assumed in these examples to be a man, no sexist implications are intended! [RPF]

information from the cards our clerk reads to some sort of scratch pad on which he can do his arithmetic:

Transfer operations

"Take Card X" = Information on card X written to pad
"Replace Card Y" = Information on pad written on card Y

All we have done is to define the instruction "take card X" to mean copying the information on card X onto the pad, and similarly with "replace card Y". Next, we want to be able to instruct the clerk to check if the location on card X was "California". He has to do this for each card, so the first thing he has to do is be able to remember "California" from one card to the next. One way to help him do this is to have *California* written on yet another card C so that his instructions are now:

Take card X (from store to pad)
Take card C (from store to pad)
Compare what is on card X with what is on card C.

We then tell him that if the contents match, do so and so, and if they don't, put the cards back and take the next ones. Keeping on taking out and putting back the California card seems to be a bit inefficient, and indeed, you don't have to do that; you can keep it on the pad for a while instead. This would be better, but it all depends on how much room the clerk has on his pad and how many pieces of information he needs to keep. If there isn't much room, then there will have to be a lot of shuffling cards back in and out. We have to worry about such things!

We can keep on breaking the clerk's complex tasks down into simpler, more fundamental ones. How, for example, do we get him to look at the "location" part of a card from the store? One way would be to burden the poor guy with yet another card, on which is written something like this:

0000 0000 0000 0000 0000 1111 0000 0000 0000 0000...

Each sequence of digits is associated with a particular piece of information on the card: the first set of zeroes is "lined up" with the salesman's name, the next with his age, say, and so on. The clerk zips through this numeric list until he hits a set of 1s, and then reads the information next to them. In our case, the 1111 is lined up with California. This sort of location procedure is actually used in computers, where you might use a so-called "bitwise AND" operation (we'll discuss this later). This little diversion was just to impress upon you the fact that we need not take any of our clerk's skills for granted – we can get him to do things increasingly stupidly.

1.2 Instruction Sets

Let's take a look at the clerk's scratch pad. We haven't yet taught the clerk how to use this, so we'll do that now. We will assume that we can break down the instructions he can carry out into two groups. First, there is a core "instruction set" of simple procedures that comes with the pad – add, transfer, and so on. These are in the hardware: they do not change when we change the problem. If you like, they reflect the clerk's basic abilities. Then we have a set that is specific to the task, say calculating a salesman's commission. The elements of this set are built out of the instructions in the core set in ways we have discussed and represent the combinations of the clerk's talents that will be required for him to carry out the task at hand.

The first thing we need to get the clerk to do is do things in the right order, that is, to follow a succession of instructions. We do this by designating one of the storage areas on the pad as a "program counter". This will have a number on it, which indicates whereabouts in the calculational procedure the clerk is. As far as the clerk is concerned, the number is an address – he knows that buried in the filing system is a special "instruction file" cabinet, and the number in the counter labels a card in that file which he has to go and get; on the card is the instruction for what he is to do next. So he gets the instruction and stores it on his pad in an area that we call the "instruction register".

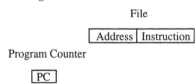

Before he carries out the instruction, however, he prepares for the next one by incrementing the program counter; he does this simply by adding one to it. Then he does whatever the instruction in the register tells him to do. Using a bracketed notation where () means "contents of" – remember this, as we will be using it a lot – we can write this sequence of actions as follows:[*]

    Fetch instruction from address PC
    PC ← (PC) + 1
    Do instruction

---

[*] The conventions adopted for such "Register Transfer Language" vary according to the whim of the author. We choose to follow the so-called "right to left" convention utilized in standard programming languages. [Editor]

The second line is a fancy way of saying that the counter *PC* "gets" the new value (*PC*) + 1. The clerk will also need some temporary storage areas on the pad; to enable him to do arithmetic, for example. These are called registers and give him a place to store something while he goes and finds some other number. Even if you are only adding two numbers, you need to remember the first until you have fetched the second! Everything must be done in sequence and the registers allow us to organize things. They usually have names; in our case, we will have four, which we call registers *A*, *B*, and *X*, and the fourth, *C*, which is special – it can only store one bit of data, and we will refer to it as the "carry" register. We could have more or fewer registers – generally, the more you have, the easier a program is to write – but four will suffice for our purposes.

So our clerk knows how to find out what he has to do and when. Let's now look at the core instruction set for his pad. The first kind of instruction concerns the transfer of data from one card to another. For example, suppose we have a memory location *M* on the pad. We want to have an instruction that transfers the contents of register *A* into *M*:

$$\text{Transfer (A) into M or M} \leftarrow \text{(A)}$$

Similarly, we might want to go the other way, and write the contents of *M* into *A*:

$$\text{Transfer (M) into A or A} \leftarrow \text{(M)}$$

M, incidentally, is not necessarily designed for temporary storage like *A*. We must also have analogous instructions for register *B*:

$$\text{Transfer (B) to M or M} \leftarrow \text{(B)}$$

$$\text{Transfer (M) to B or B} \leftarrow \text{(M)}$$

Register *X* we will use a little differently. We shall allow transfers from *B* to *X* and *X* to *B*:

$$X \leftarrow \text{(B) and B} \leftarrow \text{(X).}$$

In addition, we need to be able to keep tabs on, and manipulate, our program counter *PC*. This is obviously necessary: if the clerk shoots off to execute some multiplication, say, when he comes back, he has to know what to do next – he has to remember the number in *PC*. In fact, we'll keep it in register *X*. Thus, we add the transfer instructions:

$$PC \leftarrow \text{(X) and X} \leftarrow \text{(PC).}$$

Next, we need arithmetical and logical operations. The most basic of these is a "clear" instruction:

$$\text{Clear A, or A} \leftarrow 0.$$

This means, whatever is in *A*, forget it, wipe it out. Then we need an Add operation:

$$\text{Add B to A, or A} \leftarrow (A) + (B)$$

This means that register *A* receives the sum of the contents of *B* and the previous contents of *A*. We also have a shift operation, which will enable us to do multiplication without having to introduce a core instruction for it:

$$\text{Shiftleft A and Shiftright A}$$

The first merely moves all the bits in *A* one place to the left. If this shift causes the leftmost bit to overflow, we store it in the carry register *C*. We can also shift our number to the right; I have no use for this in mind, but it could come in handy!

The next instructions are logical ones. We will be looking at these in greater detail in Chapter 2, but I will mention them here for completeness. There are three that will interest us: AND, OR, and XOR. Each is a function of two digital "inputs" *x* and *y*. If *both* inputs are 1, then AND gives you 1; otherwise, it gives you zero. As we will see, the AND operation turns up in binary addition, and hence multiplication; if we view *x* and *y* as two digits we are adding, then (*x* AND *y*) is the carry bit: it's only one if both digits are one. In terms of our registers, *x* and *y* are (*A*) and (*B*), and AND operates on these:

$$\text{AND: A} \leftarrow (A) \wedge (B),$$

where we have used the logical symbol $\wedge$ for the AND operation. The result of acting on a pair of variables with an operator such as AND is often summarized in a "truth table" (Table 1.1).

TABLE 1.1

The Truth Table for
the AND Operator

A	B	X	
0	0	0	
0	1	0	$X = A \wedge B$
1	0	0	
1	1	1	

Our other two operators can be described in similar terms. The OR also operates on (*A*) and (*B*); it gives a one unless both (*A*) and (*B*) are zero – (*x* OR *y*) is one if *x* or *y* is one. XOR, or the "exclusive or", is similar to OR, except it gives zero if both (*A*) and (*B*) are one; in the binary addition of *x* and *y*, it corresponds to what you get if you add *x* to *y* and ignore any carry bits. A binary addition of 1 and 1 is 10, which is zero if you forget the carry. We can introduce the relevant logical symbols:

$$\text{OR } A \leftarrow (A) \vee (B)$$

$$\text{XOR } A \leftarrow (A) \oplus (B)$$

The actions of OR and XOR can also be summarized with truth tables (Table 1.2).

TABLE 1.2

The Truth Tables for the OR and XOR Operators

A	B	X
0	0	0
0	1	1
1	0	1
1	1	1

$X = A \vee B$

OR

A	B	X
0	0	0
0	1	1
1	0	1
1	1	0

$X = A \oplus B$

XOR

Two more operations that are convenient to have are the instructions to increment or decrement the contents of *A* by one:

$$\text{Increment } A, \text{ or } A \leftarrow (A) + 1$$

$$\text{Decrement } A, \text{ or } A \leftarrow (A) - 1$$

Obviously, one can go on adding instructions that may or may not turn out to be very convenient. Here, we already have more than the minimum number necessary to be able to do some useful calculations. However, we want to be able to do as much as possible, so we can bring in other instructions. One other that will be useful is one that allows us to put a data item directly into a register. For example, rather than writing *California* on a card and then transferring from card to pad, it would be convenient to be able to write *California* on the pad directly. Thus, we introduce the "Direct Load" instruction:

$$\text{Direct Load: } B \leftarrow N,$$

where N is any constant.

There is one class of instructions that it is vital we add: that of branches or jumps. A "jump to Z" is basically an instruction for the clerk to look in (instruction) location Z; that is, it involves a change in the value of the program counter by more than the usual increment of one. This enables our clerk to leap from one part of a program to another. There are two kinds of jumps, "unconditional" and "conditional". The unconditional jump we touched on above:

$$\text{Jump to (Z) or PC} \leftarrow \text{(Z)}$$

The really new thing is the conditional jump:

$$\text{Jump to (Z) } \textit{if } C=1$$

With this instruction, the jump to location (Z) is only made if the carry register C contains a carry bit. The freedom given by this conditional instruction will be vital to the whole design of any interesting machines.

There are many other kinds of jumps we can add. Sometimes it turns out to be convenient to be able to jump not only to a definite location but to one a specific number of steps further on in the program. We can therefore introduce jump instructions that add this number of steps to the program counter:

$$\text{Jump to (PC) + (Z) or PC} \leftarrow \text{(PC) + (Z)}$$

$$\text{Jump to (PC) + (Z) if } C=1$$

Finally, there is one more command that we need; namely, an instruction that tells our clerk to quit:

$$\text{Halt}$$

With these instructions, we can now do anything we want and I will suggest some problems for you to practice on below. Before we do that, let us summarize where we are and what we're trying to do. The idea has been to outline the basic computer operations and methods and indicate what is actually in a computer (I haven't been describing an actual design, but I've come close). In a simple computer, there are only a few registers; more complex ones have more registers, but the concepts are basically the same, just scaled up a bit.

It is worth looking at how we represent the instructions we considered above. In our particular case, the instructions contain two pieces: an instruction address and an instruction number, or "opcode".

Instruction address	Instruction opcode/number

For example, one of the instructions was "put the contents of memory M into register A". The computer doesn't speak English, so we have to encode this command into a form it can understand, in other words, into a binary string. This is the opcode, or instruction number, and its length clearly determines how many different instructions we can have. If the opcode is a four-digit binary number, then we can have $2^4 = 16$ different instructions, of which loading the contents of a memory address into A is just one. The second part of the instruction is the instruction address, which tells the computer where to go to find what it has to load into A; that is, memory address M. Some instructions, such as "clear A", don't require an address direction.

Details such as how the instruction opcodes are represented or exactly how things are set out in memory are not needed to use the instructions. This is the first and most elementary step in a series of hierarchies. We want to be able to maintain such ignorance consistently. In other words, we only want to have to think about the lower details once and then design things so that the next guy who comes along and wants to use your structure does not have to worry about the lower-level details.

There is one feature that we have so far ignored completely. Our machine as described so far would not work because we have no way of getting numbers in and out. We must consider input and output. One quick way to go about things would be to assign a particular place in memory, say address 17642, to be the input, and attach it to a keyboard so that someone from outside the machine could change its contents. Similarly, another location, say 17644, might be the output, which would be connected to a TV monitor or some other device, so that the results of a calculation can reach the outside world.

Now there are two ways in which you can increase your understanding of these issues. One way is to remember the general ideas and then go home and try to figure out what commands you need and make sure you don't leave one out. Make the set shorter or longer for convenience and try to understand the tradeoffs by trying to do problems with your choice. This is the way I would do it because I have that kind of personality! It's the way I study – to understand something by trying to work it out or, in other words, to understand something by creating it. Not creating it 100%, of course; but taking a hint as to which direction to go but not remembering the details. These you work out for yourself.

The other way, which is also valuable, is to read carefully how someone else did it. I find the first method best for me once I have understood the basic idea. If I get stuck, I look at a book that tells me how someone else did it. I turn the pages and then I say "Oh, I forgot that bit", then close the book and carry on. Finally, after you've figured out how to do it, you read how they did it and find out how dumb your solution is and how much more clever and efficient theirs is! But this way, you can understand the cleverness of their ideas and have a framework in which to think about the problem. When I start straight off to read someone else's solution I find it boring and uninteresting, with no way of putting the whole picture together. At least, that's the way it works for me!

Throughout the book, I will suggest some problems for you to play with. You might feel tempted to skip them. If they're too hard, fine. Some of them are pretty difficult! But you might skip them thinking that, well, they've probably already been done by somebody else; so what's the point? Well, *of course* they've been done! But so what? Do them for the *fun* of it. That's how to learn the knack of doing things when you have to do them. Let me give you an example. Suppose I wanted to add up a series of numbers,

$$1+2+3+4+5+6+7 \ldots$$

up to, say, 62. No doubt you know how to do it, but when you play with this sort of problem as a kid, and you haven't been shown the answer ... it's *fun* trying to figure out how to do it. Then, as you go into adulthood, you develop a certain confidence that you can discover things; but if they've already been discovered, that shouldn't bother you at all. What one fool can do, so can another, and the fact that some other fool beat you to it shouldn't disturb you: you should get a kick out of having discovered something. Most of the problems I give you in this book have been worked over many times, and many ingenious solutions have been devised for them. But if you keep proving stuff that others have done, getting confidence, increasing the complexities of your solutions – for the fun of it – then one day you'll turn around and discover that *nobody actually did that one!* And that's the way to become a computer scientist.

I'll give you an example of this from my own experience. Above, I mentioned summing up the integers. Now, many years ago, I got interested in the generalization of such a problem: I wanted to figure out formulae for the sums of squares, cubes, and higher powers, trying to find the sum of m things each up to the nth power. And I cracked it, finding a whole lot of nice relations. When I'd finished, I had a formula for each sum in terms of a number, one for each n, that I couldn't find a formula for. I wrote these numbers down, but I couldn't find a general rule for getting them. What was interesting was that they were integers, until you got to $n = 13$ – when it wasn't (it was something just over 691)! Very shocking! And fun.

Anyway, I discovered later that these numbers had actually been discovered back in 1746. So I had made it up to 1746! They were called "Bernoulli Numbers". The formula for them is quite complicated, and unknown in a simple sense. I had a "recursion relation" to get the next one from the one before, but I couldn't find an arbitrary one. So I went through life like this, discovering next something that had first been discovered in 1889, then something from 1921 ... and finally I discovered something that had the same date as when I discovered it. But I get so much fun out of doing it that I figure there must be others out there who do too, so I am giving you these problems to enjoy yourselves with. (Of course, everyone enjoys themselves in different ways.) I would just urge you not to be intimidated by them, or put off by the fact that they've been done. You're unlikely to discover something new without a lot of practice on old stuff, but further, you should get a heck of a lot of fun out of working out funny relations and interesting things. Also, if you read what the other fool did, you can appreciate how hard it was to do (or not), what he was trying to do, what his problems were, and so forth. It's much easier to understand things after you've fiddled with them before you read the solution. So, for all these reasons, I suggest you have a go.

Problem 1.1 (a) Go back to our dumb file clerk and the problem of finding out the total number of sales in California. Would you advise the management to hire two clerks to do the job quicker? If so, how would you use them, and could you speed up the calculation by a factor of two? You have to think about how the clerks get their instructions. Can you generalize your solution to K or even 2^K clerks?

(b) What kinds of problems can K clerks actually speed up? What kinds can they apparently not?

(c) Most present-day computers only have one central processor – to use our analogy, one clerk. This single file clerk sits there all day long working away like a fiend, taking cards in and out of the store like mad. Ultimately, the speed of the whole machine is determined by the speed at which the clerk – that is, the central processor – can do these operations. Let's see how we can maybe improve the machine's performance. Suppose we want to compare two n-bit numbers, where n is a large number like 1024; we want to see if they're the same. The easiest way for a single file clerk to do this would be to work through the numbers, comparing each digit in sequence. Obviously, this will take a total time proportional to n, the number of digits needing checking. But suppose we can hire n file clerks, or $2n$ or perhaps $3n$: it's up to us to decide how many, but the number must be proportional to n. Now, it turns out that by increasing the number of file clerks we can get the comparison time down to be proportional to $\log_2 n$. Can you see how?

(d) If you can do this compare problem, you might like to try a harder one. See if you can figure out a way of adding two n-bit numbers in "log n" time. This is more difficult because you have to worry about the carries!

Problem 1.2 The second problem concerns getting the clerk to multiply (multiplication, remember, is not included in his basic instruction set). The problem comes in two parts. First, find the appropriate set of basic instructions required to perform multiplication. Having these, let's assume we save them some place in the machine so that we don't have to duplicate them every time we want to multiply; put them, say, in locations m to $m+k$. Show how we can give the clerk instructions to use this set-up to do a multiplication and return to the right place in the program.

1.3 Summary

We have now covered enough stuff for us to go on to understand any particular machine design. But instead of looking at any particular machine in detail, we are going to do something rather different. From where we are now, we can go up, down, or sideways. What do I mean by this? Well, "up" means hiding more details of the workings of the machine from the user – introducing more levels of abstraction. We have already seen some examples of this; for example, building up new operations such as multiplication from operations in our basic set. Every time we want to multiply, we just use this multiply "subroutine". Another example worth discussing is the ability to talk about algebraic variables rather than locations in memory. Suppose you want to take the sum of X and Y, and call it Z:

$$Z = X + Y$$

X and Y are already known to the computer and stored at specific locations in memory. The first thing we have to do is assign some place in memory to store the value of Z and then ensure that this location holds the sum of the contents of the X and Y memory cells. Now we know all about Z and can use it in other expressions, such as $Z + X$. It is clearly much simpler talking about algebraic variables rather than memory locations all the time, although it is quite a job to set this up. However, up to now, we have had to know exactly where a number is located in order to make a transfer. We can now introduce a new number Z and say to the computer "I want a number Z – find a place to put it and don't bother telling me where it is!" This is what I mean by moving "up".

Of course, we already went "up" a bit when we summarized operations by instructions such as "Clear A" and so on. This sort of shorthand is introduced for our benefit, and programs written in it cannot be understood directly by the machine itself. Such "assembly language" programs have to be translated into a "machine language" that the computer can understand, and this is done by a program called an "assembler". The next level up, where we have multiplication and variables and so on, needs another program to translate these "high-level" programs into assembly language. These translation programs are called "compilers" or "interpreters". The difference between them is in when the translation is done. An interpreter works out what to do step by step, as the program runs, interpreting each successive instruction in terms of the cruder language. A compiler takes the program as a whole and converts it all into assembly or machine language before the program is run. Compilers have the advantage that, in some cases, looking at the whole "code" it is possible for them to find clever ways to simplify the required operations. This is the nub of the important field of "compiler optimization" and is becoming of increasing importance for the new types of "non-Von Neumann" parallel computers.

Clearly, one can keep going up in level, putting together new algorithms, programming languages, adding the ability to manipulate "files" containing programs and data, and so on. Nowadays, it is possible for most people to happily work at these higher levels using high-level languages to program their machines. Imagine how tedious it was – and is, for modern computer designers – to work solely in machine code!

That was "up"; now it's time to go down. How can anything be simpler than our dumb file clerk model and our simple list of instructions? What we have not considered is what our file clerk is *made of*; to be more realistic, we have not looked at how we would actually build electronic circuits to perform the various operations we have discussed. This is where we are going to go next, but before we do, let me say what I mean by moving "sideways". Sideways means looking at something entirely different from our Von Neumann architecture, which is distinguished by having a single central processing unit (CPU) and everything coming in and going out through the "fetch and execute" cycle. Many other more exotic computer architectures are now being experimented with, and some are being marketed as machines people can buy. Going "sideways", therefore, means remaining at the same level of detail but examining how calculations would be performed by machines with differing core structures. We already invited you to think of such "parallel" computers with the problem of organizing several file clerks to work together on the same problem.

2

Computer Organization

CONTENTS

2.1 Gates and Combinational Logic

We shall begin our trip downwards by looking at what we need to be able to perform our various simple operations – adds, transfers, control decisions, and so forth. We will see that we will need very little to do *all* of these things! To get an idea of what's involved, let's start with the "add" operation. Our first, important, decision is to restrict ourselves to working in base 2, the binary system: the only digits are 1 and 0, and as we shall see, these can easily fit into a computer framework: we represent them electronically by a simple "on/off" state of a component. In the meantime, we shall adopt a somewhat picturesque, and simpler, technique for depicting binary numbers: rather than just write out strings of 1s and 0s, we will envisage a binary number to be a compartmentalized strip of plastic, rather like an ice tray, with each compartment corresponding to a digit; if the compartment is empty, that means the digit is 0, but if the digit is 1, we put a pebble there. Now let us take two such strips and pretend these are the numbers to be added – the "summands". Underneath these two we have laid out one more, to hold the answer (Figure 2.1).

DOI: 10.1201/9781003358817-2

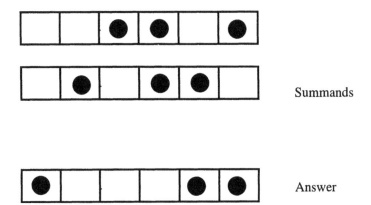

FIGURE 2.1
A pictorial depiction of binary addition.

This turns our abstract mathematical problem into a matter of real-world "mechanics". All we need to do the addition is a simple set of rules for moving the pebbles. Now instead of pebbles, which are slow and hard to handle, we could use anything else, say, wires with either a high voltage for 1 and low voltage for 0. The basic problem is the same: what are the rules for combining pebbles or voltages? For binary addition, the basic rules are:

$$0 + 0 = 0$$
$$0 + 1 = 1$$
$$1 + 0 = 1$$
$$1 + 1 = 0 \text{ plus a carry}$$

So now you can imagine giving instructions on how to move the pebbles to someone who is a complete idiot: if you have two pebbles here, one above the other, you put no pebble in the sum space beneath them, but carry one over one space to the left – and so on. The marvelous thing is, with sufficiently detailed rules, this "idiot" is able to add two numbers of any size! With a slightly more detailed set, he can graduate to multiplication. He can even, eventually, do very complicated things involving hypergeometric functions and what have you. What you tell an apparent idiot, who can do no more than shuffle pebbles around, is enough for him to tackle the evaluation of hypergeometric functions and the like. If he shifts the pebbles quickly enough, he could even do this quicker than you – in that respect, he is justified in thinking himself smarter than you!

Of course, real machines do not calculate by fiddling with pebbles (although don't forget the abacus of old!). They manipulate electronic signals. So, if we are going to implement all of our notions about operations, we have to start thinking about electric circuits. Let us ditch our ice trays and stones and look at the problem of building a real, physical adder to add two binary digits A and B. This process will result in a sum, S, and a carry, C; we set this out in Table 2.1.

TABLE 2.1

A "Truth Table" for Binary Addition

A	B	S	C
0	0	0	0
0	1	1	0
1	0	1	0
1	1	0	1

Let us represent our adder as a black box with two wires going in – A and B – and two coming out – S and C^* (Figure 2.2).

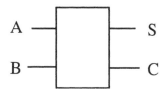

FIGURE 2.2
A black box adder.

We will detail the actual nature of this box shortly. For the moment, let us take it for granted that it works. (As an aside, let us ask how many such adders we would need to add two r-bit numbers. You should be able to convince yourself that ($2r$-1) single-bit adders are required. This again illustrates our general principle of systematically building complicated things from simpler units.)

Let us go back to our black box, single-bit adder. Suppose we just look at the carry bit: this is only non-zero if both A and B are one. This corresponds precisely to the behavior of the so-called AND gate from Boolean logic. Such a gate is itself no more than a black box, with two inputs and one output, and a "truth table" that tells us how the output depends on the inputs. This truth table and the usual pictorial symbol for the AND gate are given in Figure 2.3.

---

* This box is sometimes known as a "half adder". We will also encounter a "full adder" in this chapter, [RPF]

A	B	A AND B
0	0	0
0	1	0
1	0	0
1	1	1

FIGURE 2.3
The AND gate.

Simple enough: *A* AND *B* is 1 if, and only if, *A* is 1 *and* *B* is 1. Thus, carry and "and" are really the same thing, and the carry bit for our adder may be obtained by feeding the *A* and *B* wires into an and gate. Although I have described the gate as a black box, we do in fact know exactly how to build one using real materials, with real electronic signals acting as values for *A*, *B*, and *C*, so we are well on the way to implementing the adder. The sum bit of the adder, *S*, is given by another kind of logic gate, the "exclusive or" or XOR gate. Like the AND, this has a defining truth table and a pretty symbol (Figure 2.4).

A	B	A XOR B
0	0	0
0	1	1
1	0	1
1	1	0

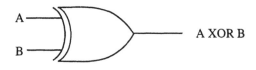

FIGURE 2.4
The XOR gate.

A XOR *B* is 1 if *A or B* is 1, but *not both*. XOR is to be distinguished from a similar type of gate, the conventional OR gate, which has a truth table and symbol shown in Figure 2.5.

A	B	A OR B
0	0	0
0	1	1
1	0	1
1	1	1

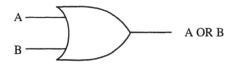

FIGURE 2.5
The OR gate.

All of these gates are examples of "switching functions", which take as input some binary-valued variables and compute some binary function.

Claude Shannon was the first to apply the rules of Boolean algebra to switching networks in his MIT master's thesis in 1937. Such switching functions can be implemented electronically with basic circuits called, appropriately enough, "gates". The presence of an electronic signal on a wire is a "1" (or "true"), the absence a "0" (or "false"). Let us continue going down in level and look in more detail at these basic gates.

The simplest operation of all is an "identity" or "do-nothing" operation. This is just a wire coming into a box and then out again, with the same signal on it. This just represents a wire (Figure 2.6).

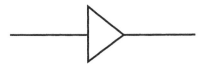

FIGURE 2.6
The identity.

In a real computer, this element would be considered a "delay": as we will see in Chapter 8, electric current actually takes time to move along wires, and this finite travel time – or delay – is something that must be taken into consideration when designing machines; with computers, even elements that do nothing on paper can do something when we build them! But let us skip this operation and look at the next simplest, namely, a box that "negates" the incoming signal. If the input is a 1, then the output will be 0, and vice versa. This is the NOT operation, with the obvious truth table (Figure 2.7).

A	NOT A
0	1
1	0

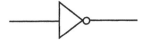

FIGURE 2.7
The NOT gate.

Diagrammatically, the NOT is just the delay with a circle at its tip. Now with a little thought, one can see that there is a relationship between OR and AND, using NOT. By playing with the truth tables, you should be able to convince yourself that *A* OR *B* is the same as NOT{(NOT *A*) AND (NOT *B*)}. This is just one example of an equivalence between operators; there are many more.* Of course, one need not express OR in terms of

---

* These relationships are actually specific instances of a general and venerable old law known as de Morgan's Theorem. [Editor]

AND and NOT; one could express AND in terms of NOT and OR, and so on. One of the nice games you can play with logic gates is trying to find out which is the best set to use for a specific purpose, and how to express other operators in terms of this best set. A question that naturally arises when thinking of this stuff is whether it's possible to assemble a basic set with which you could, in principle, build *all possible* logic functions: that is, if you invent *any* black box whatsoever (defined by assigning an output state to each possible input state), could you actually *build* it using just the gates in the basic set? We will not consider this matter of "completeness" of a set of operators in any detail here; the actual proof is pretty tough, and way beyond the level of this course. We will content ourselves with a hand-waving proof in Section 2.4. Suffice it to say that the set AND, OR, and NOT *is* complete; with these operators, one can build absolutely any switching function. To tempt you to go further with all this cute stuff, I will note that there exist *single* operators that are complete!

We now have pretty much all of the symbols used by engineers to depict the various gates. They're a useful tool for illustrating the links between their physical counterparts. For example, we can diagrammatically depict our relationship between AND, OR, and NOT (Figure 2.8).

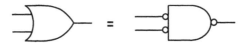

FIGURE 2.8
The relationship between AND, OR, and NOT.

Note that we have adopted the common convention of writing the NOTs as circles directly on the relevant wires; we don't need the triangles.

Let's play with these awhile. How do we make an XOR gate out of them? Now XOR only gives 1 if $A=1$ and $B=0$, or $A=0$ and $B=1$. The general rule for constructing novel gates like this is to write out the truth tables for A AND B, A OR B, A AND (NOT B), and so on, and see how you might turn the outputs of such gates into the inputs for another, in such a way that you get the desired result. For example, we can get a 1 from $A=1$ and $B=0$ if we feed A and B into an AND gate, with a NOT on the B line. Similarly, we use the same trick to get the second option, using an AND, but with the NOT on the A line. If we then feed the outputs of these two gates through a third – an OR – we end up with an XOR (Figure 2.9).

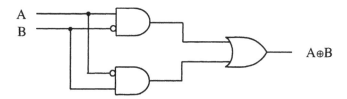

FIGURE 2.9
XOR expressed in ANDs and ORs.

(Notice the convention we are using: if two crossing wires are electrically connected, we place a dot on the crossing point. If the lines cross without connection, there is no dot.) Of course, you have to check that this combination works for the other two input sets of *A* and *B*; and indeed it does. If both *A* and *B* are 0, both AND gates give zero, and the OR gives zero; if both *A* and *B* are 1, again, both AND gates give zero, leading to zero as the final result. Note that this circuit is not unique. Another way of achieving an XOR switch is shown in Figure 2.10.

Which way should we make the XOR circuit in practice? It just depends on the details of the particular circumstance – the hardware, the semiconductor technology, and so on. We might also be interested in other issues, such as which method requires the fewest elements. As you can imagine, such stuff amounts to an interesting design problem, but we are not going to dwell on it here. All we care to note is that we can make any switch we like as long as we have a big bag of ANDs, ORs, and NOTs. We have already seen how to make a single-bit adder – the carry bit comes from an AND gate, and the sum bit from an XOR gate, which we now know how to build from our basic gates. Let us look at another example: a multiple AND, with four inputs *A*, *B*, *C*, *D*. This has four inputs but still just one output, and by extension from the two-input case, we declare that this

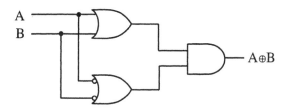

FIGURE 2.10
An alternative XOR.

gate only "goes off" – that is, gives an output of one – when all four inputs are 1. Sometimes people like to write this problem symbolically thus:

$$A \wedge B \wedge C \wedge D$$

where the symbol $\wedge$ means "AND" in propositional logic (as we mentioned earlier). Of course, when logicians write something like this, they have no particular circuit in mind that can perform the operation. We, however, can design such a circuit, built up from our primitive black box gates: to be precise, three AND gates as in Figure 2.11.

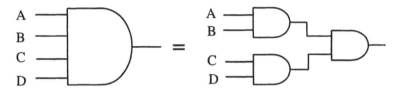

FIGURE 2.11
A multiple AND gate.

In a similar way, one can build up a multiple AND of any size.

Now the time has come to hit nearly rock bottom in our hierarchy by looking at the actual electronic components one would use to construct logic gates. We *will* actually hit rock bottom, by which I mean discussing the physics of semiconductors and the motion of actual electrons through machines, later in the course (in Chapter 8). For now, I will give some quick pointers to gate construction that should be intelligible to those of you with some grasp of electronics.

Central to the construction of all gates is the *transistor*. This is arguably the most important of all electronic components and played a critical role in the development and growth of the industry. Few electronic devices

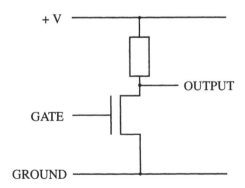

FIGURE 2.12
The transistor inverter, or NOT gate.

contain no transistors, and an understanding of the basic properties of these elements is essential for understanding computers, in which they are used as switches. Let us see how a transistor can be used to construct a NOT gate. Consider the circuit shown in Figure 2.12.

A transistor is a three-connection device: one input is connected to the gate signal, one to ground, and the other to a positive voltage via a resistor. The central property of the transistor is that if the gate has a distinctly positive voltage, the component conducts, but if the gate is zero or distinctly negative, it does not. Now look at the behavior of the output voltage as we input a voltage to the gate. If we input a positive voltage, which by convention we label a 1, the transistor conducts: a current flows through it, and the output voltage becomes zero, or binary 0. On the other hand, if the gate was a little bit negative, or zero, no current flows, and the output is the same as +V, or 1. Thus, the output is the opposite of the input, and we have a NOT gate.*

What about an AND gate? Due to the nature of the transistor, it actually turns out to be more convenient to use a NAND gate as our starting point for this. Such a gate is easier to make in a MOS environment than an AND gate, and if we can make the former, we can obtain the latter from it by using one of de Morgan's rules: that AND = NOT {NAND}. So, consider the following simple circuit in Figure 2.13.

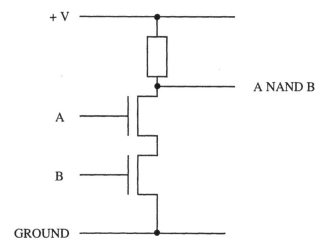

FIGURE 2.13
A transistor NAND gate.

---

* As a technical aside, we have assumed that our circuits are fabricated using MOS (Metal Oxide Semiconductor) technology. Resistors are hard to implement in this type of silicon technology, and in practice the resistor would actually be replaced by another type of MOS transistor (see Chapter 8). [RPF]

In order for the output voltage to be zero here, we need to have current flow through both A and B, which we can clearly only achieve if both A and B are positive. Hence, this circuit is indeed a "NOT AND" or NAND gate. To get an AND gate, we simply take the NAND output from Figure 2.13 and feed it in as input to the NOT gate illustrated in Figure 2.12. The resultant output is our AND.

What about an OR gate? Well, we have seen how to make an OR from ANDs and NOTs, and we could proceed this way if we wished, combining the transistor circuits above; however, an easier option (both conceptually and from the viewpoint of manufacture) results from consideration of the following, parallel circuit (Figure 2.14).

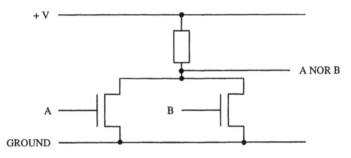

FIGURE 2.14
A transistor NOR gate.

If either A or B is positive, or both positive, current flows and the output is zero. If both A and B are zero, it is $+V$, or 1. So again, we have the opposite of what we want: this is a "NOT OR" or NOR gate. All we do now is send our output through a NOT, and all is well.

Hopefully, this has convinced you that we can make electrical circuits which function as do the basic gates. We are now going to go back up a level and look at some more elaborate devices that we can build from our basic building blocks.

2.2 The Binary Decoder

The first device that we shall look at is called a "binary decoder". It works like this. Suppose we have four wires, A, B, C, D coming into the device. These wires could bring in any input. However, if the signals on the wires are a specific set, say 1011, we want to know this: we want to receive a signal from the decoder telling us that 1011 has occurred. It is as if we have some demon scanning the four bits coming into the decoder and, if they turn out to be 1011, he sends us a signal! This is easy to arrange using a modified AND gate (and much cheaper than hiring a demon). The device in Figure 2.15 clearly only gives us an output of 1 when A, C, D are 1 and B is 0.

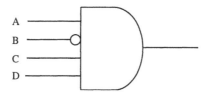

FIGURE 2.15
A simple decoder.

This is a very special type of decoder. Suppose we want a more general one, with lots of demons each looking for their own particular number amidst the many possible input combinations. Such a decoder is easy to make by connecting individual decoders in parallel. A full decoder is one that will decode every possible input number. Let us see how this works with a three-to-eight binary decoder. Here, we have three input bits on wires A, B, C giving $2^3 = 8$ combinations. We therefore have eight output wires, and we want to build a gate that will assign each input combination to a distinct output line, giving a 1 on just one of these eight wires, so that we can tell at a glance what input was fed into the decoder. We can organize the decoder as shown in Figure 2.16.

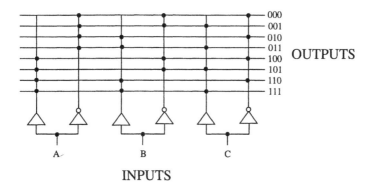

FIGURE 2.16
A binary decoder.

We have introduced the pictorial convention that three dots on a horizontal line implies a triple AND gate (see the discussion surrounding

Figure 2.11). Notice that each input wire branches into an *A* and NOT *A* signal and so on. As we have arranged things, only the bottom four wires can go off if *A* is one, and the top four if *A* is zero. The dots on the wires for *B* and *C* (and NOT *B* and NOT *C*) similarly show us immediately which of the eight output wires can go off: we have labeled each output line with its corresponding input state. Thus, we have explicitly constructed a three-to-eight binary decoder.

Now, there is a profound use to which we can put the device in Figure 2.16; one which reveals the decoder to be an absolutely essential part of the machine designer's arsenal. Suppose we feed 1s from the left into all of the horizontal input wires of the decoder. Now interpret each dot on an intersection as a two-way AND:

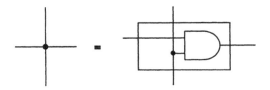

and a simple crossing as no connection:

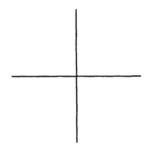

In order for the 1 input from the left to get past the first dot, the correct signal *A*=1 or NOT *A*=1, depending on the wire, must be present. Similarly for *B* and *C*. So we still have a binary decoder; nothing has changed in this regard. However, we have also invented something else, which a little thought should show you is indispensable in a functioning computer: *this device can serve as a multiple switch to connect you to a selected input wire.* The original input lines of the decoder, *A*, *B*, *C* now serve as "address" lines to select which output wire gives a signal (which may be 1 or 0). This is very close to something called a "multiplexer": multiplexing is the technique of selecting and transmitting signals from multiple input lines to a single output line. In our example, we can make our device into a true multiplexer by adding an eight-way OR gate to the eight output lines (Figure 2.17).

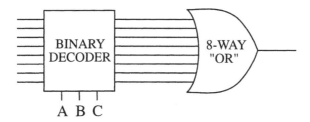

FIGURE 2.17
The multiplexer.

This rather neat composite device clearly selects which of the eight input lines on the left is transmitted, using the three-bit address code. Multiplexers are used in computers to read and write into memory, and for a whole host of other tasks.

Let me give you some problems to play with.

> **Problem 2.1** Design an eight to three encoder. In other words, solve the reverse problem to that considered earlier: eight input wires, only one of which has a signal on at any given time; three output wires that "encode" which wire had the signal on.

> **Problem 2.2** Design a simple adder using AND, OR, and NOT gates.

> **Problem 2.3** Design a one-bit full adder.

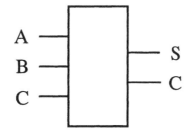

> **Problem 2.4** Make an r-bit full adder using r one-bit full adders. How many simple adders would be needed?

2.3 More on Gates: Reversible Gates

We stated earlier, without proof, that the combinational circuits for AND and NOT are sufficient building blocks to realize any switching function.

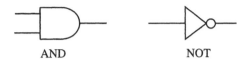

AND   NOT

Actually, there are two other elements that we added without noticing. These are the "fanout" and "exchange" operations (Figure 2.18).

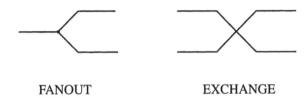

FANOUT   EXCHANGE

FIGURE 2.18
FANOUT and EXCHANGE.

In the case of wires and pulses of 1s and 0s, the presence of these "gates" is an obvious assumption; FANOUT just splits a wire into two or more and EXCHANGE just swaps over a pair of connections. If, on the other hand, the information was carried by pebbles, then a fanout into two means that one pebble has become two, so it is quite a special operation. Similarly, if the information was stored in separate boxes in distinct locations, then exchange is also a definite operation. We are emphasizing the logical necessity of including these two "obvious" operations since we will be needing them in our discussion of reversibility. The other thing we will assume we have is an endless supply of 0s and 1s; a store somewhere into which we can stick wires and get signals for as long as we want. This can have unforeseen uses. For example, we have already noted that one can in fact replace the AND and NOT set of gates with a single NAND gate (Figure 2.19).

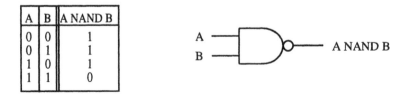

A	B	A NAND B
0	0	1
0	1	1
1	0	1
1	1	0

FIGURE 2.19
The NAND gate.

It is easy to see that this single gate is as good as having both AND and NOT. To get a NOT operation from a NAND, all we do is turn to our storehouse of 1s and connect one of the NAND inputs up to it. Now, whatever the other input *A*, the output will be NOT *A*:

Now that we have a NOT and an AND, we can clearly construct a NAND, and we have demonstrated their equivalence as a set of operators.

We want to discuss a rather different problem, which will enable us to look at some rather more exotic logic gates. Both the AND and the NAND operation – and the OR and XOR – are *irreversible* operations. By this, I simply mean that from the output of the gate you cannot reconstruct the input: information is irreversibly lost. If the output of an AND gate with four inputs is zero, it could have resulted from any one of 15 input sets, and you have no idea which (although you obviously know about the inputs if the output is one!). We would like to introduce the concept of a *reversible* operation as one with enough information in the output to enable you to deduce the input. We will need such a concept when we come to study the thermodynamics of computation later. It will make it possible for us to make calculations about the free energy – or, if you like, the physical efficiency – of computation.

The problem of reversible computers has been studied independently by Bennett and Fredkin. Our basic constructs will be three gates: NOT (N), CONTROLLED NOT (CN), and a CONTROLLED CONTROLLED NOT (CCN). Let us explain what these are. A NOT is just a NOT as before, a one-element object. A CONTROLLED NOT is a two-wire input gadget that, unlike the AND and NAND gates, has two outputs as well. It works in the following way. We have two wires, on one of which we write a circle, representing a control, and on the other a cross (Figure 2.20).

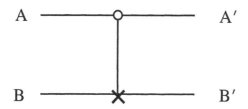

FIGURE 2.20
The CN gate.

The "**X**" denotes a NOT operation: however, this NOT is not a conventional one; it is *controlled by the input to the **O**-wire*. Specifically, if the input to the **O**-wire is 1, then the input to the **X**-wire is inverted; if the **O**-input is zero, then the NOT gate does not work, and the signal on the **X**-wire goes through unchanged. In other words, the input to the **O**-line activates the NOT gate on the lower line. The **O**-output, however, is always the same as the **O**-input – the upper line is the identity. The truth table for this gate is simple enough (Table 2.2).

TABLE 2.2

Truth Table for the CN Gate

A	B	A′	B′
0	0	0	0
0	1	0	1
1	0	1	1
1	1	1	0

Note that we can interpret B' as the output of an *XOR* gate with inputs A and B: $B'= $ XOR(A,B).

One of the most important properties of this CN gate is that it is reversible – from what comes out we can deduce what went in. Notice that we can actually reverse the operation of the gate by merely repeating it (Figure 2.21).

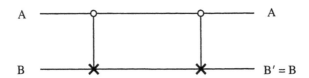

FIGURE 2.21
The identity utilizing CN gates.

We can use a CN gate to build a fanout circuit. If we set $B=0$, then we have $B'=A$ and $A'=A$. As an exercise, you might like to show how CN gates can be connected up to make an exchange operator (hint: it takes several).

Sadly, we cannot do everything with just N and CN gates. Something more is needed, for example, a CCN, or CONTROLLED CONTROLLED NOT gate* (Figure 2.22).

---

* The CNN gate is sometimes called the Toffoli gate after its inventor Tommaso Toffoli (editor).

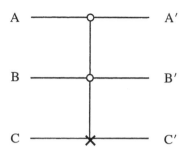

FIGURE 2.22
The CCN gate.

In this gate, we have two control lines A and B, each marked by an **O**, and as with the CN gate, the signals on this line are unchanged on passage through the gate: $A'=A$, $B'=B$. The remaining line, once again, has a NOT on it, but this is only activated if both $A=1$ and $B=1$: then, $C'=$NOT C. Notice that this single gate is very powerful. If we keep both A and B equal to one, then the CCN gate is just an N, a NOT. If we keep just $A=1$, then the gate is just a CN gate with B and C as inputs. So if we have a CCN gate and a source of 1s and 0s, we can junk both the CN and N gates. But things are even better: with this CCN gate, we can do everything! We have already seen how a CN gate can be used to produce an XOR output. We know that throwing in a NOT or two enables us to get an AND gate. As an example, the AND gate can be made by holding $C=0$ and taking the inputs to be A and B. The output, A AND B is then C', which is clearly 1 only when the NOT gate is activated to invert $C=0$, which in turn is only the case – by the property of the CCN gate – when $A=B=1$. So clearly, we can generate any gate we like with just a CCN gate: by itself, it forms a complete operator set.

The next thing we must do is show that we can do something useful with only these reversible operations. This is not difficult, as we have just shown that we can do anything with them that we can do with a complete operator set! However, we would like whatever we build to be itself reversible. Consider the problem of making a full adder:

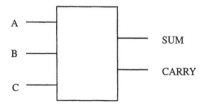

We need to add *A*, *B*, and *C* and obtain the sum and carry. Now as it stands, this operation is not reversible – one cannot, in general, reconstruct the three inputs from the sum and carry. We have decided that we want to have a reversible adder, so we need more information at the output than at present. As you can see with a little thought, reversible gates have the general property that "lines in = lines out" – this is the only way that all possible inputs can be separately "counted" at the output – and so we need another line coming out of our adder. In fact, it turns out that we need two extra lines coming out of the gate, and one extra going in, which you set to 0, say. Using N, CN, and CCN (or just the latter), we can get AND, OR, and XOR operators, and we can clearly use these to build an adder: the trick of making it reversible lies in using the redundancy of the extra outputs to arrange things such that the two extra output lines, on top of the sum and carry ones, are just the inputs *A* and *B*. It is a worthwhile exercise to work this out in detail.

Fredkin added an extra constraint on the outputs and inputs of the gates he considered. He demanded that not only must a gate be reversible, but the number of 1s and 0s should never change. There is no good reason for this, but he did it anyway. He introduced a gate performing a controlled exchange operation (Figure 2.23).

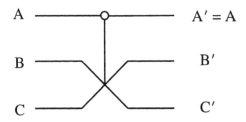

FIGURE 2.23
The Fredkin gate: a controlled exchange.

In his honor, we will call this a Fredkin gate. You should be used to the notion of control lines by now; they just activate a more conventional operation on other inputs. In this case, the operation is exchange. Fredkin's gate works like this: if *A*=0, *B* and *C* are not exchanged; *B'*=*B* and *C'*=*C*. However, if *A*=1 they are, and *B'*=*C*, *C'*=*B*. You can check that the number of 1s and 0s is conserved. As a further, and more demanding, exercise, you can try to show how this Fredkin gate can be used (perhaps surprisingly) to perform all logical operations instead of using the CCN gate.

2.4 Complete Sets of Operators

I have introduced you to the notion of reversible gates so that you can see that there is more to the subject of computer logic than just the standard AND, NOT, and OR gates. We will return to these gates in Chapter 5. I want for the moment to leave the topic of reversible computation and return to the issue of complete sets of operators. Now I've been very happy to say that with a so-called "complete set" of operators, you can do anything, that is, build any logical function. I will take as my complete set the operations AND, NOT, FANOUT, and EXCHANGE. The problem I would like to address is how we can know that this set *is* complete. Suppose we have a bunch of n input wires, which we'll label $X_1, X_2, X_3,... X_n$. For each pattern of inputs $\{X\}$, we will have some specific output pattern on a set of wires $Y_1, Y_2,..., Y_m$, where m is not necessarily equal to n. The output on Y_i is a logical function of the X_i. Formally, we write

$$Y_i = F_i\left(\{X\}\right), i = 1,...,m \tag{2.1}$$

What we want to demonstrate is that for *any* set of functions F_i we can build a circuit to perform that function on the inputs using just our basic set of gates. Let us look at a particular example, namely, the sum of the input wires. We can see how, in principle, we can do this, as follows. In our binary decoder, we had n input wires and 2^n output wires, and we arranged for a particular output wire to fire by using a bunch of AND gates. This time we want to arrange for that output to give rise to a specific signal on another set of output wires. In particular, we can then arrange for the signals on the output wires to be the binary number corresponding to the value of the sum of the particular input pattern.

Let us suppose that for a particular input set of Xs we have selected one wire. One wire only is "hot", and all the others "cold". When this wire is hot, we want to generate a specific set of output signals. This is the opposite problem to the decoder. What we need now is an *en*coder. As you should have figured out from one of the problems you were set, this can be constructed from a bunch of OR gates. So you see, we have separated the problem into two parts. The first part that we looked at before was how to arrange for different wires to go off according to the input. The answer was our decoder. Our encoder must have a lot of input wires but only one goes off at a time. We want to be able to write the number of which wire went off in the binary system. A three-bit encoder may be built from OR gates as follows (Figure 2.24):

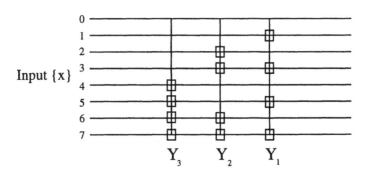

FIGURE 2.24
The three-bit encoder.

where we have used the following notation for the OR gates:

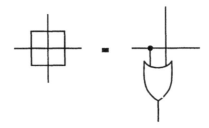

Thus, if we are not bothered about the proliferation of 2^n wires, then we can construct any logical function we wish. In general, we have an AND plane and an OR plane and a large number of wires connecting these two regions (Figure 2.25).

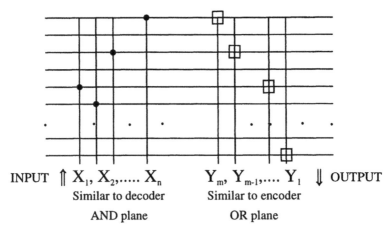

FIGURE 2.25
Construction of a general logical function.

where we have used the same notation for AND gates as in Figure 2.16. If you succeeded in solving any of the problems 2.2–2.4, which required you to construct a number of different adders, then you will have already seen simple examples of this principle at work.

Some of the logical functions we could construct in this way are so simple that using Boolean algebra we can simplify the design and use fewer gates. In the past, people used to invest much effort in finding the simplest or smallest system for particular logical functions. However, the approach described here is so simple and general that it does not need an expert in logic to design it! Moreover, it is also a standard type of layout that can easily be laid out in silicon. Thus this type of design is usually used for programmable logic arrays or PLAs. These are often used to produce custom-made chips for which relatively few copies are needed. The customer only has to specify which ANDs and which ORs are connected to get the desired functionality. For mass-produced chips, it is worthwhile investing the extra effort to do the layout more efficiently.

2.5 Flip-Flops and Computer Memory

Now I want to come onto something different, which is not only central to the functioning of useful computers but should also be fun to look at. We start with a simple question: can we store numbers? That is, can we build a computer's memory from the gates and tidbits we've assembled so far? A useful memory store will allow us to alter what we store; to erase and rewrite the contents of a memory location. Let's look at the simplest possible memory store, one which holds just one bit (a 1 or 0), and see how we might tinker with it. As a reasonable first guess at building a workable memory device, consider the following black box arrangement (Figure 2.26).

FIGURE 2.26
A black box memory store.

We take the signal on line C to represent what is in our memory. The input A is a control line, with the following properties: as long as A is 0, i.e., we are feeding nothing into our box, C remains the same. However, if we switch A to 1, then we change C: it goes from 0 to 1 or vice versa. We can write a kind of "truth table" for this (Table 2.3).

TABLE 2.3

"Truth Table" for the Memory Device

A	Present C	Next C
0	0	0
0	1	1
1	0	1
1	1	0

It is easily noticed from Table 2.3 that "Next C" is the XOR of A and the present C. So it might seem that if we get clever and replace our black box with an XOR gate with *feedback* from C, we may have a possible memory unit (Figure 2.27).

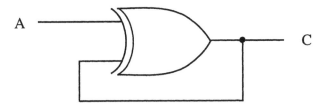

FIGURE 2.27
A plausible (but non-workable) memory device.

Will this work? Well, it all depends on the timing! We have to interrupt our abstract algebra and take note of the limitations on devices imposed by the physical world. Let's suppose that A is 0 and C is 1. Then everything is stable: so far, so good. Now change the input A to 1. What happens? C changes to 0, by definition, which is what we want. But this value is then fed back into the XOR gate, where, with $A=1$, it gives an output of 1 – so C changes back to 1. This then goes back into the XOR, where with $A=1$ it now gives an output $C=0$. We then start all over again. Our gate oscillates horribly and is of no use whatsoever.

However, if you think about it, you can see that we can salvage the gate somewhat by building in delays to the various stages of its operation; for example, we can make the XOR take a certain amount of time to produce its output. However, we cannot stop it oscillating. Even if we were prepared to build a short-term memory bank, the physical volatility of electronic components would introduce extra instabilities leading to unforeseen oscillations that make this gate pretty useless for practical

purposes. Out of interest, note what happens if we build the circuit with
an OR rather than an XOR.

Clearly, the crucial troublesome feature of this device is the element
of *feedback*. Can we not just dispense with it? The answer is yes, but this
would be at quite a cost. For reasons of economy and space, one thing we
would like our computer to be able to do is repeated calculations with the
same pieces of operating equipment. For example, if we used a certain
adder to do part of a calculation, we would like to use the same adder
to do another subsequent part of the calculation, which might involve
using its earlier output. We would not want to keep piling more and more
adders into our architecture for each new stage of a program: yet without
feedback, we would have no choice. So we will want to crack this problem!

What we want is a circuit that can hold a value, 0 or 1, until we decide to
reset it with a signal on a wire. The circuit that turns out to do the job for
us is called a *flip-flop*, schematically drawn as shown in Figure 2.28.

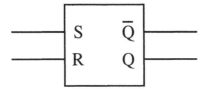

FIGURE 2.28
A flip-flop.

The flip-flop has two input wir*es* – the "set" (S) and "reset" (R) wires – and
two outputs, which we call Q and $\bar{Q}$. This latter labeling reflects the fact
that one is always the logical complement – the inverse – of the other. They
are sometimes misleadingly referred to as the 0 and 1 lines; misleading,
because each can take either value, as long as the other is its inverse.

We can actually use NOR gates (for example) to build a circuit that functions as a flip-flop (Figure 2.29).

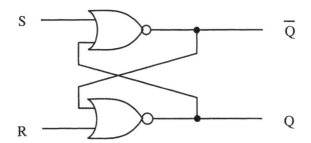

FIGURE 2.29
Gate structure of a simple flip-flop.

Note that the device incorporates feedback! Despite this, it is possible to arrange things so that the flip-flop does not oscillate, as happened with our naive XOR store. It is important to ensure that S and R are never simultaneously 1, something that we can arrange the architecture of our machine to ensure. The device then has just two output states, both of which are stable: $Q=1$ (hence $\bar{Q}=0$) and $Q=0$ (hence $\bar{Q}=1$). How does this help us with memory storage?

The way the thing works is best seen by examining its truth table (Table 2.4).

TABLE 2.4

Truth Table for a Simple Flip-Flop

Present Q	S	R	Next Q
0	0	0	0
0	0	1	0
0	1	0	1
1	0	0	1
1	1	0	1
1	0	1	0

The signal on the Q-line is interpreted as the contents of the flip-flop, and this stays the same whenever S and R are both 0. Let us first consider the case when the reset line, R, carries no signal. Then we find that, if the contents Q of the flip-flop are initially 0, setting $S=1$ changes this to 1; otherwise, the S-line has no effect. In other words, the S-line sets the contents of the flip-flop to 1, but subsequently manipulating S does nothing; if the flip-flop is already at 1, it will stay that way even if we switch S. Now look at the effect of the reset line, R. If the flip-flop is at 0, it will stay that way if we set $R=1$; however, if it is at 1, then setting $R=1$ resets it to 0. So the R line clears the contents of the flip-flop. This is pretty confusing upon first exposure, and I would recommend that you study this set-up until you understand it fully. We will now examine how we can use this flip-flop to solve our timing problems.

2.6 Timing and Shift Registers

We have now designed a device – a flip-flop – that incorporates feedback and doesn't suffer from the oscillations of naive structures. However, there is a subtle and interesting problem concerning this gadget. As I pointed out in the last lecture, the signals traveling between the various components take differing times to arrive and be processed, and sometimes the physical volatility of the components you use to build your equipment will give you freaky variations in these times in addition, which you wouldn't allow for if you assumed technology to be ideal. This means that often you will find signals arriving at gates later than they are supposed to and doing the wrong job! We have to be aware of the possible effects of this. For the flip-flop, for example, what would happen if both the outputs turned out to be the same? We have assumed, as an idealization, that they would be complementary, but things can go wrong! You can see that if this happens, then the whole business of the set and reset would go out the window.

The way around this is to introduce into the system a *clock* and have this send an "enable" signal to the flip-flop at regular intervals. We then contrive to have the flip-flop do nothing until it receives a clock signal. These signals are spaced far enough apart to allow everything to settle down before operations are executed. We implement this idea by placing an AND gate on each input wire and also feeding into each gate the clock signal (Figure 2.30).

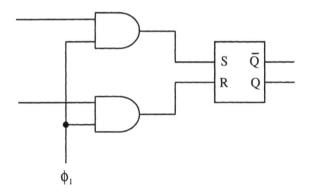

FIGURE 2.30
A clocked RS flip-flop.

This is sometimes called a *transparent latch* since all the time the clock is asserted any change of input is transmitted through the device.

We represent the signal ϕ_1, from the clock as a series of pulses (Figure 2.31).

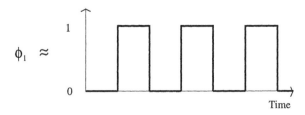

FIGURE 2.31
The clock pulse.

Clearly, whatever the input to the AND gates, it will only get through to
S and R when the signal from the clock ϕ_1 is 1. So as long as we get the
timing of the clock right, and we can be sure it does not switch the gate
on until there is no chance of the inputs playing up, we have cleared up
the problem. But of course, we have created another one! We have merely
deferred the difficulty: the output of this gate will shoot off to another,
or more than one, and we will have the same problems with travel times,
and so on, all over again. It will not help to connect everything up to our
clock ϕ_1 – far from it; one part of the system may be turning on just as
another is changing its outputs. We still have delays. So we might think,
to get around this, to try to build a machine with great precision, calculating delay times and making sure that everything comes out right. It can
be done, and the resultant system is fast and efficient, but it's also very
expensive and difficult to design. The best way to get around the problem
is to introduce another clock, ϕ_2, and not allow the next gate in the chain
to accept input from the first until *this* clock is asserted. This arrangement
is the basis for a special type of flip-flop called a *Master-Slave Flip-Flop*
(Figure 2.32).

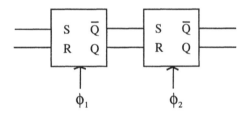

FIGURE 2.32
The master-slave flip-flop.

The signals from the two clocks should be complementary:

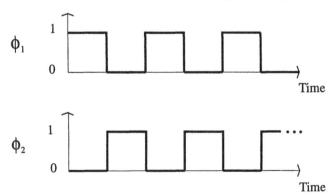

The easiest way to ensure this is to get ϕ_2 from NOT ϕ_1. We also note that we need our logical operations to be fast in comparison with the clock pulse length. Don't forget that in all this, we are using the abstractions that (1) all levels are 0 or 1 (not true: they are always changing with time. They are never exactly one or zero, but they are near saturation), and (2) there is a definite, uniform delay time between pulses: we can say that *this* happens, then *that* happens, and so on. This is a good idealization, and we can get closer to it by introducing more clock signals if we like.

It is possible to design a variety of flip-flop devices and learning how and why they work is a valuable exercise. One such device is the *D-type* flip-flop, which has the structure shown in Figure 2.33.

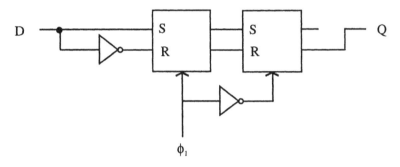

FIGURE 2.33
A "D-type" flip-flop.

It is unclear why this device is labeled a "D-type" flip-flop. One plausible suggestion is that the "D" derives from the "delaying" property of the device: basically, the output is the same as the input, but only becomes so after a clock pulse.

Let us introduce the following shorthand notation for the D-type flip-flop (Figure 2.34).

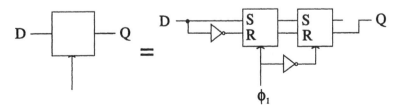

FIGURE 2.34
Simplified notation for the D-type flip-flop.

A very useful device that may be built from flip-flops, and one that we shall take the trouble to examine, is a *shift register*. This is a device that can, among other things, store arbitrary binary numbers – sequences of bits – rather than just one bit. It comprises a number of flip-flops, connected sequentially, into which we feed our binary number one bit at a time. We will just use our basic S-Rs, with a delay built in. The basic structure of a shift register is shown in Figure 2.35.

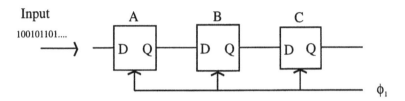

FIGURE 2.35
A shift register

Each unit of this register is essentially a stable delay device of the kind I described earlier. Note that each flip-flop in the array is clocked by the same clock ϕ_1.

The reader should have little difficulty in seeing how the device works. We start with the assumption (not necessary, but a simplifying one) that all of the flip-flops are set to zero. Suppose we wish to feed the number 101 into the device. What will happen? We feed the number in the lowest digit first, so we stick a 1 into the left-hand S-R, which I've labeled **A**, and wait until the clock pulse arrives to get things moving. After the next clock pulse, the output of **A** becomes 1. We now feed the next bit, 0, into **A**. Nothing happens until the next clock pulse. After this arrives, the next S-R in the sequence, **B**, gets a 1 on its output (the original 0 has been reset). However, the output of **A** switches to 0, reflecting its earlier input. Meanwhile, we have fed into **A** the next bit of our number, which is 1. Again, we wait for the next clock pulse. Now we find that **A** has an output of 1, **B** of 0, and **C** of 1 – in other words, reading from left to right, the very number we fed into it! Generalizing to larger binary strings is straightforward (note that each flip-flop can hold just the one bit, so a register containing n flip-flops can only store up to 2^n). So you can see that a register like this takes a sequential piece of information and turns it into parallel information; shifting it along bit by bit and storing it for our later examination. It is not necessary to go any further with them; the reader should be able to see that registers clearly have uses as memory stores for numbers and as shifting devices for binary arithmetical operations and that they can therefore be built into adders and other circuits.

3

The Theory of Computation

CONTENTS

Thus far, we have discussed the limitations on computing imposed by the structure of logic gates. We now come on to address an issue that is far more fundamental: is there a limit to what we can, in principle, compute? It is easy to imagine that if we built a big enough computer, then it could compute anything we wanted it to. Is this true? Or are there some questions that it could never answer for us, however beautifully made it might be?

Ironically, it turns out that all this was discussed long before computers were built! Computer science, in a sense, existed before the computer. It was a very big topic for logicians and mathematicians in the 1930s. There was a lot of ferment at court in those days about this very question – what can be computed *in principle*? Mathematicians were in the habit of playing a particular game, involving setting up mathematical systems of axioms and elements – like those of Euclid, for example – and seeing what they could deduce from them. An assumption that was routinely made was that any statement you might care to make in one of these mathematical languages could be proved or disproved, in principle. Mathematicians were used to struggling vainly with the proof of apparently quite simple statements – like Fermat's Last Theorem, or Goldbach's Conjecture – but always figured that, sooner or later, some smart guy would come along and figure them out.[*] However, the

---

[*] In the case of Fermat's Last Theorem, some smart guy *did* come along and solve it! Fermat's Theorem, which states that the equation $x^n + y^n = z^n$ (n an integer, $n \geq 3$) has no solutions for which x, y, and z are integers, has always been one of the outstanding problems of number theory. The proof, long believed impossible to derive (mathematical societies even offered rewards for it!), was finally arrived at in 1994 by the mathematicians Andrew Wiles and Richard Taylor, after many, many years' work (and after a false alarm in 1993). [Editor]

DOI: 10.1201/9781003358817-3

question eventually arose as to whether such statements, or others, might be inherently unprovable. The question became acute after the logician Kurt Gödel proved the astonishing result – in "Gödel's Theorem" – that arithmetic was incomplete.

3.1 Effective Procedures and Computability

The struggle to define what could and could not be proved, and what numbers could be calculated, led to the concept of what I will call an *effective procedure*. If you like, an effective procedure is a set of rules telling you, moment by moment, what to do to achieve a particular end – it is an algorithm. Let me explain roughly what this means by example. Suppose you wanted to calculate the exponential function of a number x, e^x. There is a very direct way of doing this: you use the Taylor series

$$e^x = 1 + x + \left(x^2 / 2!\right) + \left(x^3 / 3!\right) + \dots . \tag{3.1}$$

Plug in the value of x, add up the individual terms, and you have e^x. As the number of terms you include in your sum increases, the value you have for e^x gets closer to the actual value. So if the task you have set yourself is to compute e^x to a certain degree of accuracy, I can tell you how to do it – it might be slow and laborious, and there might be more efficient techniques, but we don't care: it works. It is an example of what I call an effective procedure.

Another example of an effective procedure in mathematics is the process of differentiation. It doesn't matter what function of a variable x I choose to give you, if you have learned the basic rules of differential calculus, you can differentiate it. Things might get a little messy, but they are straightforward. This is in contrast to the inverse operation, integration. As you all know, integration is something of an art; for any given integrand, you might have to make a lot of guesses before you can integrate it: Should I change variables? Do we have the derivative of a function divided by the function itself? Is integration by parts the way to go? In that none of us have a hotline to the correct answer, it is fair to say that we do not possess an effective procedure for integration. However, this is not to say that such a procedure does not exist: one of the most interesting discoveries in this area in the past 20 years has been that there *is* such a procedure! Specifically, any integral that can be expressed in terms of a predefined list of elementary functions – sines, exponentials, error functions, and so forth – can be evaluated by an effective procedure. This means, among other things, that machines can do integrals. We have to thank a guy named Risch for this ("The Problem of Integration in Finite Terms", *Trans. A.M.S.* 139 (1969), pp. 167–189).

There are other examples in mathematics where we lack effective procedures; factoring general algebraic expressions, for example. There are effective procedures for expressions up to the fourth degree, but not fifth and over. An interesting example of a discipline in which every school kid would give his eye teeth for an effective procedure is geometry. Geometrical proof, like integration, strikes most of us as more art than science, requiring considerable ingenuity. It is ironic that, like integration, there is an effective procedure for geometry! It is, in fact, Cartesian analytic geometry. We label points by coordinates, (x,y), and we determine all lengths and angles by using Pythagoras' Theorem and various other formulae. Analytic geometry reduces the geometry of Euclid to a branch of algebra, at a level where effective procedures exist.

I have already pointed out that converting questions to effective procedures is pretty much equivalent to getting them into a form whereby computers can handle them, and this is one of the reasons why the topic has attracted so much attention of late (and why, for example, the notion of effective procedures in integration has only recently been addressed and solved). When mathematicians first addressed these problems, their interest was more general than the practical limits of computation; they were interested in principle with what could be proved. The question spawned a variety of approaches. Alan Turing, a British mathematician, equated the concept of "computability" with the ability of a certain type of machine to perform a computation. Church defined a system of logic and propositions and called it effective calculability. Kleene defined certain so-called "general recursive propositions" and worked in terms of these. Post had yet another approach (see the problem at the end of this chapter), and there were still other ways of examining the problem. All of these workers started off with a mathematical language of sorts and attempted to define a concept of "effective calculability" within that language. Thankfully for us, it can be shown that all of these apparently disparate approaches are equivalent, which means that we will only need to look at one of them. We choose the commonest method, that of Turing.

Turing's idea was to make a machine that was kind of an analog of a mathematician who has to follow a set of rules. The idea is that the mathematician has a long strip of paper broken up into squares, in each of which he can write and read, one at a time. He looks at a square, and what he sees puts him in some state of mind that determines what he writes in the next square. So, imagine the guy's brain having lots of different possible states that are mixed up and changed by looking at the strip of paper. After thinking along these lines and abstracting a bit, Turing came up with a kind of machine that is referred to as – surprise, surprise – a Turing machine. We will see that these machines are horribly inefficient and slow – so much so that no one would ever waste their time building one except for amusement – but that, if we are patient with them, they can do wonderful things.

Now Turing invented all manner of Turing machines, but he eventually discovered one – the so-called universal Turing machine (UTM) – that was the best of the bunch. Anything that any specific, special-purpose Turing machine could do, the UTM could do. But further, Turing asserted that *if anything could be done by an effective procedure, it could be done by his Universal machine, and vice versa*: if the UTM could not solve a problem, there was no effective procedure for that problem. Although just a conjecture, this belief about the UTM and effective procedures is widely held and has received much theoretical support. No one has yet been able to design a machine that can outdo the UTM in principle. I will actually give you the plans for a UTM later. First, we will take a closer look at its simpler brother – the finite state machine.

3.2 Finite State Machines

A typical Turing machine consists of two parts; a tape, which must be of potentially unlimited size, and the machine itself, which moves over the tape and manipulates its contents. It would be a mistake to think that the tape is a minor addition to a very clever machine; without the tape, the machine is really quite dumb (try solving a complex integral in your head). We will begin our examination of Turing machines and what they can do by looking at a Turing machine without its tape; this is called a *finite state machine*.

Although we are chiefly interested in finite state machines (FSMs) as component parts of Turing machines, they are of some interest in their own right. What kinds of problems can such machines do, or not do? It turns out that there are some questions that FSMs cannot answer but that Turing machines can. Why this should be the case is naturally of interest to us. We will take all of our machines to be black boxes, whose inner mechanical workings are hidden from us; we have no interest in these details. We are only interested in their behavior. To familiarize you with the relevant concepts, let me give an example of a finite state machine (Figure 3.1).

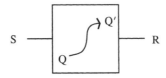

FIGURE 3.1
A generic finite state machine.

The basic idea is as follows. The machine starts off in a certain *internal state, Q*. This might, for example, simply be holding a number in memory. It then receives an *input*, or *stimulus, S* – you can either imagine the machine reading a bit of information off a (finite) tape or having it fed in in some other way. The machine reacts to this input by *changing to another state, Q'*, and spitting something out – a *response* to the input, R. The state it changes to and its response are determined by both the initial state and the input. The machine then repeats this cycle, reading another input, changing state again, and again issuing some response.

To make contact with real machines, we will introduce a discrete time variable, which sets the pace at which everything happens. At a given time *t*, we have the machine in a state *Q(t)* receiving a symbol *S(t)*. We arrange things so that the response to this state of affairs comes one pulse later, at time *(t+1)*. Let us, for notational purposes, introduce two functions *F* and *G*, to describe the FSM and write:

$$R[t+1] = F[S(t), Q(t)] \qquad (3.2)$$

$$Q[t+1] = G[S(t), Q(t)]$$

We can depict the behavior of FSMs in a neat diagrammatic way. Suppose a machine has a set of possible states $\{Q_j\}$. We represent the basic transition of this machine from a state Q_j to a state Q_k upon reception of a stimulus *S*, resulting in a response *R* (Figure 3.2).

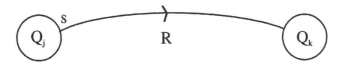

FIGURE 3.2
A graphical depiction of a state transition.

This graphical technique comes into its own when we have the possibility of multiple stimuli, responses, and state changes. For example, we might have the system shown in Figure 3.3.

This FSM behaves as follows: if it is in state Q_1 and it receives a stimulus S_1, it spits out R_1 and goes into state Q_2. If, however, it receives a stimulus S_2, it spits out R_2 and changes to state Q_3. Getting S_3, it switches to state Q_4 and produces R_3. Once in state Q_2, if it receives a stimulus S_1, it returns to state Q_1, responding with R_2, while if it receives a stimulus S_2, it stays

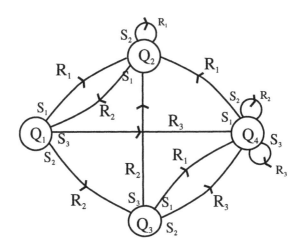

FIGURE 3.3
A complex finite state machine.

where it is and spits out R_1. The reader can figure out what happens when the machine is in states Q_3 and Q_4, and construct more complex examples for himself.

One feature of this example is that the machine was able to react to three distinct stimuli. It will suit our purposes from here on to restrict the possible stimuli to just two – the binary one and zero. This doesn't actually affect what we can do with FSMs, only how quickly we can do it; we can allow for the possibility of multiple input stimuli by feeding in a sequence of 1s and 0s, which is clearly equivalent to feeding in an arbitrary number, only in binary format. Simplifications of this kind are common in the study of FSMs and Turing machines where we are not concerned with their speed of operation.

Let me now give a specific example of an FSM that actually does something, albeit something pretty trivial – a delay machine. You feed it a stimulus and, after a pause, it responds with the same stimulus. That's all it does. Figure 3.4 shows the "state diagram" of such a delay machine.

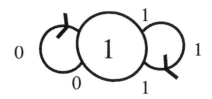

FIGURE 3.4
A delay machine.

You can hardly get a simpler machine than this! It has only two internal states and acts as a delay machine solely because we are using pulsed time and demanding that the machine's response to a stimulus at time t comes at time $t+1$. If we tell our machine to spit out whatever we put in, we will have a delay time of one unit. It is possible to increase this delay time, but it requires more complicated machines. As an exercise, try to design a delay FSM that remembers *two* units into the past: the stimulus we put in at time t is fed back to us at time $t+2$. (Incidentally, there is a sense in which such a machine can be taken as having a memory of only one time unit: if we realize that the *state* at time $t+1$ tells us the input at time t. It is often convenient to examine the state of an FSM rather than its response.)

Another way of describing the operation of FSMs is by tabulating the functions F and G we described earlier. Understanding the operation of an FSM from such a table is harder than from its state diagram, and becomes hopeless for very complex machines, but we will include it for the sake of completeness (Table 3.1).

TABLE 3.1

State Table for a Generic FSM

G	Q_0	Q_1
S_0	Q_0	Q_0
S_1	Q_1	Q_1

F	Q_0	Q_1
S_0	R_0	R_1
S_1	R_0	R_1

Now it is surprising what you can do with these things, and it is worth getting used to deciphering state diagrams so that you can appreciate this. I am going to give you a few more examples, a little more demanding than our delay machine. First up is a so-called "toggle" or "parity" machine. You feed this machine a string of 0s and 1s, and it keeps track of whether the number of 1s it has received is even or odd; that is, the parity of the string.

From the diagram in Figure 3.5, you can see that, one unit of time after you feed in the last digit, the response of the FSM tells you the parity. If

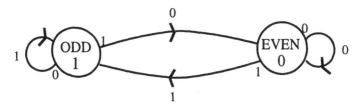

FIGURE 3.5
The parity machine.

it is a 1, the parity is odd – you have fed in an odd number of 1s. A 0 tells you that you have fed in an even number. Note that, as an alternative, the parity can be read off from the state of the machine, which I have flagged by labeling the two possible states as "odd" and "even".

Let me give you some simple problems to think about.

Problem 3.1 Suppose we feed a sequence of 1s and 0s – a binary number – into a machine. Design a machine that performs a pairwise sum of the digits, that is, one that takes the incoming digits two at a time and adds them, spitting the result out in two steps. So, if two digits come in as 00, it spits out 00; a 10 or 01 results in a 01 (1+0 = 0+1!); but a 11 results in binary 10: 1+1 = 2, in decimal, 10 in binary. I will give you a hint: the machine will require four states.

Problem 3.2 Another question you might like to address is the design of another delay machine, but this time one that remembers and returns *two* input stimuli. You can see that such a device needs four states – corresponding to the four possible inputs 00, 01, 10, and 11.

Problem 3.3 Finally, if you are feeling particularly energetic, design a two-input binary adder. I want the full works! I feed in two binary numbers, simultaneously, one bit from each at a time, *with the least significant bits first*, and the FSM, after some delay, feeds me the sum. I'm not interested in it telling me the carry, just the sum. We can schematically depict the desired behavior of the machine as follows:

Time   ⟶

Inputs   1 0 1 0 1 1

            0 1 1 0 1 0

Output = sum 1 1 0 1 0 0 (carrying 1 into the next column)

3.3 The Limitations of Finite State Machines

If you have succeeded in designing an adder, then you have created a little wonder – a simple machine that can add two numbers of any size. It is slow and inefficient, but it does its job. This is usually the case with FSMs. However, it is important to appreciate the limitations of such machines; specifically, there are many tasks they cannot perform. It is interesting to take a look at what they are. For example, it turns out that one cannot build an FSM that will multiply any two arbitrary binary numbers together. The reason for this will become clear in just a moment after we have examined a simpler example. Suppose we want to build a *parenthesis checker*. What is one of these? Imagine you have a stream of parentheses, as follows:

$$(((\ ))((\ )(\ )((\ )(\ )(\ )((\ )))(\ ))$$

The task of a parenthesis checker is to ascertain whether such an expression is "balanced": that the brackets open and close consistently throughout the expression. This is not the same as just counting the number of left and right brackets and making sure they are equal! They have to match in the correct order. This is a common problem in arithmetic and algebra, whenever we have operations nested within others. The above example, incidentally, is invalid; this one:

$$((\ )((\ )(\ ))((\ )(\ )((\ )(\ )))))$$

is valid. You might like to check in each case.

On the face of it, building a parenthesis checker seems a pretty straightforward thing to do. In many ways it is, but anything you get to implement the check would not be an FSM. Here is one way you could proceed. Starting from the left of the string, you count open brackets until you reach a close bracket. You "cancel" the close bracket with the rightmost open bracket, then move one space to the right. If you hit a close bracket, cancel it with another open bracket; if you hit an open bracket, add one to the number of open brackets you have uncanceled and move on to the next one. It is a very simple mechanism, and it will tell you whether or not your parenthesis string is OK: if you have any brackets left over after you process the rightmost one, then your string is inconsistent. So why cannot an FSM do something this simple?

The answer is that the parenthesis checker we want has to cope with *arbitrary* strings. That means, in principle, strings of arbitrary length that might contain arbitrarily large numbers of "open" brackets. Now recall that an essential feature of the machine is that it must keep track of how many open brackets remain uncanceled by closed ones at each stage of its operation; yet to do this, in the terminology of FSMs, it will need a distinct state for each distinct number of open brackets. Here lies the problem. An *arbitrary* string requires a machine with an arbitrary – that is, ultimately, infinite – number of states. Hence, no *finite* state machine will do. What will do, as we shall see, is a Turing machine – for a Turing machine is, essentially, an FSM with infinite memory capacity.

For those who think I am nitpicking, it is important to reiterate that I am discussing matters of principle. From a practical viewpoint, we can clearly build a finite state machine to check the consistency of any bracket string we are likely to encounter. Once we have set its number of states, we can ensure that we only feed it strings of an acceptable size. If we label each of its states by 32 32-bit binary numbers we can enumerate over 2^{1000} states, and hence deal with strings 2^{1000} brackets long. This is far more than we are ever likely to encounter in practice: by comparison, note that current estimates place the number of protons in the universe only of the order of 2^{200}. But from a mathematical and theoretical standpoint, this is a very different thing from having a universal parenthesis checker: it is, of course, the difference between the finite and the infinite, and when we are discussing academic matters this is important. We *can* build an FSM to add together two arbitrarily large numbers; we *cannot* build a parenthesis-checking FSM to check any string we choose. Incidentally, it is the need for an infinite memory that debars the construction of an FSM for binary multiplication.

Before getting onto Mr. Turing and his machines, I would like to say one or two more things about those with a finite number of states. One thing we looked at in detail in previous chapters was the extent to which complicated logic functions could be built out of simple, basic logic units – such as gates. A similar question arises here: is there a core set of FSMs with which all others can be built? To examine this question, we will need to examine the ways in which FSMs can be combined.

Figure 3.6 shows two machines, which I call **A** and **B**. I have linked them up in something of a crazy way, with feedback and whatnot. Don't worry if you can't see at a glance what is going on!

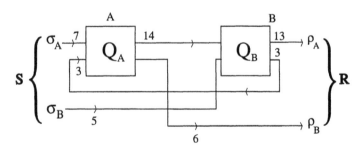

FIGURE 3.6
A composite FSM.

Let me describe what the diagram represents. In a general FSM, the input stimulus can be any binary number, as can its response. Whether the stimulus is fed in sequentially, or in parallel (e.g., on a lot of on/off lines), we can split it up into two sets. Suppose the stimulus for **A** has ten bits. We split this up into, say, a seven-bit and a three-bit stimulus. Now comes the tricky part: we take the seven-bit input to be external, fed in from outside on wire σ_A, but the three-bit input we take from the *response of machine B –* which we have also split up. In the case of **B**, we take the response to have, say, 16 bits, and three of these we re-route to **A**, the other 13 we take as output. Bear with me! What about the response of A? Again, we split this up: suppose it is 20 bits. We choose (this is all arbitrary) to feed 14 into **B** as input, and with the remaining six we bypass **B** and take them as output. The remainder of **B**'s input — whatever it may be – is fed in from outside, on wire σ_B. Let's say σ_B carries five bits.

The point of all these shenanigans is that this composite system can be represented as a single finite state machine:

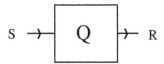

where the input stimulus is the combined binary input on wires σ_A and σ_B, and the output is the partial responses from **A** and **B**, again combined. Clearly, the machine has an input stimulus of 7+5 = 12 bits, and a response of 13+6 = 19 bits. Exactly what the thing does depends on the properties of **A** and **B**; it seems feasible that the number of internal states of this combined machine is the product of the number of states of **A** and **B**, but one must be careful about the extent to which things can be affected by feedback and the information running around the wires. What I wanted to show was how you could build an FSM from smaller ones by tying up the loose wires appropriately. You might like to see what happens if you arrange things differently – by forgetting feedback, for example. You will find that feedback is essential if you want as few constraints as possible on the size of the overall input and output bit sizes: connecting up two machines by, say, directly linking output to input not only fixes the sizes of the overall stimulus/response but also requires the component FSMs to match up in their respective outputs and inputs.

Let me return to my question: can we build any FSM out of a core set of basic FSMs? The answer turns out to be yes: in fact, we find ourselves going right back to our friends AND and NOT, which can be viewed as finite state machines themselves, and which we can actually use to build any other FSM. Let me show roughly how this is done. We will first need a bit of new notation. Let us represent a set of k signal-bearing wires by a single wire crossed with a slash, next to which we write the number k:

With this convenient diagrammatic tool, we can draw a schematic diagram of a general finite state machine (Figure 3.7).

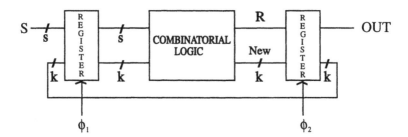

FIGURE 3.7
The general FSM.

The operation of this rather complicated-looking device is quite straightforward. It comprises two registers (such as those we constructed in Chapter 2 from clocked flip-flops) and a black box that performs certain logical functions. The input to the first register has two pieces, the stimulus S to the FSM and the state the machine is in, Q: central to our design is the fact that we can label the internal states by a binary number. In this case, the stimulus has s bits, and is fed in on s wires, while the state has k bits, fed in on k wires. (The FSM has therefore up to 2^k internal states.) Subject to timing, which I will come back to, the register passes these two inputs into the logic unit. Here is the trick. An FSM, in response to a given stimulus and being in a given state, produces a response and goes into a (possibly) new state. In terms of our current description, this simply amounts to our black box receiving two binary strings as input and producing two as output – one representing the response, the other the new state. The new state information is then fed back into the first register to prime the machine for its next stimulus. Ensuring that the FSM works is then just a matter of building a logic unit that gives the right outputs for each input, which we know is just a matter of combining ANDs and NOTs in the right way.

A quick word about timing. As we have discussed, the practicalities of circuit design mean that we have to clock the inputs and outputs of logic devices; we have to allow for the various delays in signals arriving because of finite travel times. Our FSM is no exception, and we have to connect the component registers up to two clocks as usual; the way these work is essentially the same as with standard logic circuits. The first register is clocked by ϕ_1, the second by ϕ_2, and we arrange things such that when one is on, the other is off – which we do by letting $\phi_2 = \text{NOT } \phi_1$ and hooking both up to a standard clock – and ensuring that the length of time for which each is on is more than enough to let the signals on the wires settle down. The crucial thing is to ensure that ϕ_2 is *off* while ϕ_1 is *on*, to prevent the second register sending information about the change of state to the first while it is still processing the initial state information.

Problem 3.4 Before turning to Turing machines, I will introduce you to a nice FSM problem that you might like to think about. It is called the "Firing Squad" problem. We have an arbitrarily long line of identical finite state machines that I call "soldiers". Let us say there are N of them. At one end of the line is a "general", another FSM. Here is what happens. The general shouts "Fire". The puzzle is to get all of the soldiers to fire simultaneously, in the shortest possible time, subject to the following constraints: first, time goes in units; second, the state of each FSM at time $T+1$ can only depend on the state of its next-door neighbors at time T; third, the method you come up with must be independent of N, the number of soldiers. At the beginning, each FSM is quiescent. Then the general spits out a pulse, "fire", and this acts as an input for the soldier immediately next to him. This soldier reacts in some way, enters a new state, and this in turn affects the soldier next to him, and so on down the line. All the soldiers interact in some way, yack yack yack, and at some point, they become synchronized and spit out a pulse representing their "firing". (The general, incidentally, does nothing on his own initiative after starting things off.)

There are different ways of doing this, and the time between the general issuing his order and the soldiers firing is usually found to be between $3N$ and $8N$. It is possible to prove that the soldiers cannot fire earlier than $T = 2N-2$ since there would not be enough time for all the required information to move around. Somebody has actually found a solution with this minimum time. That is very difficult though, and you should not be so ambitious. It is a nice problem, however, and I often spend time on airplanes trying to figure it out. I haven't cracked it yet.[*]

3.4 Turing Machines

Finally, we come to Turing machines. Turing's idea was to conceive of himself, or any other mathematician, as a machine, having a finite state machine in his head, and an unlimited amount of paper at his disposal to write on. It is the unlimited paper – hence effectively unbounded memory – that distinguishes a Turing machine from an FSM. Remember that some problems – parenthesis checking, multiplication – cannot be done

---

[*] See "An Optimum Solution to the Firing Squad Synchronization Problem" by Abraham Waksman, Information and Control, **9**, 66–78 (1966) if you want to see the answer. [Editor]

by finite state machines, because, by definition, they lack an unlimited memory capacity. This restriction does not apply to Turing machines. Note that we are not saying that the amount of paper attached to such a machine *is* infinite; at any given stage it will be finite, but we have the option of adding to the pile whenever we need more. Hence, our use of the word "unlimited".

Turing machines can be described in many ways, but we will adopt the picture that is perhaps most common. We envisage a little machine, with a finite number of internal states, that moves over a length of tape. This tape is how we choose to arrange our paper. It is sectioned off into cells, in each of which might be found a symbol. The action of the machine is simple, and similar to that of an FSM, it starts off in a certain state, looking at the contents of a cell. Depending on the state, and the cell contents, it might erase the contents of the cell and write something new or leave the cell as it is (to ensure uniformity of action, we view this as erasing the contents and writing them back in again). Whatever it does, it next moves one cell to the left or right, and changes to a new internal state. It might look something like Figure 3.8.

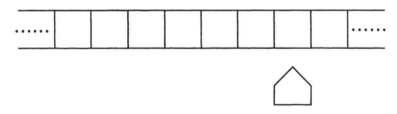

FIGURE 3.8
A Turing machine.

We can see how similar the Turing machine is to an FSM. Like an FSM, it has internal states. Reading the contents of a cell is like a stimulus, and overwriting the contents is like a response, as is moving left or right. The restriction that the machine move only one square at a time is not essential; it just makes it more primitive, which is what we want. One feature of a Turing machine that *is* essential is that it is able to move both left and right. You can show (although you might want to wait until you are more familiar with the ideas) that a Turing machine that can only move in one direction is just a finite state machine, with all its limitations.

Now we are going to start by insisting that only a finite part of the tape has any writing on it. On either side of this region, the tape is blank. We first tell the machine where to start, and this is at time T. Its later behavior, at a time $T+1$ say (Turing machines operate on pulsed time like FSMs), is specified by three functions, each of which depends on the state Q_i at time

T and the symbol S_i it has just read: these are its new state, Q_j, the symbol it writes, S_j, and the direction of its subsequent motion, D. We can write:

$$Q_j = F(Q_i, S_i)$$

$$S_j = G(Q_i, S_i) \qquad (3.3)$$

$$D = D(Q_i, S_i)$$

This list is just like the specification of an FSM but with the extra function D. The complete machine is fully described by these functions, which you can view as one giant (and finite) look-up list of "quintuples" – a fancy name for the set of five functions we have defined, two at time T (Q_i and S_i), and three at $T+1$ (Q_j, S_j, and D). All you do now is stick in some data – which you do by writing on the tape and letting the machine look at it – tell the machine where to start and leave it to get on with it. The idea is that the machine will finish up by printing the result of its calculation somewhere on the tape for you to peruse at your leisure. Note that for it to do this, you have to give it instructions as to when it is to halt or stop. This seems pretty trivial, but as we will see later, matters of "halting" hide some very important, and very profound, issues in computation.

Before giving you some concrete examples of Turing machines, let me remind you of why we are looking at them. I have said that finding an effective procedure for doing a problem is equivalent to finding a Turing machine that could solve it. This does not seem much of an insight until we realize that among the list of all Turing machines, by which I mean all lists of quintuples, there exists a very special kind, a *universal Turing machine*, which can do anything any other Turing machine can do! Specifically, a UTM is an imitator, mimicking the problem-solving activities of simpler Turing machines. (I say "a" UTM, rather than "the" UTM since, while all UTMs are computationally equivalent, they can be built in many different ways). Suppose we have a Turing machine, defined by some list of quintuples, which computes a particular output when we give it a particular set of input data. We get a UTM to imitate this process by feeding it a description of the Turing machine – that is, telling the UTM about the machine's quintuple list – and the input data, both of which we do by writing them on the UTM's tape in some language it understands, in the same way we feed data into any Turing machine. We also tell the UTM where each begins and ends.* The UTM's internal program then takes this information and mimics the action of the original machine. Eventually, it spits out the result of the calculation: that is, the output of the original

---

* The section of the UTM's tape containing information about the machine it is imitating is usually referred to as the "pseudo-tape". [RPF]

Turing machine. What is impressive about a UTM is that *all* we have to do is give it a list of quintuples and some initial data – its own set of defining quintuples suffice for it to mimic any other machine. We don't have to change them for specific cases.* Why such machines are important to us is because it turns out that, if you try to get a UTM to impersonate *itself*, you end up discovering that there are some problems that no Turing machine – and hence no mathematician – can solve!

Let us now look at a few real Turing machines. The first, and one of the simplest, is related to a finite state machine we have already examined – a parity counter. We feed the machine a binary string and we want it to tell us whether the number of 1s in the string is odd or even. This is schematically shown in Figure 3.9.

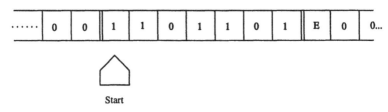

FIGURE 3.9
Input tape for the parity counter.

We begin by writing the input data, the binary string, onto the tape as shown; each cell of the tape holds one digit. The "tape head" of the machine rests at the far left of the string, on the first digit, and we define the machine to be in state Q_0. To the left of the string are nothing but zeroes, and to the right, more zeroes – although we separate these from the string with a letter E, for "end", so that the machine does not assume they are part of it.

The operation of the machine, which we will shortly translate into quintuples, is as follows. The state of the machine tells us the parity of the string. The machine starts off in state Q_0, equal to even parity, as it has not yet encountered any 1s. If it encounters a zero, it stays in state Q_0 and moves one space to the right. The state does not change because the parity does not change when it hits a zero. However, if it hits a 1, the machine erases it, replaces it with a zero, moves one space to the right, and enters a state Q_1. Now if it hits a zero, it stays in state Q_1 and moves a space to the right, as before. If it hits a 1, it erases it, putting a zero in its place, and moves to the right, this time reverting to state Q_0. You should now have an idea about what is happening. The machine works its way across the string from left to right, changing its state whenever it encounters a 1,

---
* We will actually construct a UTM later. [RPF]

and leaving a string of 0s behind. If the machine is in state Q_0 when it kills the last digit of the string, then the string has even parity; if it is in state Q_1, it is odd. How does the machine tell us the parity? Simple – we include a rule telling the machine what to do if it reads an E. If it is in state Q_0 and reads E, it erases E and writes "0", meaning even parity. In state Q_1, it overwrites E with a "1", denoting odd parity. In both cases, it then enters a new state Q_H, meaning "halt". It does not need to move to the right or left. We examine the tape, and the digit directly above the head is the answer to our question. We end up with the situation shown in Figure 3.10.

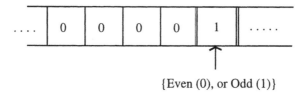

{Even (0), or Odd (1)}

FIGURE 3.10
Output tape from the parity counter.

The quintuples for this machine are straightforwardly written out (Table 3.2).

TABLE 3.2
Quintuples for the Parity Counter

Initial state	Read	New State	Write	Direction of move
0	0	0	0	R
0	1	1	0	R
1	0	1	0	R
1	1	0	0	R
0	E	H(alt)	0	-
1	E	H	1	-

Now this device is rather dumb, and we have already seen that we could solve the parity problem with a finite state machine (note here how our Turing machine has only moved in one direction!). We will shortly demonstrate the superiority of Mr. Turing's creations by building a parenthesis checker with them, something that we have seen cannot be done with an FSM, but first let me introduce some new diagrammatics that will make it easier for us to understand how these machines work without tying ourselves in knots wading through quintuple lists.

The idea is, unsurprisingly, similar to that we adopted with FSMs. In fact, the only real difference in the diagrams is that we have to somehow include the direction of motion of the head after it has overwritten a cell, and we have to build in start and halt conditions. In all other respects, the diagrams resemble those for FSMs. Take a look at Figure 3.11, which describes our parity counter:

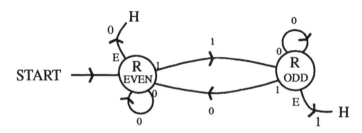

FIGURE 3.11
A Turing machine parity counter.

This is essentially the same as Figure 3.5, the FSM which does the same job. Where the FSM has a stimulus, the TM has the contents of a cell. In these diagrams, both are written at the point of contact of lines and circles. Where the FSM spits out a response, which we wrote on the arrow linking states, the TM overwrites the cell contents, what it writes being noted on the arrow. The state labels of both FSMs and TMs are written inside the circles. The major differences are that, first, we have to know where the machine starts, which we do by adding an external arrow as shown; and we have to show when it stops, which we do by attaching another arrow to each state to allow for the machine reading E, each arrow terminating in a "Halt". More subtly, we also have to describe the direction of its motion after each operation. It turns out that machines whose direction of motion depends *only* on their internal state – and not on the symbols they read – are not fundamentally less capable of carrying out computations than more general machines that allow the tape symbols to influence the direction of motion. I will thus restrict myself to machines where motion to the right or left depends solely on the internal state. This enables me to solve the diagrammatic problem with ease: just write L or R, as appropriate, inside the state box. In this case, both states are associated with movement to the right.

I have gone on at some length about the rather dumb parity machine as it is important that you familiarize yourself with the basic mechanics and notation of Turing machines. Let me now look at a more interesting problem, that of building a parenthesis checker. This will illustrate the superiority of Turing machines over finite state machines. Suppose we provide our Turing machine with a tape, in each cell of which is written a parenthesis (Figure 3.12).

....E(())()(()))(()E....

FIGURE 3.12
Input tape to the parenthesis checker.

Each end of the string is marked with a symbol E. This is obviously the simplest way of representing the string. How do we get the machine to check its validity? One way is as follows. I will describe things in words first and come back to discuss states and diagrams and so forth in a moment. The machine starts at the far-left end of the string. It runs through all the left brackets until it comes to a right bracket. It then overwrites this right bracket with an X – or any other symbol you choose – and then moves one square to the left. It is now on a left bracket. It overwrites this with an X, too. It has now canceled a pair of brackets. The key property of the Xs is that the machine doesn't care about them; they are invisible. After having canceled a pair in this way, the machine moves right again, passing through any Xs and left brackets, until it hits a right bracket. It then does its stuff with the X again. As you can see, in this way, the machine systematically cancels pairs of brackets. Sooner or later, the head of the machine will hit an E – it could be either one – and then comes the moment of truth. When this happens, the machine has to check whether the tape between the two Es contains only Xs or some uncanceled brackets too. If the former, the string is valid, and the machine prints (say) a 1 somewhere to tell us this; if the latter, the machine prints 0, telling us the string is invalid. Of course, after printing, the machine is told to halt.

If you think about it, this very simple procedure will check out any parenthesis string, irrespective of size. The functioning of this machine is encapsulated by the state diagram of Figure 3.13 (following Minsky [1967]).

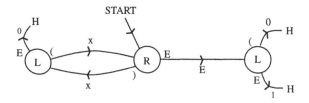

FIGURE 3.13
The parenthesis checker state diagram.

Note how the diagram differs from that for an FSM: we have to include start and stop instructions, and also direction of motion indicators. In fact, this machine, unlike the parity counter, requires two different left-moving states.

Now that you have some grasp of the basic ideas, you might like to try and design a few Turing machines for yourself. Here are some example problems to get you thinking.

Problem 3.5 Design a unary multiplier. "Unary" numbers are numbers written in base 1 and are even more primitive than binary. In this base, we have only the digit 1, and a number N is written as a string of N 1s: 1 = 1, 2(base 10) = 11(base 1), 3 = 111, 4 = 1111, and so on. I would like you to design a Turing machine to multiply together any two unary numbers. Start with the input string:

$$\overline{\quad 0\,0\,...\,E\ 1\,1\,1\,1\,....\,1\ B\ 1\,1\,1\,1\ ....\,1\,1\ E\,...\,0\,0 \quad}$$
$$\qquad\qquad\quad m\qquad\qquad\qquad\quad n$$

which codes the numbers being multiplied, m and n and separates the two numbers with the symbol B. The goal is to end up with a tape that gives you mn. It might look something like this:

$$\overline{\quad ...\,0\,0\ E\ 0\,0\,0\,0\,...\,0\ B\ X\,X\,X\,...\,X\ E\ Y\,Y\,Y\,Y\,...\,Y\ 0\,0\,... \quad}$$
$$\qquad\qquad m\qquad\qquad\quad n\qquad\qquad mn$$

where Y is some symbol distinct from 0, 1, X, E, and B. You can consider the given tape structure a strong hint as one way in which you could solve the problem!

Problem 3.6 We have discussed binary adders before. I would now like you to design a Turing machine to add two binary numbers, but only for the case where they have the same number of bits (this makes it easier). You can start with the initial tape:

$$\overline{\quad ...\,0\,0\ A\ 1\,1\,0\,1\,..\,1\ B\ 1\,0\,0\,1\,..\,0\ C\ 0\,0\,0\,... \quad}$$
$$\qquad\qquad\quad m\qquad\qquad\quad n$$

for numbers m and n with the field of the two numbers delineated by the symbols A, B, and C. I will leave it to you to decide where the machine starts, how it proceeds, what its final output looks like, where it appears, and so on.

Problem 3.7 If you're finding these problems too easy, here's one that is much harder: design a Turing machine for a binary multiplier!

Problem 3.8 This last problem is neat: design a unary to binary converter. That is, if you feed the machine a string of 1s representing a unary number, it gives you that number converted to binary. The secret to this problem lies in the mathematics of divisors and remainders. Consider what we mean when we talk of the binary form of an n-bit number $N = N_n N_{n-1}...N_1 N_0$. By definition we have:

$$N = N_n.2^n + N_{n-1}.2^{n-1} + ... + N_1.2 + N_0$$

We start with N written in unary – i.e., a string of N 1s – and we want to find the coefficients N_i, the digits in binary. The rightmost digit, N_0, can be found by dividing N by two, and noting the remainder, since:

$$N = 2.X + N_0$$

with X easily ascertained. To find N_1, we get rid of N_0, and use the fact that:

$$X = 2.Y + N_1$$

That is, we divide X by two and note the remainder – N_1. We just keep doing this, shrinking the number down by dividing by two and noting the remainder, until we have the binary result. Note that, since N is an n-bit number, by definition N_n *must be 1*.

If we are given the number N in unary form, we can simulate the above procedure by grouping the 1s off pairwise and looking at what is left. Let us take a concrete example. Use the number nine in base ten, or 111111111 in unary. Pair up the 1s:

$$(11)\ (11)\ (11)\ (11)\ 1$$

Clearly, this is just like dividing by two. There is an isolated digit on the right. This tells us that N_0 is 1. To find N_1, we scratch the right-hand 1 and pair up the pairs in the remaining string:

$$(11\ 11)\ (11\ 11).$$

This time, there is no remainder: N_1 is 0. Similarly, we find that N_2 is 0. We have now paired up all our pairs and pairs of pairs, and the only thing left to do is tag a 1, for N_3, to the left of the number, giving us 111111111 (unary) = 1001 (binary).

I will leave it up to you to implement this algorithm with a Turing machine. You have to get the thing to pair off digits, mark them as pairs, and check the remainder; and then come back to the beginning and mark off pairs of pairs and so on. Marking pairs is probably best done by starting at the left end of the string and going to the right, striking out every other digit and replacing it with an X symbol. When the machine gets to go through the string again, it ignores the Xs and strikes out every other 1 again. This method, suitably refined, will work! I leave it to you to figure out the details. Don't forget that you have to get the machine to start, perform the conversion, write its output, and then stop.

3.5 More on Turing Machines

I would now like to take a look at a fairly complicated Turing machine that bears on a different aspect of computing. Earlier in these lectures, I pointed out that computers were more paper pushers than calculators, and it would be nice to see if we can build a Turing machine that performs filing, rather than arithmetic, functions. The most primitive such function is looking up information in a file, and that is what we are going to examine next. We want a machine that first locates a file in a file system, then reads its contents, and finally relays these contents to us.[*]

We will employ the following Turing "filing system", or tape (Figure 3.14), which we are to feed into our machine:

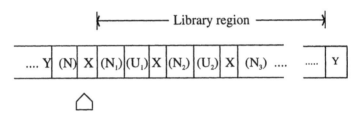

FIGURE 3.14
Input tape to the locating machine.

This is a bit schematic. The X-symbols this time play the role of segregating various file entries; there is one entry between each pair of Xs. Each entry comprises a name (or address), "N", and contents, "U", both of which we take to be binary strings, one digit per tape square as usual. We have attached to the left-hand end of the tape the name of a file that we want the machine to read for us, and denoted the left end of the tape by a symbol Y. To the left of this is a string of zeroes; the same is true at the right-hand end of the tape. The machine is to start where marked to the right of the name N of the file we want to find.

The first task confronting the machine is that of locating the right file. It does this by systematically comparing each file name in the list with the target name, working from left to right, until it finds the correct one. How should it do this? For ease of understanding, suppose we have the following filing tape (Figure 3.15):

---
[*] Our discussion closely follows Minsky [1967]. [RPF]

... 0 0 0 Y 101 X 001 011 X 101 110 X 111 000 Y 0 0 0 ...

N   U

FIGURE 3.15
A sample filing tape.

For convenience, we are taking both the name strings and the data strings to be of the same length – three bits. We want to read the contents of file (101), which we'll call the target file. Now it might seem that the best thing to do is the following: assign to each possible target a distinct state of the Turing machine. This will give us at most eight states. The machine starts in the state 101 dictated by the target file name and goes to the first file from the left and looks at the name. If there is a match, all well and good. If not, it goes to the next file on the right, checks that, and so on. In this way, the machine smoothly moves from left to right until it hits the correct address. However, the problem with such a machine is that it has only eight states and will only be any good for three-bit filing systems: it has no universality of application. We want a single machine that can handle any size of filename. To achieve this, the machine must compare each filename with the target on a sequential, digit-by-digit basis, laboriously shuttling between the two until a mismatched digit is found, in which case it goes onto the next filename, or until a complete match is found, when (say) it returns to its starting point. To keep track of those parts of the tape it has already considered, the machine would do the now-familiar trick of overwriting digits with symbols that it subsequently ignores, just as we did with the parenthesis checker. By assigning different symbols to 0s and 1s – As and Bs, say – we can keep track of which were 0s and 1s; if we wanted to come along tomorrow and use the file again, we could, only we would find it written in a different alphabet. We could then reconstruct the entire original file by overwriting the new symbols with 0s and 1s.

Minsky's solution for a locating Turing machine is shown in Figure 3.16.

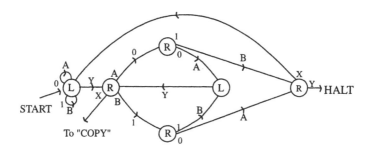

FIGURE 3.16
The locating machine.

There is a loose end in this diagram, pointing to "copy". This represents the stage at which the machine has located the correct filename and is wondering what to do next. We will shortly show how we are going to get it to copy the information in the file to a point of our choosing on the tape.

For the moment, let us stick with our location machine and look in more detail at how it works. The head starts on the first X to the right of the target number. As the loop instructions show, the machine then heads left, changing the 0s and 1s in the target to As and Bs, respectively. This may seem a little bizarre, but there is a point to it, as we will see. Eventually, the machine hits the Y. It then goes into a new state, and as is clear from the diagram, it will start moving right. It will first encounter one of the As or Bs it has just written: it overwrites this with the original digit (this *definitely* seems bizarre, but it will make sense!), a 0 or 1, and moves right again. It now enters one of two states in which it will only recognize a 0 or 1: not an A or B. If it hits an A or B, it will ignore it, keeping on moving right – in other words, it is going to pass right through the remainder of the rewritten target string, having in a sense "noted" the first digit of the string. This is why we overwrote the 0s and 1s of the string with As and Bs. It will also pass straight through the X it encounters and go on to the first filename to be checked.

Now comes the crucial sequence of operations. The machine is going to hit either a 1 or a 0, and how it reacts depends on how it has been primed – i.e., on the state it is in as a result of reading the first target digit. There are two possibilities. First, if the digit it hits is different from the first target digit, so the filenames do not match from the outset, the machine overwrites the digit as appropriate, and then moves to the right until it hits the next X, denoting the end of the file. It then starts to move to the left, overwriting the contents of the rest of the file with As and Bs. It passes through the leftmost X, zips through the target filename (As and ns are invisible to it), changes the first digit to an A or B, and hits Y. This is a cue for the whole process to start again: only now it goes to the next filename. Sooner or later, the first target digit and that of the checked filename will match, and this is the second possibility we must consider.

When a match occurs, the machine overwrites the matching digit, and enters a state in which it moves back left until it encounters the Y. Then, it goes forwards, overwrites the second target symbol with the correct digit, and then moves on to the files. It checks the second digit for a match, and so it goes on. Working through the machine diagram, you should be able to convince yourself that the tape above would ultimately be converted into the tape of Figure 3.17.

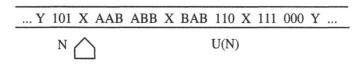

FIGURE 3.17
Output tape from the locating machine.

Note that the head has returned to its starting point, and the effect of its activities has been to change all 0s and 1s between the start and the end of the desired filename (but not the contents of the file) to *A*s and *B*s. (There is the important possibility that the target filename cannot be found, because we have typed it in wrongly, say, and in this case, the machine head will end up on the *Y* at the far right; as the diagram indicates, at this juncture, it is instructed to "Halt".)

As I have said, there is a "loose wire" on our diagram, representing a feed to a copy machine: we have our file, now we want to know what to do with it! True to the spirit of Turing machines, we are going to copy it slowly and laboriously to another part of the tape. That is, *you* are: the copying machine is shown in Figure 3.18, and its input tape is the output tape of the location machine. Have fun figuring out how it works!

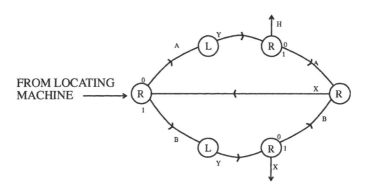

FIGURE 3.18
The copy machine.

A cute feature of this machine is that it copies the contents of the file into the block containing the target filename on the original tape; that is, the target string is overwritten. (We can do this because we chose to have filenames and contents the same size.) The end result of this machine operating on its tape is the tape in Figure 3.19.

... Y BBA X AAB ABB X BAB BBA X B11 000 Y ...

U

FIGURE 3.19
Output tape from the copy machine.

I will finish this section by giving you a couple more Turing problems.

Problem 3.9 Make a Turing machine that starts with a blank tape and ends up with all the binary numbers written on it in succession, separated by "Ys", with the restriction that after you write the terminating Y, you never change the number again. An additional restriction you might impose is that the machine does not even look at terminated numbers.

Problem 3.10 Design a machine that recognizes only, and all, sequences of the form

$$1011001110001111 0000.....1^n 0^n.$$

That the machine has "accepted" such a tape is indicated by its halting and leaving the tape blank after its machinations. More generally, we define an arbitrary sequence as "acceptable" by a Turing machine if the machine eventually halts with a blank tape. We can extend this notion to cover finite state machines. Design a Turing machine that accepts exactly the set of sequences accepted by any FSM. (Hint: use the FSM functions F and G to make Turing quintuples.)

3.6 Universal Turing Machines and the Halting Problem

Let us return to the reason why we are studying Turing machines. I said earlier that if you had an effective procedure for doing some computation, then that was equivalent to it being possible in principle to find a Turing machine to do the same computation. It is useful to talk in terms of functions. Suppose we start with a variable x, and we take a function of that variable, $F(x)$. We say that $F(x)$ is *Turing computable* if we can find a Turing machine $\mathbf{T}_F$ which, if fed a tape on which x is written, in some representation – binary, unary, whatever – will eventually halt with $F(x)$ printed on the tape. Every other effective procedure that anyone else has been able to cook up has turned out to be equivalent to this – the general recursive functions are Turing computable, and vice versa – so we can take "Turing computable" to be an effective synonym for "computable".

Now it may be the case that for some values of x, the Turing machine might not halt. This is weird behavior, but it might happen. Many functions – such as x^2 – are called "complete", meaning that for all values of x we plug into our machine, it will halt with the value of the function written on the tape. Functions for which this is not true are called "partial". In such cases, we have to alter our operational definition of the function as follows: if, for a value x, the machine stops, we define the value of the function to be $F(x)$; if the machine does not stop, we *define* the value of the function to be *zero*. This does *not* mean that if we put x into F we get zero, in the way that putting $x = 3$ in the function $(x–3)$ gives us zero. Here, "zero" is just a useful label we attach to $F(x)$ when our Turing machine does not quit its computing. This redefined function is complete in the sense that we can assign *some* numerical value to it for *any* x.

A question naturally arises: can we say, in advance, which values of x might cause our machine to hang up? In some cases, the answer is yes. For example, there may be times when the machine goes into a recognizable infinite loop, perhaps shuttling between a couple of states and not achieving anything, and we can then say for sure that it will never stop. But in general, we cannot say in advance when a particular value of x is going to give us trouble! *Put another way, it is not possible to construct a computable function that predicts whether or not the machine* $\mathbf{T}_F$ *halts with input x.* In seeing why this is so, we shall appreciate the power of Mr. Turing's little machines.

I have flagged what is to follow in the penultimate sentence of the previous paragraph. I have raised the question of whether there is a computable function that will tell us whether or not $\mathbf{T_F}$ halts for input x. But, if there is such a function, by definition, it must be describable by a Turing machine. This concept, of Turing machines telling us about other Turing machines, is central to the topic of *universal* Turing machines to which we now turn.

We can pose the question we have set ourselves in the following way. Suppose we have a machine that we call **D**. As input, **D** takes a tape that contains information about $\mathbf{T_F}$ and $\mathbf{T_F}$'s initial tape (that is, information about x). Machine **D** is required to tell us whether $\mathbf{T_F}$ will halt or not: yes or no. Importantly, **D** must always write the answer and halt, itself. What we now do is introduce another machine **Z** that reacts to the output from **D** in the following way:

> If $\mathbf{T_F}$ halts (**D** says "yes"), then **Z** does not.
> If $\mathbf{T_F}$ does not halt (**D** says "no"), then **Z** does.

We then get **Z** to operate on itself and find a contradiction! Let us expand on this argument.

To begin our quest for **D**, we first need to look at how we get one Turing machine to understand the workings of another. We need to characterize a given machine **T**, and its tape *t*; there are several ways of doing this. We will choose a description in terms of quintuples (Table 3.3).

TABLE 3.3

Quintuple Description of a Turing Machine

Initial		Final		
State	Read	State	Write	Move
Q	S	Q'	S'	d (= L or R)

We want to build a universal machine that is capable of imitating any **T**. In other words, if we feed it information about **T** and about **T**'s tape *t*, our universal machine spits out the result of **T** acting on *t*. We will characterize our universal machine – call it **U** – in terms of quintuples in a similar fashion to T. Let these quintuples for **U** be written $(q, s; q', s', d')$ and note that they must suffice for all possible machines **T** that we want **U** to imitate: q, s, etc. must not depend on the specifics of **T**. A constraint we shall impose on our machines is that the tape symbols S, S', s, s' must be binary numbers. An arbitrary Turing machine **T** will come with an arbitrary set of possible symbols, but with thought you should be able to see that we can always label the distinct symbols by binary numbers and work with these (e.g., if we had eight symbols, each could be redescribed by a three-bit binary string).[*]

The basic behavior of **U** is simple enough to describe. (Our discussion again closely follows Minsky [1967].) We need **U** to imitate **T** step by step, keeping a record of the state of **T**'s tape at each stage. It must note the state of **T** at each point, and by examining its simulated T-tape, it can inform itself what **T** would read at any given stage. By looking at the description it has of **T**, **U** can find out what **T** is supposed to do next. Minsky nicely relates this process to what *you* would do when using a quintuple list and a tape to figure out what a Turing machine does. The universal Turing machine **U** is just a slower version of you!

Let us supply **U** with the tape shown in Figure 3.20.

---

[*] In fact, as an exercise, examine how you would reprogram a Turing machine **T** that operated with 2^n symbols to become a machine **T**' operating on 0 and 1. Hint: where **T** had to read one symbol at a time, **T**' has to read n. [RPF]

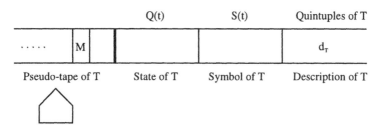

		Q(t)	S(t)	Quintuples of T
. . . . .	M			d_T
Pseudo-tape of T		State of T	Symbol of T	Description of T

FIGURE 3.20
Input tape to the universal Turing machine.

The infinite "pseudo-tape" on the left is **U**'s working space, where **U** keeps track of what **T**'s tape looks like at each stage of its simulation. Choosing to have it infinite only to the left is not essential but simplifies things. The marker M tells **U** where the tape head of **T** currently is on t. To the right of this working space is a segment of tape containing the state of **T**; then, next right is a segment containing the symbol just read by **T**; and finally, to the right again, is a region containing the description of **T**. This description of **T**, which we denote as d_T, comprises a sequential listing of the quintuples of **T**, written as a binary sequence (Figure 3.21).

Quintuple			Quintuple			Quintuple		
X 0 0 0 0 0 0 0 X	· ·	·	0 0 0 0 0 0 0 X	· ·	X	0 0 0 0 0 0 0 0 Y		
1 1 1 1 1 1 1			1 1 1 1 1 1 1			1 1 1 1 1 1 1		

$Q_1 \; S_1 \; Q_{11} \; S_{11} D_{11}$ $Q_i \; S_j \; Q_{ij} \; S_{ij} D_{ij}$

FIGURE 3.21
The description d_T of T for U's tape.

Each quintuple is segregated from the next by the symbol X. To start **U** off, we need to tell it **T**'s initial state Q_0 and the symbol S_0 it reads first. Let us assume that **U**'s tape head is initially over the leftmost X as shown in Figure 3.22.

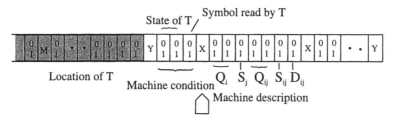

State of T / Symbol read by T

0 0 M 0 0 0 0 0 Y	0 0 0 X	0 0 0 0 0 0 0 X	0 0 X	· ·	Y
1 1 1 1 1 1	1 1 1	1 1 1 1 1 1 1	1 1		

Location of T
Machine condition $Q_i \; S_j \; Q_{ij} \; S_{ij} D_{ij}$
Machine description

FIGURE 3.22
Starting position for the tape head of U.

Essential to the operation of **U** are the locating and copying operations we described earlier. In general terms, **U** operates as follows. First, **U** looks in the section of the tape describing the action of **T** for a given Q and S, exactly as we did with the locating machine: the set (Q,S) can be regarded as the filename of the file containing the relevant quintuple. As with the locator, on its way this operation changes all the 0s and 1s it encounters to As and Bs. After it finds the relevant pair (and changes them to As and Bs), it returns to the leftmost X.

The next stage involves the copy machine. **U** moves to the right until it hits the first set of 0s and 1s; because of the way we have set up the tape, these represent the three remaining parts of the quintuple specified by Q and S. These are the new state of **T**, the symbol it writes (on the pseudo-tape in position M) and its subsequent direction of motion. The machine then copies As and Bs representing both the new Q and new S into the machine condition region in the middle of the tape. It remembers the direction of motion d (L or R, represented as A or B). The machine now heads left until it reaches M. Once there, it erases M and temporarily overwrites it with the direction d (A or B). It then moves right, changing all As and Bs to 0s and 1s on the way (leaving an A or B in M's old location). Finally, it moves to the immediate left of the leftmost X, erases the symbol S that is there (but remembers it) and prints the special symbol V in its place (this is all that V is used for).

The machine now enters its final phase. It shifts left until it encounters the A or B that we stored in M; this represents the direction d in which **T** should next move. The machine overwrites the A or B with the S it has remembered, and then moves left or right depending on the instruction d. It reads the symbol of the square it is now on, remembers it and prints an M in its place. It then shifts right until it reaches the V, which it replaces with the remembered symbol. Now the sequence starts all over again.

What the machine has done is simulate one cycle of **T**'s operation: it has started off in a certain state Q and a given input symbol S; it has then changed state, written a new symbol, and moved on to the next symbol dictated by **T**. **U** continues like this until it has mimicked **T** completely. Importantly, **U** has a halt state: it recognizes when **T** has halted and proceeds to stop itself.

The description of **U** given above, due to Minsky, requires U to have eight symbols and 23 states. So that you can appreciate the beauty of his machine, we reproduce it in full in Figure 3.23. You should not find it too hard to break it down into its constituent sub-machines.

It is possible to build a UTM with the same number of symbols but just six states. If one wants to get tricky, there are ways of using the same state

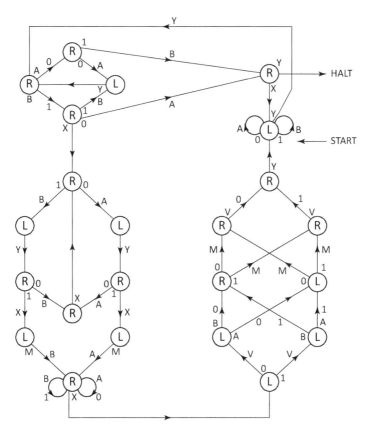

FIGURE 3.23
A universal Turing machine.

for more than one purpose – to minimize the number of states required. A UTM can be built with just two states and lots of symbols, or two symbols and lots of states. It is surprising that such a general-purpose machine should require so few parts for its description; surely a machine that can do *everything* should be enormously complicated? The surprising answer is that it's not! How efficient one can make a UTM is an entertaining question but has no deep significance.

Let us now turn to the real reason why we have been interested in demonstrating the existence of a UTM. We have asked whether it is possible to build a machine that will tell us whether a Turing machine **T** with tape *t* will halt, for all **T** and *t*. We can clearly rephrase this as a halting problem for a universal machine **U**. Let us define a new machine **D**, which is just **U** with the added property that it tells us whether or not **T** halts with tape *t*, and that it can do this for all machines **T** and all tapes *t* (Figure 3.24).

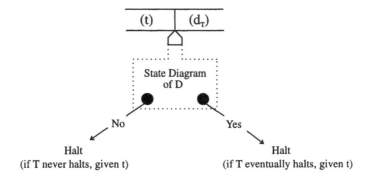

FIGURE 3.24
Universal machine D with tape t and d$_T$.

In other words, **D** *always* halts with an answer. Can such a machine exist? The answer is no! We can actually show that **D** is an impossible dream, and we do this by picking a machine **T** and a tape *t*, for which **D** cannot do what it is supposed to.

Information about **T** and *t* are fed into a universal machine in the form d_T, the quintuple description of **T**, and the information on the tape *t* (see Figure 3.24). Now, for no apparent reason, let us see what happens if we let the tape *t* contain the description d_T. We now enhance our machine **D** slightly and introduce another machine **E**. This new machine only requires as input a tape containing d_T; it then copies d_T onto a blank part of the tape and now behaves like machine **D** with an input tape containing $t = d_T$ and d_T. **E** will now behave the same way as **D** and halt giving the answer "yes" if **T** halts when reading its own description: otherwise, **E** will answer "no" (Figure 3.25). Whatever the case, **E** always halts.

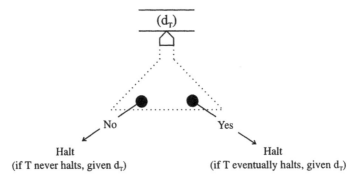

FIGURE 3.25
Universal machine E with input tape d$_T$.

Now we introduce a modified version of **E** that we shall call **Z**. Our new machine **Z** has two extra states that are used to prevent **Z** from halting if **E** takes the "yes" route (Figure 3.26).

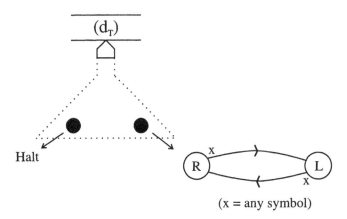

$(x = \text{any symbol})$

FIGURE 3.26
Universal machine Z.

Thus, **Z** has the property that, if **E** spits out the answer "yes", it does not halt; whereas if **E** spits out "no", it also gives us a "no" and does halt (i.e., **Z** = **E** in this case). So, **Z** halts when we feed it d_T, if **T** applied to d_T does *not* halt, but does not halt if **T** applied to d_T *does*. Now comes the crucial step. Let us write a description d_z for **Z**, and substitute **Z** for **T** in the foregoing argument. We then deduce that:

Z applied to d_z halts if and only if **Z** applied to d_z does not halt.

This is a clear contradiction! Going back through our argument, we find that it is our assumption that **D** exists that is wrong. *So there are some computational problems (e.g., determining whether a UTM will halt) that cannot be solved by any Turing machine.* This is Turing's main result.

3.7 Computability

There must be many uncomputable functions. How many are there? We can gain some insight into this by considering a counting argument. Consider computable real numbers: by which we mean those whose binary expansions can be printed on a tape, whether the machine halts or not. We can show that there are many more real numbers than computable real numbers since the latter are countable, while the former are not. We call a set "countable" if we can put its elements in one-to-one correspondence with elements of the set of positive integers; that is, if we can label each set member by a unique integer. Two examples of countable sets are the even and rational numbers:

Even numbers	0	2	4	6	8	10....
	0	1	2	3	4	5

Rational numbers	(1/2)	(1/3	2/3)	(1/4	2/4	3/4) ....
	1	2	3	4	5	6

The real numbers, however, are not countable. We can supply a neat proof of this as follows. Let us suppose the opposite. Then we would be able to pair off the reals with the integers in some way, say as follows:

Integer	Real
1	0.1̲24
2	0.01̲5
3	0.536̲92
4	0.8003̲444
5	0.33410̲5011
6	0.3425...

The exact assignment of real numbers to integers, and we have chosen a weird one here, is arbitrary; as long as we have one real number per integer, and all the reals are accounted for, we are OK. However, this cannot be so! To see why, we will find a real number that cannot be on our list. In the above list, I have underlined certain digits: the first digit of the first number, the second digit of the second, the third of the third, and so on. We define a new number using these: all we require is that the nth digit of this number *differs* from the nth digit in our list. The real number:

$$0.22741...$$

going on forever is just such a number. We have obtained this by adding one to each of the underlined digits. (We can include the rule "9+1=0" to make this a consistent procedure or we can use other procedures entirely to generate new real numbers.) What have we achieved? By construction, the above number differs from the mth number in our correspondence list in its mth digit, and this is true for all m – that is, for all integers. Hence, we have found a real number that cannot appear on our list. So by "diagonalization" as it is called (referring to the "diagonal" line we can draw through all of the underlined numbers above) we have shown that the real numbers are not countable.

Turing machines, however, *are* countable. To see this, consider the tape description d_T of a machine **T**. We can consider this to be a string of binary symbols unique to the machine if we ignore the spacings between quintuple listings. The resulting binary number serves to uniquely label the machine by an element of the set of integers. On the other hand, if we define a function $f(n)$ to be 1 if the nth Turing machine halts and 0 otherwise, then clearly this function is not computable, as we have seen from the Halting Problem. There are many other examples.

Let us return to the subject of effective procedures and make a few comments. Although we have tended to portray effective procedures as algorithms that enable us to calculate things, in reality, many such procedures are of little practical use – they might require too much tape for their execution, for example, or some other extravagant use of resources. A procedure might take the age of the universe to complete yet still be technically "effective". In practice, we want procedures that are not just effective but also efficient. The word "efficient", of course, is not easy to define precisely and so we end up leaving the clean and unambiguous world of logic and entering that of the real world of the comparatively dirty and vague – or exciting and interesting – depending on your viewpoint! Many problems in "artificial intelligence", such as face recognition, involve effective procedures that are not efficient – and in some cases, they are not even very effective!

Sometimes we do not strictly need effective procedures at all. It might be the case, for example, that you can ask a question and, while I cannot give you a sure answer, I can answer it with a probability of correctness of $(1–10^{-20})$. You might be quite happy with such good odds. There is nothing particularly bad about uncertainty. An obvious, and rather uninteresting, example of this would be if you asked me whether a given number x was divisible by some other number y. I could simply say "no", and if y is big enough, the odds are in my favor that I am right: to be precise, the odds are 1 in y that a randomly chosen number is divisible by y. The principle here is that you can know a lot more than you can prove! Unfortunately, it is also possible to think you know a lot more than you actually know. Hence the frequent need for proof.

A related, but more interesting problem, is the question of whether or not a given number n is prime. An effective procedure for this might involve taking all prime numbers up to $n^{1/2}$ and seeing if any divide n; if not, n is prime. This is fine, and rather neat, for small n, but when we get to the big numbers it becomes impractical. A better test is a probabilistic one. This uses one of Fermat's famous theorems:

$$a^p = a \mod p \qquad (3.4)$$

What this means is that, for any number a and prime p, if we divide a^p by p, we get the remainder a. So, for example, we write:

$$3^5 = 243 = (48 \times 5) + 3$$

The idea behind the method is to take a large value of a, and calculate a mod p. For large p, the odds are good that p is not a prime and that this quantity does not equal a since there are so many possible remainders. (The actual odds are not simple to calculate, but you get the idea.) However, if p is huge – something of the order of 10^{200}, say – how do we calculate a^p? Well, we don't actually need this number: we only need the remainder after division by p. Why this is so I will leave as an exercise for you! (Don't worry about the general case: do it for a nine-digit p.)

Another similar problem deals with factorization: I give you a number m and tell you that it is the product of two primes, $m=pq$. You have to find p and q. No efficient algorithm exists for this problem as yet, and it in fact forms the basis of a coding system. It is possible to build our ignorance of the general solution of this mathematical problem into ciphering a message. The moment some clever guy cracks it[*] – and people have gotten up to 72-digit m's so far – the code is useless, and we'd better find another one.

Before leaving the subject of computability, I want to make some remarks about the related topic of "grammars". In mathematics, as in linguistics, a grammar is basically a set of rules for combining the elements of a language, only the language is a mathematical one (such as arithmetic or algebra). It is possible to misapply these rules. Consider the following statements:

$$(a+b)c \qquad a+b(c$$

Within the context of arithmetic, only the first of these makes sense. The second, however, does not: the parenthesis is wrongly, even meaninglessly, placed. An interesting general question in computing is whether we can build machines that will test mathematical (and other) expressions for their grammatical correctness. We have seen one example: the parenthesis checker. This checked a very simple grammar involving ")" and "(" and the only grammatical rule was that strings of parentheses balanced. But remember it took a Turing machine to do this: a finite state machine was not up to it. Now there are certain classes of grammar that FSMs can check – for example, strings of ones, 1111111... where valid strings have to have even numbers of digits – but the abilities of this type of machine are limited. We can actually draw up a table relating types of grammars to the machines required for their analysis (Table 3.4).

---

[*] Or someone builds a sufficiently large quantum computer and uses Peter Shor's factorization algorithm – see Chapter 7. [Editor]

TABLE 3.4

General Grammars and Their Machine Implementation

Language	Description	Example	Machine required
Finite enumerable	A list of acceptable expressions	ab, abc	Memory (table look-up)
Regular language	Regular expressions built with *, ∨, ∧, ()	ab^*c, * = any no. of repetitions, incl. none. $a(b \vee d)^*c$	Finite state machines (a theorem)
Context free	Language generated by production rules that admit recursion	$a^n b^n$ (not $a^n b^m$ where $n \neq m$)	An in-between machine: a push-down automaton. Has one "stack" inside – a pile of paper with a spring underneath, can only take off the top one
General recursive	Computable functions	$a^n b^m c^q$	Turing

It is sad that Turing machines are so easy to make that we have to leap over all this pretty theory. Nevertheless, in the design of compilers (which involve the interpretation of languages), the use of such theory is so fundamental that you might find further study of it worthwhile.

We will finish our look at computability with an interesting problem discovered by Post as a graduate student in 1921. Consider a binary string, say 10010. It is arbitrary. Given the string, play with it according to the following rules: read the first three digits; if the first is 0, delete all three and add 00 to the end of the string; if the first is 1, delete all three and add 1101 to the end. So with our string, we would have:

10010

--- 101101

--- 1011101

The question is this: does this process go on forever, stop, or go on periodically? The last I heard, all tested sequences had either stopped or gone into a loop, but that this should be so generally had not been proved. It is an interesting issue because it has been shown that a so-called "Post machine" – one that takes a string g and writes a result $h(g)$ depending on the first digit g_1, of the string – can act as a universal machine and do anything a Turing machine can do!

4

Coding and Information Theory

CONTENTS

In this chapter, we move from abstract considerations of computation to the more concrete realities of computer structure. I want to examine the limitations on machines resulting from the *unreliability* of their component parts. A typical machine may be built from millions of logic gates and other bits and pieces and if these components have a tendency to malfunction in some way, the operation of the machine could be seriously affected.

Components can let us down chiefly in two ways. First, they may contain faults: these can arise during manufacture and are obviously of extreme importance. For example, when making a memory chip from silicon, flaws can be anywhere – where there was a bit of dirt in the material, or where the machine making it made a mistake – and the smallest fault can screw up an element of memory. If your memory system is such that all the cells have to work or the whole thing is useless, then just one tiny mistake can be very costly. A neat way to resolve this problem is to design systems that work around such flaws, spotting them and, perhaps, sealing them off from further usage. However, I will look at the physical structure of components later.

What I want to focus on now is a second way in which an element can let us down. This is when it fails *randomly*, once in a while, perhaps because of the random Brownian motion of atoms or just irregular noise in a circuit.

DOI: 10.1201/9781003358817-4

Any such glitch can cause a component to fail, either temporarily or permanently. Now, the odds against a particular element failing in such a way may be a million to one, but if we have billions of such elements in our machine, we will have thousands failing, all over the place, at any one time. When the earliest Von Neumann machines were in operation, they were constructed from relays and vacuum tubes whose failure rate was very high (of the order of one in a thousand), and the problem of unreliability was acute: with a million components one could expect a thousand of them to be acting up at any one time! Now it has turned out that as we have developed better and better systems with transistors, the failure rate has been going down for almost every machine we build. Indeed, until recently, the problem has ceased to be considered very serious. But as we manufacture computers with more and more parts, and get them to work faster and faster, and particularly as we miniaturize things more and more, this might not remain true. There are something like 10^{11} atoms in a modern transistor, but if we try to get this number down, to build switching devices with, say, a thousand atoms, the importance of noise and random failure becomes very great. So, with one eye on future developments, it is wise to examine the matter of unreliability in some detail. Besides, it is an interesting subject, and that should be reason enough to study it!

4.1 Computing and Communication Theory

We begin our discussion of unreliability by considering the aspect of computers for which it is most problematic, that of memory storage. For example, suppose that we have some data stored somewhere for a long time, and at some point, the system makes a mistake and switches a bit somewhere – a one gets changed to a zero, for example. This sort of error can occur elsewhere in the machine, in its CPU for example, but this is less likely than it happening in the memory, where the number of transistors and elements is so much larger. To examine this situation, I am going to draw a useful analogy with another area of engineering; namely, *communication theory*. In a communication system, you send out a bunch of bits at one end, the transmitter, and at the other end, the receiver, you take them in. This is just sending a message. In the process, errors can creep in: noise could affect the message in transit, reception could be bad, or we might get glitches. Any of these could mean that the message we receive differs from the one sent, this is the so-called "communication problem in the presence of noise". Now, this isn't exactly the same situation as with memory – which is like sending a message through time rather than space – but you can see the similarities. We store something in memory and at a later time we read it back out – in the interim, the stored "message" is subject to noise. When it comes to the reliability of stored memory and sent messages, there are important practical differences. It is possible in principle, for example, to continually check on the contents of our memory, whereas, if NASA sends a radio communication to a Jupiter probe, there is no way of checking its contents while in transit. Nevertheless, the analogy is strong enough to make a look at communication theory worthwhile. We will start with a look at how we might go about detecting errors, an essential step before we can correct them.

4.2 Error Detecting and Correcting Codes

From now on I am going to use the language of communications and will generally leave it to the reader to make the connection with memory systems in machines. Let us suppose we have a transmitted message, which we take to be some sequence of symbols,* and we are going to be doing the receiving. Obviously, we would like to know how trustworthy the received message is, and this brings us to our first matter, that of error detection. Is there some way in which we could know whether the message we have received is correct? Clearly, all we have to work with is the message: calling up the sender for confirmation defeats the object! Is there some way of building a check into the message itself that will enable us to confirm it? The answer is yes, as we will shortly see.

4.2.1 Parity Checking

We first assume that the probability of an error arising in the message is very small; in fact, so small that we never have to worry about more than one error turning up. Furthermore, we will only consider errors in individual bits and not, for example, errors spread out across several neighboring, or related bits (such as "error bursts" caused by scratches on disks). Suppose the chance of an error in a symbol is one in 10,000, and our message is ten symbols long. The chances of an error in the message are about one in 1000. However, the odds against two errors are of the order of a million to one, and we shall consider this negligible. We will only bother trying to detect single errors, assuming doubles are too rare to worry about.

Here is a very simple scheme for checking for single errors, known as a *parity checking* scheme. Suppose we are sending the following ten-bit message:

1101011001

---

* From here on, we restrict these symbols to be binary digits, 1s and 0s. [RPF]

What we do is tag another bit onto the end of this string, which tells us the parity of the string – the number of 1s it contains, or, the same thing, the sum of its digits modulo 2. In other words, the extra digit is a 1 if the original message has an odd number of 1s: otherwise, the above message would have a 0 attached. This is an example of *coding* a message; that is, amending its basic structure in some way to allow for the possibility of error correction (or, of course, for reasons of security). When we receive the message, we look at the parity digit and if the number of 1s is wrong, then we have obviously received a faulty message. Note that this simple check actually enables us to detect any odd number of errors, but not any even number (although as we have ascribed vanishing probability to anything more than a single error, these are assumed not to occur).[*]

There are two main shortcomings of this procedure that we should address. First, we might get an error in the parity bit! We would then be mistaken if we thought the message itself was wrong. Clearly, the longer the message, the less likely the error is to occur in the parity bit itself. Second, at best, the check only tells us whether an error exists – it does not tell us *where* that error might be. All we can do on finding a mistake is have the message sent again. In our case, where we are using a computer, we might simply reboot the machine and go back to square one. Another minor shortcoming of this particular approach is that it leads to a certain inefficiency of communication – in this case, a 10% inefficiency, as we had to send 11 bits to communicate a message of ten.

The obvious next question to ask is: can we construct a method for not only detecting the existence of an error but actually locating it? Again, the answer is yes, and the method is quite ingenious. It is a generalization of the simple parity check we have just examined. What we do is imagine that the data in the message can be arranged into a rectangular array, of (say) m rows and n columns (Table 4.1).

TABLE 4.1

A Rectangular Data Array

$$
\begin{array}{c|ccc}
 & \mathbf{n} & & \\
\hline
 & 1\ 1\ 0\ 1 \dots 0\ 1 & & \\
 & 0\ 0\ 0\ 1 \dots 1\ 1 & & \\
m & \dots & \dots & \dots \\
 & \dots & \dots & \dots \\
 & 1\ 1\ 0\ 0 \dots 0\ 1 & &
\end{array}
$$

---

[*] Note that a simple machine that could do this checking would be the parity FSM discussed in Chapter 3. [RPF]

Of course, the data would not be sent in this form: it would be sent as a binary sequence and then arranged according to some predefined rule, such as breaking the message into m blocks of n symbols and placing them one over the other. To check for errors, what we do is include at the end of each row a digit giving the parity of the row, and at the base of each column, a parity digit for the column. These parity digits can be seeded into a sequential message without difficulty. An error in the array will then lead to a parity mismatch in both the row and the column in which the error appears, enabling us to pinpoint it precisely. In principle, this scheme can detect any number of message errors, as long as they occur in different rows and columns.

We have to be careful, however, about errors occurring among the parity check bits. A particularly nasty instance would be a double error where a message digit and the parity digit for the row (say) both switch. We would know that there was an error, due to the column parity being wrong, but we might be inclined to think that it was a single column parity bit that was at fault – as we would have no confirming row parity error. However, we can safeguard against this ambiguity by placing another parity check in the array, this time at its lower right corner. This bit gives us the parity of the whole message (i.e., of the row and column totals), and using it we can detect – but not locate – such a double error. The end result of this double error detection may well be the same as with our single error detector – we go back to square one and send the message again – but it is still an improvement.

A useful way to quantify the efficiency of a coding method like this is by calculation of a quantity called the *redundancy R*:

$$R = \frac{no. \ of \ bits \ used \ in \ full}{no. \ of \ bits \ in \ message} \tag{4.1}$$

The bigger R the less efficient our code. The quantity $R-1$ is usually known as the "excess redundancy". For our first, single-error-detecting code, the redundancy is $(n+1)/n$. For the rectangular array above, we are using mn bits to send a message that is only $(m-1)(n-1)$ bits long. So we have the following result:

$$R = \frac{mn}{(m-1)(n-1)} \tag{4.2}$$

This quantity is a minimum, and hence the code most efficient, when the array is a square, i.e., $m=n$. You might be tempted to say, "Well, I can get the redundancy down to near one by just taking m and n very large – let's just send our message in blocks and rows with not ten, but 10,000 bits!" The problem with this is that there is a certain probability of each bit being in error, and if the number you are sending gets too big, the chance of multiple errors begins to creep up.

4.2.2 Hamming Codes

I will now take a look at another single-error-correcting (and double-error-detecting) coding method based on parity checking, which is both more efficient and a lot more subtle than the rectangular type. Actually, it is a kind of higher-dimensional generalization of the array method. In any message we send, some of the bits will be defined by the message itself, and the rest will be coding symbols – parity bits and the like. For any given message, we can ask the question: "How many check bits do I need to not only *spot* a single error but also to *correct* it?" One clever answer to this question was discovered by Hamming, whose basic idea was as follows. The message is broken down into a number of subsets of digits, which are not independent, over each of which we run a parity check. The presence of an error will result in some of these checks failing. We use a well-defined rule to construct a binary number, called the "syndrome", which is dependent on the outcome of the parity checks in some way. If the syndrome is zero, meaning all parity checks pass, there is no error; if it is non-zero, there is an error, and furthermore, the *value of the syndrome tells us the precise location of this error*. For example, if the syndrome reads 101, that is decimal 5, then the error is in bit five of the message. If, on the other hand, it reads 110010, then the error is in the 50th bit. The trick is to implement this idea.

We can straightaway make some statements about how many check bits we will need. Suppose our syndrome is m bits long so that we have m check bits. If we decide that a vanishing syndrome is to represent no error, that leaves at most (2^m-1) message error positions that can be coded. However, errors can occur in the syndrome as well as the original message we are sending. Hence, if n is the length of the original message, we must have:

$$2^m - 1 \geq (n+m) \qquad (4.3)$$

or

$$n \leq 2^m - m - 1 \qquad (4.4)$$

For example, if we wanted to send a message 11 bits in length, we would have to include a syndrome of at least four bits, making the full message 15 bits long. This does not seem particularly efficient (efficiency = 11/15 or about 70%). However, if the original message was, say, 1000 bits long, we would only need ten bits in our syndrome ($2^{10} = 1024$), which is a considerable improvement!

Let us now see precisely how this syndrome idea works. As an example, let us continue with our problem of sending a message 11 bits long. As we saw, we will need four check bits. Each such bit will be a parity check run over a subset of the bits in the full 15-bit message. Just as with the simple parity check method, we will select a few specific bits in the message, calculate their overall parity, and adjust the corresponding check bit to make the total parity of the (subset plus check) zero. If there is an error in this subset, the parity check will fail. The clever thing about the Hamming code is that each message bit is in more than one subset and hence contributes to more than one parity check, but not to all of them. By seeing which parity checks fail and which pass, we can home in on the error uniquely. We assign to each parity check a one if it fails and a zero if it passes, and arrange the resulting bits into a binary number, the syndrome. This directly indicates the error position. It is pretty much arbitrary, but we will construct the syndrome by reading the parity checks from left to right in the message.

For the moment, we will assume that the parity check bits are placed in some order throughout the message, although we will not mind where for the moment. We will first identify the subsets each covers. To do this, it will help to list the four-digit binary representations of the positions within the message:

1	0001
2	0010
3	0011
4	0100
5	0101
6	0110
7	0111
8	1000
9	1001
10	1010
11	1011
12	1100
13	1101
14	1110
15	1111

Let us look at the rightmost parity check, the far-right digit of the syndrome. Suppose this is non-zero. Then there will be a parity failure in a position whose binary representation ends in a one: that is, one of positions 1, 3, 5, 7, 9, 11, 13, or 15. This is our first subset. To get the second, look at the second digit from the right. This can only be non-zero for numbers 2, 3, 6, 7, 10, 11, 14, and 15. Note what is happening. Suppose we have found that both of these parity checks failed, i.e., we assign a 1 to each. The error must be in a position that is common to both sets, i.e., a binary number of the form $ab11$. This can only be 3, 7, 11, or 15; we have narrowed the possible location choices. To find out which, we have to do the remaining parity checks. The third check runs over digits 4–7 and 12–15. The final check covers digits 8 through 15. Suppose both of these are zero, that is, the parity checks out; we hence put two zeroes in our syndrome. Then there is no error in positions 4-7, 12–15, and 8–15 (which obviously overlap). But there is an error somewhere in 3, 7, 11, 15. The only one of these latter four that is not excluded from the previous sets is position 3. That must be where the error lies. Of course, in binary, 3 is 0011 – the syndrome calculated from the parity checks.

Let us pick a real example to illustrate these ideas in a more concrete manner. Suppose we want to send the 11-bit message 10111011011. We first have to decide where to stick in our parity bits. There is nothing in what we have said so far that tells us whereabouts in the message these must go, and in fact, we can put them anywhere. However, certain positions make the encoding easier than others, and we will use one of these. Specifically, we place our check bits at positions 1, 2, 4, and 8. We now have:

Codeword	a	b	1	c	0	1	1	d	1	0	1	1	0	1	1
Position	1	2	3	4	5	6	7	8	9	10	11	12	13	14	15

Importantly, the check bits here, read from left to right, will give the *reverse* of the syndrome: that is, the first leftmost digit of the syndrome, when written out, would be read from the parity check of d, not a, and the last rightmost read from a, not d. Again, this is a matter of calculational simplicity.

We can now work out a, b, c, and d. Bit a is the parity of the odd positions: 1, 3, 5, 7, 9… This is 1. Hence, a is 1. Bit b is found by summing the parities of positions 2, 3, 6, 7, 10, 11, 14, and 15. This gives zero. Bit c comes from the parity of positions 4 through 7 and 12 through 15, giving $c=1$. Finally, we get d by doing a check on 8 through 15, giving $d=1$. Note, incidentally, how this placing of the check bits leads to straightforward encoding, that is, calculation of $abcd$. If we had chosen an apparently more straightforward option, such as placing them all at the left end of the string occupying positions 1 to 4, we would have had to deal with a set of simultaneous equations in a,b,c,d. The important feature of our choice is that the parity positions 1, 2, 4, 8 each appear in only one subset, giving four independent equations. The completed message is thus:

10110111011011.

Let us see what happens when an error occurs in transmission. Suppose we receive the message:

101101011011011.

Where is the mistake? The sum over the odd-placed digits is 1 – a failure. We assign a 1 to this in our syndrome, this being the rightmost digit. The second parity check also gives 1, another failure, and our syndrome is now $xy11$, with x and y to be determined. Bit y comes from our next check and is 1 again – another failure! The syndrome is now $x111$. To find x, we check the parity of places 8 through 15 and find that this is zero – a pass. We, therefore, assign a zero to x, giving our syndrome as 0111. This represents position 7, and indeed, if you compare the original and corrupted messages, position 7 is the only place you will find any difference.

An interesting feature of the Hamming code is that the message and code bits are on the same footing – an error in a code bit is located the same way as in a message bit. We can extend this code to detect double errors quite simply. We tag on to the end of the message yet another check bit, this time representing the parity of the whole thing. For the (uncorrupted) message we gave above, the parity is 1, so we attach a 1 at the rightmost end, to give us zero overall parity as usual. Now, if there is a single error in the message, the parity of the 15-bit message will change, and this will show up as a mismatch with the 16th bit. However, if there is a double error, the parity of the 15-bit message will not change, and all will look normal in the 16th bit; yet the parity checks within the former will fail, and this indicates a double error. Observe that if the 15-bits check out, but the overall parity does not, this indicates an error in the overall check bit. Note that the cost of these benefits is almost 50% inefficiency – five check bits for an 11-bit message. As we pointed out earlier however, the inefficiency drops considerably as we increase the message length. For a 1000-bit message, the inefficiency is a tiny 1% or so.

It is worth examining the practical usefulness of error-detecting codes like this by looking at how the consequences of message failure become less drastic for quite small losses of efficiency. Let us suppose that we are sending a message in separately coded batches of about 1000 bits apiece, and the probability of an error in a single one of the bits is 10^{-6} or 10^{-3} per batch. We take these errors to be random and independent of each other. We can use Poisson's Law to get a handle on the probabilities of multiple errors occurring when we send our batches. If the mean number of errors expected is m, then the actual probability of k errors occurring is given by:

$$\frac{1}{k!}m^k\exp(-m). \tag{4.5}$$

The expected number of errors per batch is, as we have said, 10^{-3}, Hence, the probability of a double error in a batch is $(1/2) \times 10^{-6}$ to high accuracy (we can ignore the exponential), and the probability of a triple error is $(1/6) \times 10^{-9}$. Now suppose we have no error detection or correction, so we are expending no cash on insurance. If an error occurs, we get a dud message; the system fails. On average, we should only be able to send 1000 batches before this happens, which is pretty miserable. Suppose now that we have single error detection, but not correction, say a simple parity check. Now, when the system detects an error, it at least stops and tells me, and I can try to do something about it. This still happens once in every 1000 or so batches, but it is an improvement that we gain at the cost of just one-tenth of a percent message inefficiency – one parity bit per 1000 message bits. This system will fail whenever there are two errors, which occur roughly once every two million batches. However, suppose we have our 1% gadget, with single error correction and double error detection. This system will take care of it itself for single errors, and only stop and let me know there is a problem if a double error occurs, once in every million or so batches. It will only fail with a triple error, which turns up in something like every six billion batches. Not a bad rate for a 1% investment!

Issues of efficiency and reliability are understandably central in computer engineering. An obvious question to ask is: "How long should our messages be?" The longer the better as regards the efficiency of coding, but the more likely an error is to occur, and the longer we will have to wait to find out if an error has occurred. On the other hand, we might be prepared to sacrifice efficiency for security, sending heavily coded but brief messages so that we can examine them and regularly feel that we can trust what we receive. An example of the latter kind that is worth noting is in the field of communications with deep solar system spacecraft such as the *Voyager* series. When your spacecraft costs billions of dollars, and you have to send radio messages across millions of miles and be as certain as possible that it arrives uncorrupted – we don't want the cameras pointing at the Sun when they should be looking at Jupiter or Saturn – efficiency goes out the window. Spacecraft communications rely on a kind of voting technique, referred to as "majority logic decisions". Here, each bit in the source message is sent an odd number of times, the idea being that most of these will arrive unchanged – i.e., correct – at the receiver. The receiver takes as the message bit whichever bit appears most in each bit-batch, in the best democratic fashion. How many copies you send depends on the expected error rate. Anyway, this is just a little example from communication, so I won't dwell on it. It doesn't seem to have too much relevance for computing (except perhaps for those in the habit of backing up their files a dozen times).

4.2.3 An Aside on Memory

Let me briefly discuss one interesting way in which the Hamming coding technique can be applied to computer memory systems. With the advent of parallel processing, it has become necessary to load and download information at an increasingly rapid rate, as multiple machines gobble it up and spit it out faster and faster. This information is still stored on disks, but individual disks are simply not fast enough to handle the required influx and outflow of data. Consequently, it is common practice to use "gang-disk" systems, where lots of disks share the load, simultaneously taking in and spewing out data. Such systems are obviously sensitive to errors on individual disks: if just one disk screws up, the efforts of the whole bunch can be wasted. Every manufacturer would like to build the perfect disk, one that is error-free – sadly, this is impossible. In fact, it is fair to say that the probability of any given floppy, and certainly any hard disk, on the market being free of flaws (e.g., bits of dirt, scratches) is virtually nil.* The reason you do not usually notice this is that machines are designed to spot flaws and work around them: if a computer locates a bad sector on a disk, it will typically seal it off and go on to the next good one. This all happens so fast that we don't notice it. However, when a disk is working alongside many others in a parallel processing environment, the momentary hang up as one disk attends to a flaw can screw everything up.

The Hamming method can come to our rescue. Let's suppose we have 32 disks working together. We take 26 of these to be loaded with information, and six to be fake. We only have 26 worth of data, but 32 lines coming into the system. With each click of the clock, we get one bit fed in from each disk. What we do is run a parity check for each input of 32 bits coming into the system, one per disk, according to Hamming's method – hence six parity bits for 26 of message – and correct the single errors as they come along. Note that in this sort of set-up the odds against double errors occurring are enormous; that would need two or more disks to have errors in the same disk locations. It is a possibility, of course, but even if it happens, we can soup up our system to detect these double errors and have it grind to a halt temporarily so that we can fix things. The flaws we are talking about here, in any case, are not really random: they are permanent, fixed on the disk, and hence will turn up in the same place whenever the disk is operating. We can avoid double errors of this kind by running the system, debugging it, and throwing away any disks that have coincidentally the same error spots. We then buy new ones. We do this until no more

---

* Readers might like to contact disk manufacturers and try to get some figures on the flaw rate on their products. You will have a hard time getting anywhere! [RPF]

double errors are found. The use of this Hamming coding method saved the whole idea of gang disks from going down the drain.

Here are some problems for you to look at.

Problem 4.1 Devise a Hamming-type code for a message alphabet with a number (a) of elements (for binary, $a=2$). Show that, if the number of code symbols is r, and the total message length is N (so the original message is $N-r$) we must have

$$1+(a-1)N = a^r$$

Work out a simple example.

Problem 4.2 This is an interesting mathematical, but not overly practical, exercise. We define a *perfect* code to be one for which:

(a) each received coded message can be decoded into some purported message, and

(b) the purported message is correct if there are less than a specific number of errors.

Codes can correct up to e errors (thus far we have only considered $e=1$). Can you construct a perfect code for binary symbols ($a=2$) with $e=3$, i.e., triple error correction? Hint: stick to $N=23$, with 12 bits of message and 11 of code (or syndrome). There is also a solution to this problem for a tertiary alphabet ($a=3$). For this case, try $N=11$, with six data digits and five of code.

4.3 Shannon's Theorem

We have asked lots of fundamental questions so far in this book. Now it is time to ask another. In principle, how far can we go with error correction? Could we make a code that corrects two, three, four, five, six, seven... errors and so on, up to the point where the error rate is so low that there is no point in going any further? Let's set the acceptable chance of us getting a failure at 10^{-30}. If you don't like that, you can try 10^{-100}: any number will do, but it must be non-zero, or you'll get into trouble with what follows. I reckon 10^{-30} will do.

Suppose we're sending a message of length M_C, which is the length of the full coded message, containing original data and coding bits. As usual, we're working in binary. The length of the data message we call M. Let's assume that the probability of any single bit going wrong is q. We want to design a coding scheme that corrects single, double, triple errors, and so on, until the chances of getting more errors in M_C are less than our chosen number, 10^{-30}. How many code bits are we going to need? How much of M_C is going to be message? What's left?

Claude Shannon has shown that the following inequality holds for M and M_C:

$$M / M_C \le f(q) = 1 - \left(q \log_2 \left[1/q \right] + \left[1 - q \right] \log_2 \left[1/(1-q) \right] \right) \quad (4.6)$$

Given this, says Shannon, if no limit is placed on the length of the batches M_C, *the residual error rate can be made arbitrarily close to zero.* In other words, yes, we can construct a code to correct n-tuple errors to any accuracy we choose. The only restriction in principle is the inequality (4.6). In practice, however, it might require a large batch size and a lot of ingenuity. However, in his extraordinarily powerful theorem, Shannon has told us we can do it. Unfortunately, he hasn't told us *how* to do it. That's a different matter!

We can construct a table with a few values for q to illustrate the upper limit Shannon has placed on coding efficiencies (Table 4.2).

TABLE 4.2

Shannon's Coding Efficiency Limits

q	M/M_C	M_C/M
1/2	0	∞
1/3	0.082	12.2
1/4	0.19	5.3
0.1	0.53	1.9
0.01	0.919	1.09
0.001	0.988	1.012

Note that if q is 0.5 – that is, there is a 50–50 chance that any bit we receive might be in error – then we can get no message through. This obviously makes sense. As the error rate drops, the upper limit on the efficiency increases, meaning that we need fewer code bits per data bit. For any given q, however, it is very difficult to reach Shannon's limit.

The actual proof of this theorem is not easy. I would like to first give a hand-waving justification of it that will give you some insight into where it comes from. Later I will follow a geometrical approach, and after that prove it in another way that is completely different and fun, and involves physics, and the definition of a quantity called information. But first, our hand-waving. Let us start with the assumption that M_C is very, very large. This will enable us to make some approximations. If the probability of a single-bit error is q, then the average number of errors we would expect in a batch is:

$$k = qM_C. \tag{4.7}$$

This isn't exactly true – the actual number may be more or less – but this is the average error rate we expect, and it will do as a rough guess. We have to figure out how much coding we need to dispose of this number of errors. The number of ways this number of errors could be distributed through a batch is given by simple combinatorics:

$$\frac{M_C!}{k!(M_C - k)!} \tag{4.8}$$

Let us assume that we have m code bits. Such a number of bits can describe 2^m things. This number of bits must be able to describe at least the M bits of the data message plus the exact locations (to give us error *correction*, not just detection) of each possible distribution of errors, of which we are saying there are $M_C!/k!(M_C-k)!$ It is clear that $m \leq M_C - M$ (since some bits could be redundant), so we have the inequality:

$$2^{M_C - M} \geq \frac{M_C!}{k! \ (M_C - k)!} \tag{4.9}$$

We now take the logarithm of both sides. The right-hand side we work out approximately, using Stirling's formula:

$$n! = \sqrt{(2\pi n)} \ n^n \ e^{-n} \ \exp\left[(1/12n) - (1/360n^3) + \ldots\right] \tag{4.10}$$

for large n. Hence:

$$\log n! \approx (1/2) \log n + n \log n - n + O(1/n) \tag{4.11}$$

(Here, $\log x = \log_e x$.) The last term simply represents a lot of junk that gets smaller as n gets bigger (tending to zero in the limit), plus terms like $\log 2\pi$. We can in fact get rid of the first term, namely $(1/2)\log n$, as this is small compared with the next two when n is large. We thus use:

$$\log n! \approx n \log n - n \tag{4.12}$$

and with this, the right-hand side of the inequality becomes:

$$M_C \log M_C - M_C \log(M_C - k) + k \log(M_C - k) - k \log k \tag{4.13}$$

which, using $k = qM_C$ and a little algebra, is:

$$M_C\left[q\log_2\left(1/q\right)+\left(1-q\right)\log_2\left(1/\left(1-q\right)\right)\right].\tag{4.14}$$

We have here converted the natural logarithm to base two, which simply introduces a multiplicative factor on both sides, which cancels. Taking the logarithm to base two of the left-hand side of the inequality, and dividing both sides by M_C, we end up with Shannon's inequality.

This inequality tells us that, if we want to code a message M, where the bit error rate is q, so that we can correct k errors, the efficiency of the result cannot exceed the bounds in (4.6). Of course, k is not arbitrary; we have taken it to be the mean number of errors, $k=qM_C$. The question we would like to have answered is whether we can code a message to be sure that the odds against more than a certain number of errors, say k', occurring is some number of our choosing, such as 10^{-30}. Shannon's actual Theorem says that we can do this; let us take our "proof" a little further to see why this might be so.

The number of errors that can occur in the message is not always going to be k but will be k within some range and probability. In fact, the distribution of errors will follow a binomial distribution, with mean qM_C ($=k$) and standard deviation $\sigma = \sqrt{[M_C.q(1-q)]}$. It is a standard result that, for M_C large and q small, we can approximate this with a Gaussian (or Normal) distribution with mean qM_C ($=k$) and standard deviation $\sigma = \sqrt{[M_C.q(1-q)]}$; that is, the same as before. Now to ask that our error rate be less than a number N (e.g., 10^{-30}) is equivalent to demanding that the number of errors we have to correct be less than k', where:

$$k' = k + g\,\sigma \tag{4.15}$$

for some finite number g dependent on N. For the Gaussian distribution, the probability that the number of errors lies within one standard deviation from the mean k is 68%; within two it is 96%; within three 99%. So, for example, if we wanted to be 95% certain that there would be no errors in our message, we would have to demand not that we be able to correct k errors but:

$$k + 2\sigma \tag{4.16}$$

errors. In this case, $g=2$. For any level of probability we pick, we can find a value for g. As a rule, the probability of finding errors g standard deviations from k goes like:

$$\exp\left(-g^2/2\right) \tag{4.17}$$

and we can see how incredibly rare we can make errors for relatively small g. If g is 20, for instance, this factor is exp(–200), or about 10^{-100} A heck of a lot smaller than our 10^{-30}! For our choice of number, we get g to be about six.

To make use of this, we simply amend (4.9) by replacing k by k', the new number of errors we want to correct. If we can still find a code to do this, then we know that the odds of errors occurring in its transmission are less than our 10^{-30} or whatever. I leave it as an exercise for the reader to put:

$$k' = k + g\sigma = k + g\sqrt{k(1-q)} \qquad (4.18)$$

into the inequality, and show that, in the limit, Shannon's result emerges as before.

What Shannon has given us is an upper limit on the efficiency. He hasn't told us whether or not we can find a coding method that reaches that limit. Can we? The answer is yes. I'll explain in more detail later, but the technique basically involves picking a random coding scheme and then letting M_C get larger and larger. It's a terrific mathematical problem. Provided M_C is big enough, we can reach the upper efficiency whatever the coding scheme. However, the message length might have to be enormous. A nice illustration of how big we have to take M_C in one case is from satellite communication. In sending messages from Earth to Jupiter or Saturn, it is not unusual for an error rate q of the order of a third to come through. The upper limit on the efficiency for this, from our table, is 8%; that is, we would have to send about 12 code bits for each data bit. However, to do this would require a prohibitively long M_C, so long that it is not practical. In fact, a scheme is used in which about *150* code bits are sent for each data bit!

4.4 The Geometry of Message Space

I am now going to look at Shannon's Theorem from another angle, this time using geometry. In doing so I will introduce the useful idea of "message space". Although this is primarily of importance in communications, and we are doing computing, I have found the idea interesting and useful, and you might too.

Message space, simply, is a space made up of the messages that we want to transmit. We are used to thinking of a space as something that can be many-dimensional, either continuous or discrete, and whose points can be labeled by coordinates. Message space is a multi-dimensional discrete space, some or all of whose points correspond to messages. To make matters a little more concrete, consider a three-bit binary code, with acceptable words:

$$000, 001, 010, 011, 100, 101, 110, 111.$$

These are just the binary numbers zero to seven. We can consider these numbers to be the coordinates of the vertices of a cube in three-dimensional space, as shown in Figure 4.1.

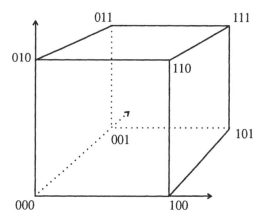

FIGURE 4.1
A simple message space.

This cube is the message space corresponding to the three-bit messages. The only points in this space are the vertices of the cube – the space between them in the diagram, the edges, and whatnot are not part of it. This space is quite tightly packed, in that every point in it is an acceptable message; if we change one bit of a message, we end up with another. There is no wastage, everything is significant. We could easily generalize to a four-bit message, which would have a message space that was a 16-vertex "hypercube", which unfortunately our brains can't visualize! An m-bit message would require an m-dimensional space.

What happens if there is an error in transmission? This will change the bits in the sent message and correspond to moving us to some other point in the message space. Intuitively, it makes sense to think that the more errors there are, the "further" we move in message space; in the above diagram, (111) is "further" from (000) than is (001) or (100). This leads us to introduce a so-called "distance function" on the message space. The one we shall use is called the *Hamming distance*. The Hamming distance between two points is defined to be the number of bits in which they differ. So, the Hamming distance from 111 to 000 is three, while from 001 to 000 it is just one. According to this definition, in a 4D space, 1110 is as far from 1101 as 0100, and so on. This makes sense. The notion of distance is useful for discussing errors. Clearly, a single error moves us from one point in message space to another a Hamming distance of one away; a double error puts us a Hamming distance of two away, and so on. For a given number of errors e, we can draw about each point in our hypercubic message space a "sphere of error", of radius e, which is such that it encloses all of the other points in message space that could be reached from that point as a result of up to e errors occurring. This gives us a nice geometrical way of thinking about the coding process.

Whenever we code a message M, we rewrite it into a longer message M_C. We can build a message space for M_C just as we can for M; of course, the space for M_C will be bigger, having more dimensions and points. Clearly, not every point within this space can be associated one-on-one with points in the M-space; there is some redundancy. This redundancy is actually central to coding. e-Error correction involves designing a set of acceptable coded messages in M_C such that if, during the transmission process, any of them develops at most e errors, we can locate the original message with certainty. In our geometrical picture, acceptable messages correspond to certain points within the message space of M_C; errors make us move to other points, and to have error correction we must ensure that if we find ourselves at a point that does not correspond to an acceptable message, we must be able to backtrack, uniquely, to one that does. A straightforward way to ensure this is to make sure that, in M_C, all acceptable coded message points lie at least a Hamming distance of:

$$d = 2e + 1 \qquad\qquad (4.19)$$

from each other. We can see why this works. Suppose we send an acceptable message M and allow e errors to occur in transmission. The received message M' will lie at a point in M_C e units away from the original. How do we get back from M' to M? Easy. Because of the separation of $d = 2e + 1$ we have demanded, M is the closest acceptable message to M'! All other acceptable messages must be at a Hamming distance of at least $e+1$ from M'. Note that we can have simple error *detection* more cheaply; in this case, we can allow acceptable points in M_c to be within $2e$ of one another. The demand that points be $(2e+1)$ apart enables us to either correct e errors or detect $2e$.

Pictorially, we can envisage what we have done as mapping the message space of M into a space for M_C in such a way that each element of M is associated with a point in M_C such that no other acceptable points lie within a Hamming distance of $2e+1$ units. We can envisage the space for M_C as built out of packed spheres, each of radius e units, centered on acceptable coded message points (Figure 4.2). If we find our received message to lie

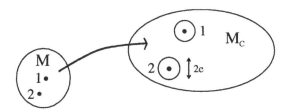

FIGURE 4.2
A message space mapping.

anywhere within one of these spheres, we know exactly which point corresponds to the original message.

To make this idea a little more concrete, let us return to the three-bit cube we drew earlier. We can consider this the message space M_C corresponding to a parity code that detects, but does not correct, single errors for the two-bit message system M comprising:

$$00, 01, 10, 11.$$

This system has the simple two-dimensional square message space:

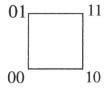

The parity code simply tags an extra digit onto each message in M, clearly resulting in a 3D cubic space for M_C. The acceptable messages are 000, 011, 101, and 110. This leaves four vertices that are redundant:

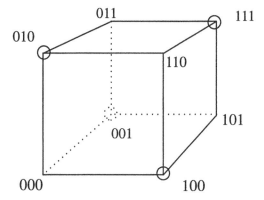

Any error in transmission will put us on one of these vertices, telling us that an error has occurred, but not where. Note that each false vertex lies within a Hamming distance of 1 from a genuine one. If we wanted single error correction for this system, we would have to use a space for M_C of four dimensions.

So if our coding system works, we should be able to move each of our message points somewhere in the message space of M_C such that they are sufficiently separated. Every now and again we will be forced to allow some overlap between spheres of error, but this is not usually a problem.* We can now quickly see how this geometrical approach offers another proof of Shannon's Theorem. Use M and M_C to denote the dimensions of the original and coded message spaces, respectively – this is just a fancy way of describing the lengths of the message strings. The number of points in M is 2^M, and in M_C, 2^{M_C}. To correct k errors, we need to be able to pack M_C with spheres of error of radius k, one for each point in M. We do not want these to overlap. Using this, we can obtain an inequality relating the volume of M_C to that of the spheres. Now in a discrete space of the kind that is message space, the volume of a sphere is defined to be the number of points contained within it. It is possible to show that, for an M_C-dimensional space, the number of points spaced a unit length apart that lie within a radius k units of a point is:

$$\frac{M_c!}{k!\,(M_c - k)!} \tag{4.20}$$

By noting that the volume of M_C must be greater than or equal to the number of points in each error sphere multiplied by the number of spheres, that is, the number of points in M, we ultimately recover the inequality (4.9). There is no need to go through the subsequent derivations again; you should be able to see how the proof works out.

Problem 4.3 Here is a nice problem you can try to solve using message space (that's how I did it). By now you are familiar with using a single parity bit at the end of a message to detect single errors. One feature of this technique is that you always need just one check bit irrespective of how long the message is; it is independent of M_C. The question is, can we also set up a method for detecting double errors that is M_C-independent? We are not interested in correcting, just detecting. We want a finite number of bits, always the same, and it would seem pretty likely that we could do it with only two bits! Recall that for the Hamming code we could correct a single error and detect doubles with a check bit for

---

* "Perfect" codes, which we introduced in a problem earlier, are actually those for which the error spheres "fill" the message space without overlapping. If the spheres have radius e, then every point in the space lies within e units of one, and only one, message point. [RPF]

overall parity and the syndrome, but the number of check bits contributing to the syndrome depended on the message length. You should find that it is actually impossible to detect doubles without bringing in the length of the message.

4.5 Data Compression and Information

In a moment, I am going to look at Shannon's Theorem in yet another way, but first I would like you to let me wander a bit. The first direction I want to wander in is that of data compression, and I'd like to explain some of the ideas behind this. Consider a language like English. Now this has 26 letters, and if we add on commas, full stops, spaces, and whatnot, we have about 30 symbols for communication. So, how many things can I say in English if I have ten symbols at my disposal? You might say, well, 30 to the power ten. Not true. If I wrote the following string for you:

cpfajrarfw

it wouldn't be English. In real, interpretable English, you can't have everything; the number of acceptable words is limited, and the ordering of the letters within them is not random. If you have a "T", the odds are the next letter is an "H". It's not going to be an "X", and rarely a "J". Why? The letters are not being used uniformly, and there are very much fewer messages in English than seem available at first sight.

Perhaps, almost certainly, each one of you has parents back home who suffer from the fact that you never get around to writing letters. So they send you a card, all addressed and everything, which has on the back a lot of squares. And one of the squares says "I'm having a good time and enjoying Caltech. Yes/No". Next square says "I met a beautiful girl or handsome boy" or something, "Yes/No". The next message says "I have no more laundry to send", or "I'm sending it at such-and-such a time", and so on. What the poor parents are doing is converting long English sentences into single bits! They are actually trying to shame you into writing, but failing that, they are producing a method for you to communicate with them that is more efficient bitwise than you going to all that trouble. (Of course, you still have to post the card!) Another example of improving efficiency, and this is something you've all probably done, is to clip a long-distance telephone company by prearranging to ring a friend and let the phone ring a set number of times if you're going to his party, another number if you're not, or maybe signal to him with three rings that you'll be at his place in your car in five minutes and he should go outside to wait, or what-have-you. You're calling long distance but getting the message through costs you nothing because he doesn't pick up the phone.

A related question is: how inefficient is English? Should I send my ten symbols directly, from the English alphabet? Or should I perhaps try a different basis? Suppose I had 32 rather than 30 symbols, then I could represent each element as a five-bit string. Ten symbols hence become 50 bits, giving a possible 2^{50} messages. Of course, as I've said, most messages won't make sense. Whenever a Q appears, we expect it to be followed by a U. We, therefore, don't need to send the U; except for awkward words like Iraq, of course, which we could deal with by just sending a double Q. So we could exploit the structure of the language to send fewer symbols, and this packing, or compression, of messages is what we're going to look at now.

A good way to look at the notion of packing is to ask, if we receive a symbol in a message, how surprised should we be by the next symbol? I mean, if we receive a T, then we would not be surprised if we got an I next, but we would if we got an X or a J. If you pick a T, the chances of a J next are very small; in English, you don't have the freedom to just pick any letter. What we want to guess is how much freedom we have. Here is an interesting way to do it. Try this experiment with your friends. Take an English text and read it up to a certain point. When you stop, your friend has to guess the next letter or symbol. Then you look at the next one and tell him whether he's right. The number of guesses he has to make to get the next letter is a handy estimate of the freedom, of the possibilities for the next letter. It won't necessarily be accurate – people's guesses will not be as good as a mechanical linguistic analysis – but it'll give you an idea. But in any case, people will be able to guess the next letter at a much better rate than one in 30. In fact, the average number of possible letters following a letter in English is not 26 but about five. You can work this out from your experiment by doing it often and averaging the number of guesses. And that gives you an idea of how English can be compacted. It also introduces us to the notion of how much *information* is in a message. We will return to this.

Another way of considering this problem of compression is to ask: if you had N symbols of English, with 32 possibilities for each, how many messages could you send? If you like, what is the greatest amount of information you could convey? As we have discussed, you could not send the full 32^N, as most would not make sense. Suppose the number of potentially sendable messages[*] is n. We can label each of these, code them if you like, by a number. We'll take this to be in binary. Now both guys at

---

[*] Strictly speaking, we mean the number of equally likely messages that can be sent. In reality, some messages will be more likely than others. However, we are not being rigorous in what follows, and will not worry about this for the time being. [RPF]

each end of the message can have a long list, telling them which number corresponds to which message, and instead of communicating by sending a full message, they just send numbers. This is exactly analogous to you with the cards your parents send – "Yes, I need a haircut"; "No, I didn't do my homework"; and so on. You cannot get a more efficient sending method than this, compressing a whole message down to one number. We can work out how many bits we will need to send to cover all the messages. If we need *I* bits, then we have:

$$2^I = n, \text{ or } I = \log_2 n. \tag{4.21}$$

This number, the number of bits we minimally need to send to convey as much as we possibly could have conveyed in the *N* bits of English (or whatever other system being used) is called the *information* content, or just the *information* being sent. It is important to stress that the meaning of the word "information" here differs from that in ordinary usage – it is not totally distinct from this, but generally "information" in our sense tells us nothing about the usefulness of or interest in a message and is strictly an academic term. There are lots of words like this in science: the meaning of the words "work" in physics and "function" in mathematics bears little relationship to their colloquial meanings. We will return to the concept of information and define it more rigorously later. For the moment, just bear in mind the fundamental idea: that we are coding messages into a binary system and looking at the bare minimum of bits we need to send to get our messages across.

It is possible to give a crude definition of the "average information per symbol" using the notions we have developed. Suppose we have a number of symbols that is not *N*, but twice *N*. What number of possible messages will correspond to this? For figurative purposes we can split the 2*N*-symbol string into two *N*-symbol strings:

2*N*

.... / ....

N   *N*

As a rough guess, we may expect that the number of potentially sendable messages will be the number of messages in each string multiplied together, or n^2. In general, of course, the precise answer will be horribly difficult to find. For example, there will be what we might call "edge effects" – the possibility of words being formable at the join of the two strings, crossing from one into the other – since the two *N*-symbol strings are not actually separated. There can also be "long-range correlations" where parts of the string influence others some distance away. This is true in English, where the presence of a word at one point in a message typically affects what other words can appear near it. As we are not being rigorous yet, we will not worry about such problems. In fact, you can see that if we let N get bigger and bigger, our rough guess gets more accurate. If we have $2N$ symbols, we get about n^2 messages; if we have $3N$, about n^3; and generally, if we have xN symbols, the number of messages will be about n^x. If we write the information content from N symbols as $I(N)$, we have:

$$I(xN) \approx \log_2\left(n^x\right) = x\log_2 n,$$ (4.22)

and we see that the ratio:

$$r = \frac{I(xN)}{xN} \approx \frac{\log_2 n}{N} = \frac{I(N)}{N}$$ (4.23)

is independent of x. So for large enough N, as long as our approximation gets better, it tends to a constant. We call this ratio the *information per symbol* of our system, an interpretation that seems clear from the right-hand side of (4.23).

Let us return to the notion of information and try to get a better idea of what it means. In a sense, the amount of information in a message reflects how much surprise we feel at receiving it. Consider, say, receiving a printed communication from a bookshop, such as: "We are pleased to tell you that the book you ordered is in stock"; or its opposite: "We are sorry to inform you that … is not in stock"; these long messages contain many more symbols but no more information than the simple "Yes" or "No" you could elicit from a shopworker if you called the bookshop direct. Most of the symbols in the printed communications are redundant in any case: you only have to spot the words "pleased" and "sorry" to figure out what they are saying. In this respect, information is as much a property of your own knowledge as anything in the message.

To clarify this point, consider someone sending you two duplicate messages: a message, then a copy. Every time you receive a communication from him, you get it twice. (This is not for purposes of error detection; it's just a bad habit!) We might say, well, the information in the two messages must be the sum of that in each (remember, $I(n_1 n_2) = \log_2(n_1 n_2) = \log_2 n_1 + \log_2 n_2$). But this would be wrong. There is still only one message, the first, and the information only comes from this first half. This illustrates how "information" is not simply a physical property of a message: it is a property of the message and your knowledge about it.

Perhaps the best way to demonstrate the difference between our definition of information and the everyday term is to consider a *random* message, that is, an N-bit binary string with random bits. If all possible strings are allowable messages, and all are equally likely (which will happen if each bit is equally likely to be 0 or 1), then the information in such a message will be:

$$I = \log_2\left(2^N\right) = N \tag{4.24}$$

This is actually the most information you can get with this particular choice of symbols. No other type of message will reach $I=N$. Now, surely this doesn't make sense – how can a random string contain *any* information, let alone the maximum amount? Surely, we must be using the wrong definition of "information"? But if you think about it, the *N-bit* strings could each label a message, as we discussed earlier, and receiving a particular string singles out which of the 2^N possible messages we could get that we are actually getting. In this sense, the string contains a lot of "information". Receiving the message changes your circumstance from not knowing what it was to now knowing what it is; and the more possible messages you could have received, the more "surprised", or enlightened, you are when you get a specific one. If you like, the difference between your initial uncertainty and final certainty is very great, and this is what matters.

4.6 Information Theory

We have defined the information in a message to be:

$$I = \log_2 n, \tag{4.25}$$

where n is the number of *equally likely* messages we might receive. Each message contains this same amount of information. In the general case, some messages will be more likely than others, and in this case, the greater the likelihood, the less information contained. This makes sense, given our claim that the information in a message represents the "surprise" we experience at receiving it. In this section, we come on to the topic of *information theory* proper, which will enable us to both generalize and make more rigorous our previous considerations.

We'll take a simple example first. Suppose our message is built from an alphabet of symbols. There could be any number of these, such as the four bases of DNA, or whatever: we certainly do not want to restrict ourselves to the letters of English. Let the number of symbols be i, and label them:

$$a_1, a_2, \ldots\ldots a_i.$$

Messages in this language are long strings of these symbols, say of length N. Now, before we go any further, we have to make some assumptions about the way these symbols are distributed throughout messages. We assume first that we can assign a probability, p_i, to each symbol, which is the probability that any given symbol in the message is the symbol a_i. The quantity p_i tells us the frequency of occurrence of a_i. We also assume that each symbol in the message is independent of every other; that is, which symbol appears at a given position does not depend on symbols at other positions, such as the one before. This is actually quite an unrealistic assumption for most languages. We will consider cases for which it is not true shortly.

How much information is carried by a given message? A simple way in which we can work this out is as follows. Suppose the message we have is length N. Then we would expect to find symbol a_1 turn up Np_1 times on average, a_2 Np_2 times, ... a_i Np_i times. The bigger N is, the better these guesses are. How many different messages do we have? Combinatorics comes to our rescue, through a standard formula. If we have N objects, m of one type, n of another, p of another, ..., and $m+n+p+... = N$, then the number of possible arrangements of the m, n, $p..$, is given by:

$$\frac{N!}{m!n!p!...}. \tag{4.26}$$

On average, then, we can say that the number of different messages in N symbols is

$$\frac{N!}{(Np_1)!(Np_2)!...}. \tag{4.27}$$

We earlier defined information to be the base two logarithm of the number of possible messages in a string. That definition was based on the case where all messages were equally likely, but you can see that it is a good definition in the unequal probability case too. We, therefore, find the expected information in a message, which we write as , by taking the $\log_2$ of (4.27). Assuming N to be very large, and using Stirling's approximation, with which you should be familiar by now, we find:

$$\langle I \rangle = N \sum_{i=1}^{M} \left(-p_i \log_2 p_i \right). \tag{4.28}$$

We can therefore obtain the *average information per symbol*:

$$I / N = \sum_{i=1}^{M} \left(-p_i \log_2 p_i \right). \tag{4.29}$$

This derivation appeals to intuition but it is possible to make it more rigorous. Shannon defined the information in a message to be the base two logarithm of the probability of that message appearing. Note how this ties in with our notion of information as "surprise": the less likely the message to appear, the greater the information it carries. Clearly, the information contained in one particular symbol a_n is:

$$-\log_2 p_n, \qquad (4.30)$$

and if a message contains n_1 a_1s, n_2 a_2s, and so on, its information will be:

$$I = -\log_2\left[\left(p_1^{n_1}\right)\left(p_2^{n_2}\right) \ldots \left(p_M^{n_M}\right)\right] \qquad (4.31)$$

which is:

$$-\left(n_1\log_2 p_1 + n_2\log_2 p_2 + \ldots \ n_M\log_2 p_M\right). \qquad (4.32)$$

Incidentally, this shows that if we place two messages end to end, the total information they convey is twice that in the messages individually, which is satisfying. Check this for yourselves. Now the average information in a message is calculated in standard probabilistic fashion; it is just:

$$Average \ \text{information} = \Sigma \ information \ in \ symbol \ a_i$$
$$\bullet (expected \ number \ of \ appearances \ of \ a_i) \qquad (4.33)$$

$$= -\Sigma\left(\log_2 p_i\right)\times\left(N p_i\right)$$

which is our previous result. Incidentally, Shannon called this average information the "entropy", which some think was a big mistake, as it led many to overemphasize the link between information theory and thermodynamics.[*]
  Here is a nice, and slightly different, illustration of these ideas. Suppose we work for a telegraph company, and we send, with unequal probabilities, a range of messages – such as "Happy birthday", "Isn't Christmas wonderful", and so on. Each one of these messages has a probability P_m of being requested by a customer (m= 1 to M, say). We can define two types of information here. There is that calculated by the person who receives the telegram – since it's my birthday it is not very surprising I get a "happy birthday" message, so there is not much information there. There is also

---

[*] Legend has it that Shannon adopted this term on the advice of the mathematician John Von Neumann, who declared that it would give him "a great edge in debates because nobody really knows what entropy is anyway". [RPF]

the information as seen by the telegraphist who gets requested to send the message. It's interesting to look at the second type. To work this out, we would have to look at the operation of the telegraphy business for some time and calculate the proportions of each type of message that are sent. This gives the probability P_m of message m being sent and we can treat each message as the symbol of some alphabet, similar to labeling each one, rather like the parent-student card we looked at earlier. We can hence calculate the information from the viewpoint of the telegraphist.

4.7 Further Coding Techniques

Let me now return to the topic of coding and describe a couple of popular techniques for coding messages to show you some of the wonderful and ingenious ways in which this can be done. These codes are unlike those we've considered so far in that they are designed for messages in which the symbol probabilities vary.

4.7.1 Huffman Coding

Consider the following system of eight symbols, where, by each, I have written the probability of its occurrence (I have arranged the probabilities in descending order of magnitude):

E	0.50
THE	0.15
AN	0.12
O	0.10
IN	0.04
R	0.04
S	0.03
PS	0.02

The probabilities add to one, as they should. A sample message in this system might be:

ANOTHER

which has probability $0.12 \times 0.10 \times 0.15 \times 0.04$. Now we notice that the symbol E appears much more often than the others: it turns up 25 times as often as the symbol PS, which takes twice as much effort to write. This symbol system doesn't look very efficient. Can we write a new code that improves it? Naively, we might think: "Well, we have eight symbols, so let's just use a three-bit binary code". But since the E occurs so often, would it not be better to describe it, if we can, by just *one* bit instead of three? We might have to use more bits to describe the other symbols, but as they're pretty rare maybe we might still gain something. In fact, it is possible to invent a *non-uniform* code that is much more efficient, as regards the space taken up by a message, than the one we have. This will be an example of compression of a code. Morse had this idea in mind when he assigned a single "dot" to the common E but "dash dash dot dash" to the much rarer Q.

The idea is that the symbols will vary in their lengths, roughly inversely according to their probability of appearance, with the most common being represented by a single symbol, and with the upshot that the typical overall message length is shortened. We will actually replace our symbols with binary strings. The technique I will outline for you – I will let you figure out for yourselves why it works – is due to Huffman. It is quite a popular method, although I believe frequently costly to implement. It is a two-stage process and is best understood by considering the following tree diagram, where initially the symbols are arranged in ascending order of probabilities (Figure 4.3).

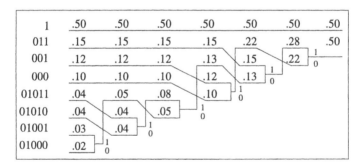

FIGURE 4.3
Huffman coding tree.

Begin by coalescing the two lowest probability symbols into one, adding the probabilities. We can now pretend that we have a source alphabet consisting of the original symbols, less the lower two, plus a new "joint" symbol with probability of occurrence (in this case) of 0.05. Redraw the column, placing the joint symbol at its appropriate point in the probability hierarchy, as shown in Figure 4.3. Now iterate. Coalesce the next two to shrink the list further. Continue in this vein until we reach the right hand of the tree, where we have an "alphabet" of two symbols, the original maximally probable one, plus a summed "joint" symbol, built from all the others.

To figure out the actual assignment of coding symbols, we now retrace a path back through the tree. The rule is straightforward: at each branch in the path required to get back to the original symbol, you add a digit to its code. If you follow the upper path at the branch, you add a one; a lower branch gives you a zero (this is purely a matter of convention). You move from right to left across the tree, but the code you write out from left to right as you go. What is happening is shown in Figure 4.4.

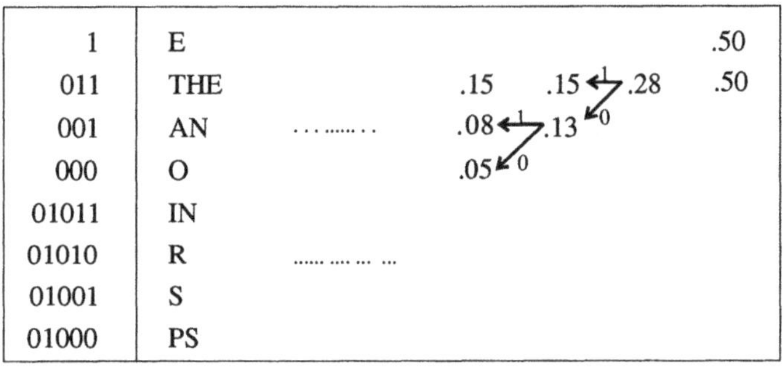

1	E				.50
011	THE		.15	.15 � .28	.50
001	AN	·· ······ ··	.08 ↔ .13 ↖0		
000	O		.05 ↙0		
01011	IN				
01010	R	······ ···· ··· ···			
01001	S				
01000	PS				

FIGURE 4.4
"Trellis" for Huffman coding.

Let us look at the code for "THE". To get to it, we have to start with a 0. We follow the upper path from the first branch, giving us 01 so far. Then, again, we have to follow the upper path from the next branch. We end up with 011, which is the code for THE. The other codes are as shown above. It is worth pointing out that other Huffman codes can be developed by exploiting the ambiguity that occasionally arises when a joint probability at some point equals one of the original probabilities. Do we put the joint above its equal in the table or beneath? You might like to think about this.

We can easily calculate the length saving of this code in comparison with a straight three-bit code. With three bits, the average length of a symbol is obviously three! With this Huffman code, the average symbol length is:

$$(1 \times 0.5) + 3 \times (0.15 + 0.12 + 0.10) + 4 \times (0.04 + 0.04 + 0.03 + 0.02) = 2.13.$$

which is a saving of nearly a third!

There is a nice subtlety in constructing non-uniform codes that the Huffman method takes care of nicely. It has the property that *no code word is the prefix of the beginning of any other code word*. A little thought shows that a code for which this is not true is potentially disastrous. Suppose we had the following set of symbols:

1, 01, 10, 101, 010, 011.

Try and decode this message: 011010110. You can't do it! At least, not uniquely. You do not know whether it is 01-1-01-01-10 or 011-01-01-10 or 01101-01-10 or another possibility. There is an ambiguity due to the fact that the symbols can run into each other. A good, uniquely decodable symbol choice is necessary to avoid this, and Huffman coding is one way forward. You can check that the code choice we have found for our symbols leads to unique decoding.

Problem 4.4 Huffman coding differs from our previous coding methods in that it was developed for compression, not error correction. The technique gives us nicely packed codes, but they are quite sensitive to errors. If we have the following message:

$$00100001101010 \; (= \text{ANOTHER})$$

then a single glitch can easily result in swapped symbols. For example, an error in position 2 would give us THEOTHER*. This throws up an interesting question that you might like to address. For general non-uniform coding, what is the greatest number of symbol errors that can be generated by a one-bit mistake? You are used to thinking of one error – one bit, but with non-uniform coding that might not be true. For example, might it not be possible that a single error might change one symbol to one of a different length, and that this will affect the next symbol, and the next, and so on, so that the error propagates throughout the entire message string? It turns out not. Can you figure out why?

4.7.2 Predictive Coding

Thus far, I have only considered situations in which the probabilities of symbols occurring in a message are independent: symbols exert no influence across the message. However, as I have stressed by the example of English, such dependence is extremely common. The full mathematical treatment of source alphabets comprising varying probabilities of appearance and inter-symbol influence is quite complex and I will not address it here. However, I would like to give you some flavor of the issues such influence raises, and I will do this by considering *predictive coding*. This is another way of compressing codes, rather than correcting them.

Let us suppose that we have a source alphabet that is such that, if we know the contents of a message up to a point, we can predict what the next symbol will be. This prediction will not typically be certain – it will be probabilistic – and the details of how we make it do not matter. The method we use might require the knowledge of only the previous symbol,

---

* It could be said that too many glitches would drive us 01011010010011. [RPF]

or the previous four, or even the whole message. It does not matter. We just have some rule that enables us to predict the next symbol.

Imagine now that we do our predicting with a predictor, a black box that we stick next to the message to be sent, which contains some formula for making predictions. Here are how things work. The predictor is fully aware of the message that has been sent so far (which we illustrate by feeding a message line into it), and on the basis of this, it makes a prediction of what symbol is to be sent next. This prediction is then compared with the actual source symbol that comes up. If the prediction is right, then we send a zero along the transmission channel. If the prediction is wrong, we send a one. The easiest way to implement this is to bitwise add the source symbol and the prediction and ignore any carry. Schematically, we have, at the transmission end (Figure 4.5):

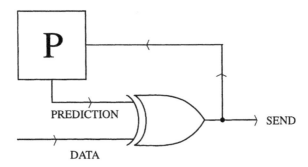

FIGURE 4.5
A predictive encoder.

Note that we have incorporated a feedback loop to let the predictor know whether its prediction was correct or not. A good predictor will produce a long string of zeroes, with the occasional one interspersed where it made a mistake: a random predictor, one that just guesses, will have ones and zeroes split 50–50, if that is the base-rate in the source. It's not difficult to see how, if we send this string, we can reconstruct the original message at the other end by using an identical predictor as a decoder. It simply works backwards. This is all very nice, of course, but what is the point of this rather exotic procedure? Well, if the first predictor is good, making pretty accurate predictions, then it will generate far more zeroes than ones. Interspersed between the ones will be long runs of zeroes. The key is this – when sending the message, we do not send these runs: *instead, we send a number telling us how many zeroes it contained*. We do this in binary, of course. If there is a run of 22 zeroes before the next one digit, we don't send out:

00000000000000000000000

but rather its binary equivalent:

10110.

That's some saving of transmission space! All we have to do is get the guy at the receiving end to break the binary numbers down into strings of zeroes and use his predictor to figure out what we were saying. Predictive coding enables us to compress messages to quite a remarkable degree.

Problem 4.5 An interesting problem with which you can entertain yourself is how to compress things still further. The average length of the runs of zeroes is dependent on how good the predictor is. Once we know how good it is, we can work out the probability that a run will have a given length. We can then use a Huffman technique to get an even tighter code! Work out the details if we are sending an equally likely binary code, and the probability of the predictor being wrong in its prediction is q. You can get pretty close to Shannon's limit using compression of this sort.

4.8 Analog Signal Transmission

I would like to discuss one more coding problem before leaving the subject. This is the question of how we can send information that is not naturally in the form of bits; that is, an analog signal. Ordinarily, information like the oil pressure in a car, the torque in a drive shaft, the temperature variation on the Venusian surface, is *continuous*: the quantities can take any value. If we only have the capacity to transmit bits, how do we send information of this kind? This is not a matter of fundamental principle; it is actually a practical matter. I will say a few words about it despite the fact it is somewhat peripheral. You could say that the whole course is somewhat peripheral. You just wait!

Let us suppose for starters that our continuous quantity – S, say – is restricted to lie between 0 and 1:

$$0 \leq S \leq 1 \tag{4.34}$$

The secret of sending the value of S is to approximate it. The most important question to ask is with what accuracy we want to send data. Suppose we want S to within 1%. Then, all we need do is split the interval [0,1] up into 100 slices (usually referred to as "bins") and transmit information about which slice the value of S is in; in other words, a number between 0 and 99. This is easy. However, as we prefer to use binary, it is better to split the range of S into 128 slices (=2⁷) and send the S value as a seven-bit number. Similarly, if we want to send S to an accuracy of one part in 1000, we would send a ten-bit number, having split [0,1] into 1024 bins.

What happens if the variable S is unbounded? This is not uncommon. Usually, such a variable will have values that are not evenly distributed. In other words, it will be more likely to have some values rather than others (very, very few physical quantities have flat probability distributions). We might have a variable with a probability distribution such as that shown in Figure 4.6.

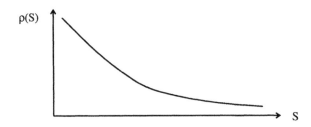

FIGURE 4.6
A sample probability distribution for a physical variable.

The probability density $\rho(S)$ has the usual definition: if we make a measurement of S, the probability of finding its value to lie between S_1 and S_2 is:

$$\int_{S_1}^{S_2} \rho(S)\, dS \qquad (4.35)$$

or, if S_1 and S_2 lie infinitesimally close to one another, $S_2 = S_1 + \delta s$:

$$\rho(S_1).\delta s \qquad (4.36)$$

The basic idea for transmitting S in this general case is the same. We divide the effective range of S into a number of bins with the important difference that we size these bins so that they are all of equal probability (Figure 4.7).

Clearly, the bins are of different widths, but they are chosen to have the same area when viewed probabilistically. They are defined by the formula:

$$\int_{S_i}^{S_{i+1}} \rho(S)\, dS = 1/128 \qquad (4.37)$$

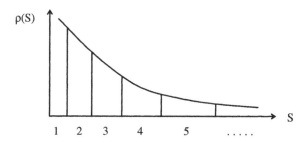

FIGURE 4.7
Division of ρ(S) into equal volume bins.

where i runs from 0 to 127, and the ith bin corresponds to the S-values S_i to S_{i+1}.

Alternatively, we can make a change of variables. For each value s of S we can define the function $P(s)$ by:

$$P(s) = \int_0^s \rho(S) \, dS \tag{4.38}$$

$P(s)$ is just the cumulative probability function of S, the probability that $S < s$. It clearly satisfies the inequality $(0 \le P \le 1)$. One well-known statistical property of this function (as you can check) is that its own distribution is *flat*: that is, if we were to plot the probability distribution of $P(s)$ as a function of s in Figure 4.6, we would see just a horizontal line. A consequence of this is that if we make equal volume bins in P, they will automatically be of equal width. That takes us back to the first case.

A different, but related, problem is that of transmitting a function of time (Figure 4.8).

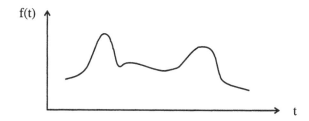

FIGURE 4.8
A typical function of time.

Consideration of such a problem will bring us to the famous *Sampling Theorem*, another baby of Claude Shannon. The basic idea here would be to sample the function at certain regular time intervals, say τ, and send the value of the function at each time digitally. The receiver would then attempt to reconstruct the original signal from this discrete set of numbers. Of course, for a general function, the receiver will have to smooth out the set, to make up for the "gaps". However, for certain types of continuous function, it is actually possible to sample in such a way as to encode *completely* the information about the function: that is, to enable the receiver to reconstruct the source function exactly! To understand how it is possible to describe a continuous function with a finite number of numbers, we have to take a quick look at the mathematical subject of Fourier analysis. I will cover this because I think it is interesting; those without a mathematical background might wish to skip it!

It turns out that if the "Fourier transform" of the function $g(\omega) = 0$ for all $|\omega| \geq \nu$, and we sample at intervals of $\tau = \pi/\nu$, then these samples will completely describe the function. What does this mean? Recall that, according to Fourier theory, any periodic function $f(t)$ can be written as a sum of trigonometric terms. For a general function of time, $f(t)$, we have:

$$f(t) = (1/2\pi) \int_{-\infty}^{\infty} g(\omega) e^{-2\pi i \omega t} \, d\omega \qquad (4.39)$$

where $g(\omega)$ is the *Fourier transform* of $f(t)$. What we have effectively done here is split $f(t)$ up into component frequencies, suitably weighted. Now the typical function (signal) that is encountered in communication theory has a limited *bandwidth*; that is, there is an upper limit to the frequencies that may be found in it (for example, the channel through which the signal is sent might not be able to carry frequencies above a certain value). In such a case, the limits of integration in (4.39) become finite:

$$f(t) = (1/2\pi) \int_{-W(\upsilon)}^{W(\upsilon)} g(\omega) e^{-2\pi i \omega t} \, d\omega, \qquad (4.40)$$

where W is the bandwidth, and ν is now the highest frequency in the Fourier expansion of $f(t)$.[*]

It is possible (the math is a bit tough) to show that this expression reduces to the infinite sum over the integers:

$$f(t) = \sum_{n=-\infty}^{\infty} f(n\pi/\upsilon).[(\sin(\upsilon t - n\pi)]/(\upsilon t - n\pi) \qquad (4.41)$$

---

[*] Conventionally, the bandwidth W is given by $W = \upsilon/2\pi$. [RPF]

This is the Sampling Theorem. If you look at this expression, you will see that as long as we know the values of the function $f(t)$ at the times:

$$t = n\pi / \upsilon, \tag{4.42}$$

where n is an integer, then we can work it out at all other times, as a super-position of terms weighted by the signal samples. This is a subtle consequence of using the well-known relation:

$$(\sin x)/x \to 1 \text{ as } x \to 0 \tag{4.43}$$

in (4.41): setting $t = n\pi / \upsilon$ in the summand, we find that all terms except the nth vanish and the nth is just unity multiplied by the value of f at $t = n\pi/\upsilon$. In other words, if we sampled the function at times spaced (π/υ) time units apart, we could reconstruct the entire thing from the sample! This finding is of most interest in the physically meaningful case when the function $f(t)$ is defined only over a finite interval $(0,T)$. Then, the sum (4.41) is no longer infinite and we only need to take a finite number of sample points to enable us to reconstruct $f(t)$. This number is $(T\upsilon/\pi)$.

Although I have skated over the mathematical proof of the Sampling Theorem, it is worth pausing to give you at least some feel for where it comes from. We are sampling a function $f(t)$ at regular intervals, τ. The graph for the sampled function arises from multiplying that of the continuous $f(t)$ by that of a spikey "comb" function, $C(t)$, which is unity at the sample points and zero elsewhere (Figure 4.9).

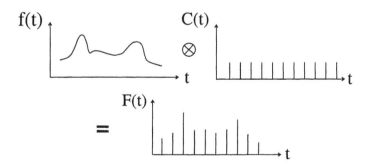

FIGURE 4.9
The sampled function.

Now, corresponding to $f(t)$ is a Fourier transform $\varphi(\omega)$. $C(t)$ also has an associated transform, $\chi(\omega)$, another comb function (Figure 4.10).

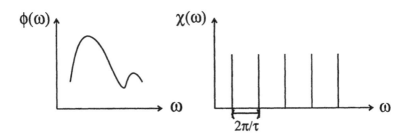

FIGURE 4.10
Fourier transforms of f(t) and C(t).

The transform χ is actually a set of equally spaced delta functions $(2\pi/\tau$ apart). The Fourier transform of the sampled function, $F(t)$, is obtained by the process of "convolution", which in crude graphical terms involves superposing the graph of φ with that of χ. We find that the transform of $F(t)$ comprises copies of the transform of $f(t)$, equally spaced along the horizontal axis, but scaled in height according to the trigonometric ratio in (4.41) as in Figure 4.11.

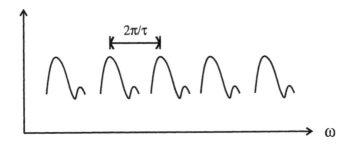

FIGURE 4.11
The Fourier transform of the sampled function.

Look closely at this graph. What it is telling us is that information about the whole of $f(t)$ could, in principle, be extracted from $F(t)$ alone. There is as much information in one of the Fourier-transformed bumps in Figure 4.11 as there is in the whole of Figure 4.9! As the former transform comes solely from the sampled function $F(t)$, we can see the basic idea of the Sampling Theorem emerging.

An interesting subtlety occasionally arises in the sampling process. The Sampling Theorem tells us that, if a signal has an upper-frequency limit of ν (i.e., a bandwidth of $\nu/2\pi$), then we need at least $(T\nu/\pi)$ sample points to enable us to reconstruct the signal. If we take more points than this, all well and good. However, if we take fewer (and this can arise by accident if the function $f(t)$ has "tails" that lie outside the interval (0,T)), our sampling will be insufficient. Under such circumstances, we get what is known as *aliasing*. The sampling interval will be too coarse to resolve high-frequency components in $f(t)$, instead, mapping them into low-frequency components – their "aliases". A familiar example of this phenomenon can be found in movies. Movies, of course, are samples – 24 times a second, we take a snapshot of the world, creating the illusion of seamless movement to our eyes and brains. However, evidence that sampling has occurred often shows up. Maybe the best known is the behavior of wagon wheels in old westerns. As kids, we all noticed that, when a wagon started moving, at first, the spokes in the wheels seemed to go around the right way. Then, as things sped up, they appeared to stop rotating altogether. Finally, as things sped up still further, the wheels appeared to be going the wrong way around! The explanation for this phenomenon lies in inadequate sampling. Another example of aliasing occurs when we inadequately sample audio data and end up with frequencies that we cannot ordinarily hear being aliased into ones we can. To avoid aliasing, we would need to filter out of the signal any unduly high frequencies before we sampled. In the case of movies, this would mean taking pictures with a wider shutter, so that the picture is just a blur, or smoothed out.

It is now possible to send sound *digitally* – 16 bits, 44.1 kHz reproduces perfectly, and is pretty resistant to noise; and such a method is far superior to any analog technique. Such developments will transform the future. Movies will be cleaned up, too – optical fibers, for example, are now giving us overcapacity. The soap ad will appear with absolute clarity. It seems that the technological world progresses but real humanistic culture slides into the mud!

5

Reversible Computation and the Thermodynamics of Computing

CONTENTS

I would now like to take a look at a subject that is extremely interesting, but almost entirely academic in nature. This is the subject of the energetics of computing. We want to address the question: *how much energy must be used in carrying out a computation?* This doesn't sound all that academic. After all, a feature of most modern machines is that their energy consumption when they run very fast is quite considerable, and one of the limitations of the fastest machines is the speed at which we can drain off the heat generated in their components, such as transistors, during operation. The reason I have described our subject as "academic" is because we are actually going to ask another of our fundamental questions: what is the *minimum* energy required to carry out a computation?

To introduce these more physical aspects of our subject I will return to the field covered in Chapter 4, namely the theory of information. It is possible to treat this subject from a strictly physical viewpoint, and it is this that will make the link with the energy of computation.

DOI: 10.1201/9781003358817-5

5.1 The Physics of Information

To begin with, I would like to try to give you an understanding of the physical definition of the information content of a message. That physics should get involved in this area is hardly surprising. Remember, Shannon was initially interested in sending messages down real wires, and we cannot send messages of any kind without some interference from the physical world. I am going to illustrate things by concentrating on a particular, very basic physical model of a message being sent.

I want you to visualize the message coming in as a sequence of boxes, each of which contains a single atom. In each box, the atom can be in one of two places, on the left or right side. If it's on the left, that counts as a 0 bit, if it's on the right, it's a 1. So the stream of boxes comes past me, and by looking to see where each atom is, I can work out the corresponding bit (Figure 5.1).

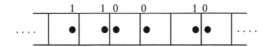

FIGURE 5.1
A basic atomic message.

To see how this model can help us understand information, we have to look at the physics of jiggling atoms around. This requires us to consider the physics of gases, so I will begin by taking a few things I need from that. Let us begin by supposing we have a gas containing N atoms (or molecules) occupying a volume V_1. We will take this gas to be an exceptionally simple one; each atom, or molecule, within it (we take the terms to be interchangeable here) is essentially free – there are no forces of attraction or repulsion between each constituent (this is actually a good approximation at moderately low pressures). I am now going to shrink the gas, pushing against its volume with a piston, compressing it to volume V_2. I do all this isothermally: that is, I immerse the whole system in a thermal "bath" at a fixed temperature T, so that the temperature of my apparatus remains constant. Isn't it wonderful that this has anything to do with what we're talking about? I'm going to show you how. First, we want to know how much work, W, it takes to compress the gas (see Figure 5.2).

FIGURE 5.2
Gas compression.

Now, a standard result in mechanics has it that if a force F moves through a small distance δx, the work* done δW is:

$$\delta W = F\delta x \tag{5.1}$$

If the pressure of the gas is p, and the cross-sectional area of the piston is A, we can rewrite this using $F = pA$ and letting the volume change of the gas $\delta V = A\delta x$ so that:

$$\delta W = p\,\delta V. \tag{5.2}$$

Now we draw on a standard result from gas theory. For an ideal gas at pressure p, volume V and temperature T, we have the relation:

$$p\,V = NkT \tag{5.3}$$

where N is the number of molecules in the gas and k is Boltzmann's constant (approximately 1.381×10^{-23} J K^{-1}). As T is constant – our isothermal assumption – we can perform a simple integration to find W:

$$W = \int_{V_1}^{V_2} \frac{NkT}{V} dV = NkT \log \frac{V_2}{V_1}. \tag{5.4}$$

(Here, $\log x = \log_e x$.) Since V_2 is smaller than V_1, this quantity is negative, and this is just a result of the convention that work done on a gas, rather than by it, has a minus sign. Now, ordinarily when we compress a gas, we heat it up. This is a result of its constituent atoms speeding up and gaining kinetic energy. However, in our case, if we examine the molecules of the gas before and after compression, we find no difference. There is the same number, and they are jiggling about no more or less energetically than they were before. There is no difference between the two at the molecular level. So where did the work go? We put some in to compress the gas, and conservation of energy says it had to go somewhere. In fact, it *was* converted into internal gas heat but was promptly drained off into the thermal bath, keeping the gas at the same temperature. This is actually what we mean by isothermal compression: we do the compression slowly, ensuring that at all times the gas and the surrounding bath are in thermal equilibrium.

From the viewpoint of thermodynamics, what we have effected is a "change of state", from a gas occupying volume V_1 to one occupying

---

* Another one of those awkward words, like "information". Note that, with this definition, a force must move through a distance to perform work; so it does not take any of this kind of "work" to hold up a suitcase – only to lift it! [RPF]

volume V_2. In the process, the total energy of the gas, U, which is the sum of the energies of its constituent parts, remains unchanged. The natural thermodynamical quantities with which such changes of state are discussed are the *free energy* F and the *entropy* S, which are related by:

$$F = U - TS. \tag{5.5}$$

The concept of free energy was invented to enable us to discuss the differences between two states even though there might be no actual mechanical differences between them. To get a better feel for its meaning, look at how expression (5.5) relates small variations at constant temperature:

$$\delta F = \delta U - T \delta S. \tag{5.6}$$

For the change under consideration, the total gas energy remains constant, so $\delta U = 0$ and $\delta F = -T \delta S$. δF is just the "missing" heat energy siphoned off into the heat bath, $NkT \log(V_1/V_2)$, and we use this to write (5.6) as an entropy change:

$$\Delta S = Nk \log \frac{V_2}{V_1}. \tag{5.7}$$

Note that as we are dealing with a finite change here, we have replaced the infinitesimal δ with a finite Δ.

Entropy is a rather bizarre and counter-intuitive quantity and I am never sure whether to focus on it or on the free energy! For those who know a little thermodynamics, the general equation $\delta S = -\delta F/T$ is a variant of the standard formula $\delta S = \delta Q/T$ for the infinitesimal change in entropy resulting from a thermodynamically reversible change of state where, at each stage, an amount of heat δQ enters or leaves the system at absolute temperature T. For an irreversible process, the equality is replaced by an inequality, ensuring that the entropy of an isolated system can only remain constant or increase – this is the Second Law of Thermodynamics. I'll say a little more about entropy in a moment.

Now we take a bit of a leap, and it is not obvious that we can do this, but we can. We consider the case where our gas contains only one molecule. That is, we put $N=1$ into our formulae. Now it's difficult to get a feeling for concepts like temperature, pressure, and volume, never mind free energy and entropy, when you only have one molecule! However, these concepts make sense as long as we consider them to be time averaged, smoothing out the irregularities of this one particle as it bounces back and forth. Indeed, our formulae actually work with $N=1$, as long as there is this hidden smoothing. The situation is more fun, too!

Let us suppose that we are halving the volume occupied by the molecule: $V_2 = V_1/2$. We then find that the free energy and the entropy of the particle change by:

$$+kT\log 2 \ and -k\log 2 \qquad (5.8)$$

respectively. What does this mean? Pictorially, the situation has changed from:

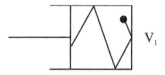

to:

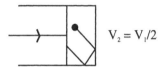

The physical state of the molecule before and after the compression appears to be the same – its actual (kinetic) energy has not changed, for example – yet, for some reason, we have a change in these quantities F and S. What has happened, and this is very subtle, is that my *knowledge of the possible locations of the molecule has changed*. In the initial state, it could be hiding anywhere in volume V_1: after the compression, it must be somewhere within V_2. In other words, there are fewer places it can be in.

   This concept of "knowledge" is extremely important, and central to the concept of entropy, so I will dwell on it awhile. It arises from the deeply statistical nature of thermodynamics. When doing the mathematics of vast numbers of particles that make up gases, we cannot practically follow the paths and momenta of every molecule in the gas, so we are forced to turn to probability theory. Concepts such as temperature and pressure of a gas are essentially defined to be statistical averages. We assign certain physical properties to each molecule, assume particular distributions for these molecules, and calculate the average by a weighting process: so many molecules will move this fast, corresponding to one temperature; so many will move that fast, giving another temperature; and we just average over everything. The entropy of a gas is defined statistically, indeed, this is its core definition, but in a different way to quantities such as temperature and energy. Unlike these, it is not a macroscopic property that arises from a sum of microscopic properties. Rather, it is directly related to the *probability that the gas be in the configuration in which it is found*. By "configuration", I mean a particular arrangement, or cluster of arrangements, of positions and momenta for each of the N constituent molecules (or, if you want to be fancy, a particular point or region in "phase space"). The existence of such a probability should not come as too much of a surprise: if you look at any given gas, it is far less likely at the outset that you will find all the molecules moving in the same direction or paired up and dancing than you will find them shooting all over the place at random. Entropy quantifies this notion. Loosely speaking, if the probability of a particular gas configuration is W, we have:

$$S \approx k \log W. \tag{5.9}$$

The bigger W, the bigger the entropy, and, like all probabilities, the Ws add, so we can straightforwardly calculate the chances of being in some range of configurations. The gas with molecules going all one way has a W much less than that of the one with a more random – or *more disordered* – structure, and hence has a lower entropy. What has all this got to do with our knowledge of a system? Simply, the less we know about the configuration of a gas, the more states it could be in, and the greater the overall W – and the greater the entropy. This gives us a nice intuitive feel of what is happening when we compress a gas into a smaller volume. Working isothermally, the momenta of the molecules within the container remains the same (δU=0), but each molecule has access to fewer possible spatial positions. The gas has therefore adopted a configuration with smaller W, and its entropy has decreased. As an aside, the Second Law of Thermodynamics tells us that in any isolated system:

$$\delta S \approx k \, \delta W / W \geq 0, \tag{5.10}$$

i.e., the entropy never decreases. The fact that the entropy of our compressed gas has dropped is a reminder that the system is not isolated – we have been draining heat into a heat bath. The heat flow into the bath increases its entropy, preserving the Second Law. Generally speaking, *the less information we have about a state, the higher the entropy.*

  As the definition of entropy is essentially statistical, it is perfectly all right to define it for a gas with a single molecule, such as the one we have been considering, although there are a few subtleties (which we will avoid). You can see that if we compress the volume by a factor of 2, then we halve the number of spatial positions, and hence the number of configurations that the molecule can occupy. Before, it could be in either half of the box: now, it can only be in one half. You should be able to see in our probabilistic picture how this leads to a decrease in entropy by an amount:

$$\delta S = k \log 2 \qquad (5.11)$$

This is the same as we obtained with our work and free energy considerations.

We can now return to the topic of information and see where all this weird physics fits in. Recall the atomic tape with which we opened this section, in which the position of atoms in boxes tells us about the binary bits in the message. Now, if the message is a typical one, for some of these bits we will have no prior knowledge, whereas for others we will – either because we know them in advance, or because we can work them out from correlations with earlier bits that we have examined. We will *define* the information in the message to be *proportional to the amount of free energy required to reset the entire tape to zero*. By "reset to zero", we mean compress each cell of the tape to ensure that its constituent atom is in the "zero" position.

Straightaway, we note what seems to be an obvious problem with this definition, namely, that it introduces an unnatural asymmetry between 0 and 1. If an atom is already in the zero part of the compartment, then surely the reset operation amounts to doing nothing, which costs no free energy. Yet, if it is in the one position in the compartment, we have to do work to move it into the zero position! This doesn't seem to make sense. One would expect to be able to introduce an alternative definition of information for which the tape is reset to one – but then we would only seem to get the same answer if the message contained an equal number of ones and zeroes! But there is a subtlety here. Only if we *do not know* which side of the compartment the atom is in do we expend free energy. It is only in this circumstance that the phase space for the atom is halved, and the entropy decreases. If we know the atom's position, then we expend *no* energy in resetting, irrespective of where the atom starts out. In other words, as one would hope, the information in the message is contained in the surprise bits. Understanding why this is so is worth dwelling on, as it involves a style of argument often seen in the reversible computing world. It seems a bit counter-intuitive to claim that the energy required to reset a one to a zero is no more than leaving a zero alone – in other words, nothing.

To clear this point up, I first have to stress the idealized nature of the setup we are considering. Although I have talked freely about atoms in "boxes", these boxes are not real boxes made of cardboard and strung together, with mass and kinetic and potential energy. Moreover, when I talk about "energy", I certainly don't mean that of the tape! We are only interested in the content of the message, which is specified by the positions of the atoms. Let us suppose we have a message bit that we know is a one – the atom is on the right-hand side – so we have the following picture:

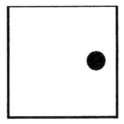

We can show that resetting this to zero costs no energy in several ways. One pretty abstract way is to first slip in a little partition to keep the atom in place, All I have to do now is *turn the box over*. The end result is that we now have a zero on the right-hand side (Figure 5.3).

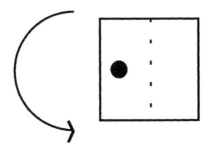

FIGURE 5.3
A simple reset procedure.

This is abstract because it might seem odd to be able to insert pistons and turn boxes without expending energy. In the real world, of course, you can't – but we are dealing with abstractions here and, as I have said, we are not interested in the kinetic energy or weight of the "boxes". Given our assumptions, it is possible to do so, although the downside is that we would have to take an eternity to do it! (We will return to this sort of argument in Section 5.2.) Another way, perhaps a little less abstract, would be to introduce two pistons, one on each side of the box, and push the atom over with one, while drawing the other out (Figure 5.4).

Now the bombardment on the left is equal to that on the right, and any work put in at one end will be taken out at the other, and so is recovered. One could even join the pistons by an up-and-over rod, and you should be

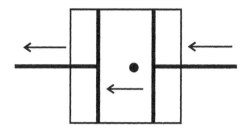

FIGURE 5.4
A more "realistic" reset.

able to see that the tiniest touch on one piston will send the whole thing coasting over to its final position. So, if you do it slowly enough – "infinitesimal in the limit" – no work is done in resetting. Clearing, or resetting, the tape is what occurs when we don't know what compartment the atom is in. Then we must perform a compression and this will take free energy, as we discussed earlier, as we are lessening our ignorance of the atom's position.

Another way of looking at these ideas is due to Bennett, who suggests using a message tape as *fuel*, and relates the information in the tape to its fuel value – that is, to the amount of energy we can get from it. His idea, which is quite subtle, goes as follows. We suppose we have a machine, in contact with some kind of heat bath, which takes in tapes at one end, and spits them out at the other. We assume to begin with, that the tape the machine eats is blank, that is, all of its atoms are in the zero state. We will show how such a tape can be used to provide us with useful work, which we can use to power our machine.

What we do is incorporate a piston into the system. As each cell comes in, we bring the piston into it, up to the halfway position in each box (Figure 5.5).

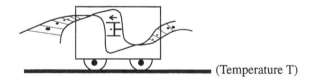

(Temperature T)

FIGURE 5.5
An information-driven engine.

We now let the heat bath warm the cell up. This will cause the atom in the cell to jiggle against the piston, isothermally pushing it outwards as in Figure 5.6.

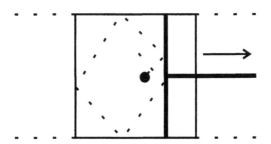

FIGURE 5.6
Work generation mechanism in the engine.

This is just the opposite process to the compression of a gas we considered at the beginning of this section. The net result is that work is done on the piston, which we can subsequently extract, in other words, we can get our tape to do work for us. You should be able to see that for a tape of n bits, this work is equal to $nkT\log 2$, the free energy, where T is the temperature of the heat bath. An important consequence of our procedure is that the tape that the machine spits out has been *randomized*: after the piston has been pushed out, the atom that did the pushing can be anywhere in that cell, and we have no way of knowing where, short of performing a measurement.

We now generalize the argument by assuming that our piston is maneuverable. This allows us to extract work from tapes that have a 1 in them. If we get a 1, we switch the piston to the other side of the cell, bring it up to the edge of the 1 half, and proceed as before. Again, we get $kT\log 2$ of useful work given out, and again the tape that emerges from the machine is randomized. What is crucial here is that we *know* what bit is about to enter the machine. Only then can we ready the piston to ensure that it does work for us. Obviously, if we left the piston in the 0 position, and we got a 1 in, we would actually have to do work to shift the atom into the 0 cell, and when the atom expands back into the full cell, we would get that work back: that is, no useful work would be done. Clearly, *a random tape has zero fuel value*. If we do not know what bit is coming in next, we do not know how to set our piston. So we would leave it in one position, and just push it in and hope, push it in and hope, boom, boom, boom. Sometimes we would get lucky, and find an atom pushing our piston out again, giving us work; but equally likely, for a truly random message, we have to do work on the atom. The net result is zero work to power our machine.

Clearly, Bennett's tape machine seems to do the opposite of our reset process. He uses a message tape to extract work, ending up with a random tape: we took a random tape and did work on it, to end up with a tape of standard zeroes. This inverse relationship is reflected in the definition of information within Bennett's framework. Suppose we have a tape with N bits. We *define* the information, I, in the tape by the formula:

$$Fuel\ value\ of\ tape = (N - I).kT\ \log 2. \tag{5.12}$$

From this, we see that a tape giving us a full fuel-load – that is, $kT\log 2$ per bit – carries zero information. This is what we would expect since such a tape must have completely predictable contents. There is a nice physical symmetry between these two approaches. If we run a message tape through the machine, we will be able to extract a certain energy E from it: this energy E will be precisely what we need to reset the newly randomized tape to its original form. It is, of course, up to you which picture you prefer to adopt when thinking about these things. I opt for the erasure picture partly because I do not like having to subtract from N all the time to get my information!

You might like to contemplate some problems on Dr. Bennett's machine.

Problem 5.1 Suppose we have two tapes: an N-bit random tape, and an exact copy. It can be shown that the fuel value of the two tapes combined is $NkT\log 2$. See if you can design a machine that will be able to extract this amount of energy from the two tapes. (Hint: you have to expand one tape "relative" to the other.)

Problem 5.2 We have a tape in which three bits are repeated in succession, say 110110110110... For a $3N$-bit tape, what is the fuel value? How do you get it out?

5.1.1 Maxwell's Demon and the Thermodynamics of Measurement

Those of you who wish to take your study of the physics of information further could do no better than check out many of the references to a 19th-century paradox discovered by the great Scottish physicist James Clerk Maxwell. *Maxwell's Demon*, as it is known, resulted in a controversy that raged among physicists for a century, and the matter has only recently been resolved. In fact, it was the contemplation of Maxwell's demon that partly led workers such as Charles Bennett and Rolf Landauer to their conclusions about reversible computing, the energy of computation, and clarified the link between information and entropy. Importantly, such research has also shed light on the role of *measurement* in all this. I will not go into the matter in great detail here but supply you with enough tidbits to at least arouse your interest.[*]

---

[*] A full discussion of the demon and of the attempts to understand it can be found in *Maxwell's Demon: Entropy, Information, Computing* by H.S. Leff and A.F. Rex (Adam Hilger, 1990). [Editor]

With Maxwell, we imagine that we have a small demon sitting on a partitioned box, each half of which is filled by a gas of molecules with a random distribution of positions and velocities (Figure 5.7).

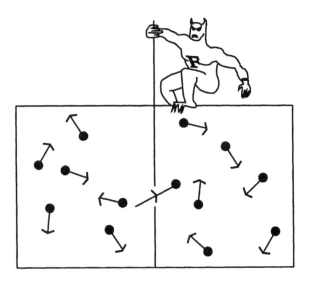

FIGURE 5.7
Maxwell's demon at work.

The demon has a very simple task. Set into the partition is a flap, which he can open and shut at will. He looks in one half of the box (say, the left) and waits until he sees a fast-moving molecule approaching the flap. When he does, he opens the flap momentarily, letting the molecule through into the right side, and then shuts the flap again. Similarly, if the demon sees a slow-moving molecule approaching from the right side of the flap, he lets that through into the side the fast one came from. After a period of such activity, our little friend will have separated the fast- and slow-moving molecules into the two compartments. In other words, he will have separated the hot from the cold, and hence created a temperature difference between the two sides of the box. This means that the entropy of the system has decreased, in clear violation of the Second Law!

This seeming paradox, as I have said, caused tremendous controversy among physicists. The Second Law of Thermodynamics is a well-established principle in physics, and if Maxwell's demon appears to be able to violate it, there is probably something fishy about him. Since Maxwell came up with his idea in 1867, many people have tried to spot the flaw in his argument. Somehow, somewhere, in the process of looking for molecules of a given type and letting them through the flap, there had to be some entropy generated.

Until recently, it was generally accepted that this entropy arose as a result of the demon's *measurement* of the position of the molecules. This did not seem unreasonable. For example, one way in which the demon could detect fast-moving molecules would be to shine a demonic torch at them; but such a process would involve dispersing at least one photon, which would cost energy. More generally, before looking at a particular molecule, the demon could not know whether it was moving left or right. Upon observing it, however this was done, his uncertainty, and hence entropy, would have reduced by half, surely accompanied by the corresponding generation of entropy in the environment.

In fact, and surprisingly, Bennett has shown that Maxwell's demon can actually make its measurements with zero energy expenditure, providing it follows certain rules for recording and erasing whatever information it obtains. The demon must be in a standard state of some kind before measurement, which we will call S: this is the state of uncertainty. After it measures the direction of motion of a molecule, it enters one of two other states – say L for "left-moving", or R for "right-moving". It overwrites the S with whichever is appropriate. Bennett has demonstrated that this procedure can be performed for no energy cost. The cost comes in the next step, which is the *erasure* of the L or R to reset the demon in the S state in preparation for the next measurement. This realization, that it is the erasure of information, and not measurement, that is the source of entropy generation in the computational process, was a major breakthrough in the study of reversible computation.

5.1.2 Energy and Shannon's Theorem

Before leaving physics and information, I would like to return to something we studied in Chapter 4, namely, the limits on sending information down a channel. It will come as no surprise to you that we can revisit Shannon's Theorem with our physical tools too! Let us combine our study of the physics of information with our earlier work on errors. An interesting question is: How does the occurrence of an error in a message affect its information content? Let's start off with a message with all its M bits perfectly known, containing information N, and suppose that we want to send it somewhere. We're going to send it through a noisy channel: the effect of this is that, in transit, each bit of the message has a probability q of coming through wrong. Let us ask a familiar question: what is the minimum number of bits we have to send to get the information in the M bits across? We will have to code up the message, and in keeping with our earlier look at this question, we'll say the coded message has length M_c. This is the number of bits we actually send. Now we have said that to clear the tape, assuming we know nothing about its contents, we need to expend the following amount of free energy:

$$M_C kT \log 2. \tag{5.13}$$

However, some of this energy is taken up in clearing errors. On average, using our earlier derivations, this amount will be:

$$M_C kT \log 2\left[-q \log_2 q - (1-q)\log_2(1-q)\right] = \left[1 - f(q)\right]M_C kT \log 2. \tag{5.14}$$

This energy we consider to be wasted. This leaves us with the free energy:

$$M_C kT\log 2 - \left[1 - f(q)\right]M_C kT \log 2 = f(q)M_C kT\log 2 \tag{5.15}$$

to expend in clearing the message. By conservation of energy, then, and using our relationship between free energy and information, the greatest amount of information I can send through this channel will be:

$$M_c\left[q \log_2(1/q) + (1-q)\log_2(1/(1-q))\right] \tag{5.16}$$

You can see how this kind of physical argument now leads us to Shannon's result.

5.2 Reversible Computation and the Thermodynamics of Computing

It has always been assumed that any computational step required energy.[*] The first guess, and one that was a common belief for years, was that there was a minimum amount of energy required for each logical step taken by a machine. From what we have looked at so far, you should be able to appreciate the argument. The idea is that every logical state of a device must correspond to some physical state of the device, and whenever the device had to choose between 0 and 1 for its output – such as a transistor in an AND gate – there would be a compression of the available phase space of the object from two options to one, halving the phase-space volume. Therefore, the argument went, a minimum free energy of $kT\log 2$ would be required per logical step.[†] There have been

---

[*] Detailed accounts of the history of this subject can be found in the papers "Zig-zag Path to Understanding", R. Landauer, Proceedings of the Workshop on Physics and Computation Physcomp '94, and "Notes on the History of Reversible Computation", C.H. Bennett, IBM J. Res. Dev. 32(1), pp. 16–23 (1988). [Editor]

[†] This is actually a lower limit far beneath anything practically realizable at present. Conventional transistors dissipate on the order of $10^8 kT$ per step! [RPF]

other suggestions. One focused on the reliability of the computational step. The probability of an error, say q, was involved and the minimum energy was supposed to be $kT\log q$. However, recently this question has been straightened out. The energy required per step is less than $kT\log q$, less than $kT\log 2$, in fact, less than any other number you might want to set – provided you carry out the computation carefully and slowly enough. Ideally, the computation can actually be done with *no* minimal loss of energy. Perhaps a good analogy is with friction. In practice, there is always friction, and if you take a look at a typical real-world engine, you will see heat energy dissipated all over the place as various moving parts rub against one another. This loss of energy is ordinarily large. However, physicists are very fond of studying certain types of idealized engines, so-called Carnot heat engines, in which heat energy is converted into work and back again, for which it is possible to calculate a certain maximum efficiency of operation. Such engines operate over a *reversible* closed cycle: that is, they start off in a particular state and, after one cycle of operation, return to it. The Second Law ensures that this cannot be done for zero energy cost but it is theoretically possible to operate such machines in such a way as to achieve the maximum efficiency, making the losses due to friction, for example, as small as possible. Unfortunately, they have to be run infinitesimally slowly to do this! You might, for example, want to drain heat from the engine into a surrounding reservoir to keep everything at thermal equilibrium, but if you operate the machine too quickly, you will not be able to do this smoothly and will lose heat to parts of the engine that will simply dissipate it. But the point is that, in principle, such engines could be made, and physicists have learned much about thermodynamics from studying them. The crucial requirement is reversibility. Now it turns out that a similar idea works in computers. If your computer is reversible, and I'll say what I mean by that in a moment, then the energy loss could be made as small as you want, provided you work with care and slowly – as a rule, infinitesimally slowly. Just as with Carnot's engines, if you work too fast, you will dissipate energy. Now you can see why I think of this as an academic subject. You might even think the question is a bit dopey – after all, as I've said, modern transistors dissipate something like $10^8 kT$ per switch – but as with our discussion of the limits of what is computable, such questions are of interest. When we come to design the ultimate computers of the far future, which might have "transistors" that are atom-sized, we will want to know how the fundamental physical laws will limit us. When you get down to that sort of scale, you really have to ask about the energies involved in computation, and the answer is that there is no reason why you shouldn't operate below kT. We shall look later at problems of more immediacy, such as how to reduce the energy dissipation of modern computers involving present-day transistors.

5.2.1 Reversible Computers

Let me return to the matter of "reversible computing". Consider the following special kind of computation, which we draw as a black box with a set of input and output lines (Figure 5.8):

FIGURE 5.8
A reversible computation.

Suppose that for every input line there is one, and only one, output, and that this is determined by the input. (In the most trivial case, the signals simply propagate through the box unchanged.) In such circumstances, the output carries no more information than the input – if we know the input, we can calculate the output and, moreover, the computation is "reversible". This is in sharp contrast to a conventional logic gate, such as an AND (Figure 5.9).

FIGURE 5.9
The AND gate.

In this case, we have two lines going in but only one coming out. If the output is found to be zero, then any one of three possible states could have led to it. I have irretrievably lost information about the input so the AND gate is irreversible. So too, is the OR gate (but not the NOT!). In other words, the phase space of the inputs has shrunk to that of the output, with an unavoidable decrease in entropy. This must be compensated by heat generation somewhere. The mistake everyone was making about energy dissipation in computers was based on the assumption that logical steps were necessarily like AND and OR – irreversible. What Bennett and others showed was that this is not necessarily the case. The fact that there is no gain in information in our abstract "computation" above is the first clue that maybe there's no loss of entropy involved in a reversible computation. This is actually correct: reversible computers are rather like Carnot engines, where the reversible ones are the most efficient. It will turn out that the only entropy loss resulting from operating our abstract machine comes in resetting it for its next operation.

We can consider a "higher" kind of computer that is reversible in a more direct sense: it gives as its output the actual result of a computation plus the original input. That is, it appends the input data to the output data printed on its tape (say). This is the most direct way of making a computation reversible. We will later show that, in principle, such a calculation can be performed for zero energy cost. The only cost is incurred in resetting the machine to restart, and the nice thing is that this does not depend on the complexity of the computation itself but only on the number of bits in the answer. You might have billions of components whirring away in the machine, but if the answer you get out is just one bit, then $kT\log 2$ is all the energy you need to run things.

We actually studied some reversible gates earlier in the course. NOT is one, as I've said. A more complicated example we looked at was Toffoli's CONTROLLED CONTROLLED NOT gate (Figure 5.10).

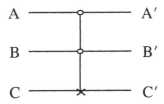

FIGURE 5.10
The CCN gate.

The lines A and B act as control lines, leaving C as it is unless both are one, in which case C becomes NOT C. This is reversible in the sense that we can regain our input data by running the output through another CCN gate (see Section 2.3).

I would now like to take a look in more detail at some reversible computations and demonstrate the absence of a minimum energy requirement. I'll start with a computation that you might not ordinarily think of as a computation: the act of copying (recall our discussion of Turing copying machines, Section 3.5). This seems like a dumb sort of computation, as you're not getting anywhere, but it is a useful introduction to some of the ideas underlying issues of energy dissipation. It's not at all obvious that you can copy information down from one place to another without expending at least some energy, even in principle. Having said this, it is easy to suggest why it might not cost any energy. We can consider a set of data and its copy as two messages on tape, both identical. Either we know what the original message is, or we don't. In the first case, no free energy is expended in clearing the tape, and none need be for the copy tape: we just turn it over when necessary, as we discussed previously. In the second case, clearing the tape will cost free energy, but not for the copy: knowing what the first tape says, we can use this information to clear the copy by turning bits over again. Simply, there is no more information in the (data plus copy) set than is in just the single data set. Clearing the system should not, therefore, require more free energy in the first case than the second. This is a common type of argument in the reversible computing world.

5.2.2 The Copy Computation

Let us make these ideas a little more concrete. In a moment, I will examine a copying machine found in nature, namely the RNA molecule found in living cells. But first, I will take a look at two rather artificial examples of copying machines. Our discussion follows Bennett.

   We start with a very general copy process. We will have an original object, which we'll call the model, which can somehow hold a zero or one. It's some kind of bistable physical device. We want another object, which we'll call the copier, which can also hold a zero or one. An example of a bistable device would be one that could be modeled by the following potential well (Figure 5.11).

FIGURE 5.11
A potential well.

I will give one possible physical realization of this shortly. What this rather abstract diagram means is that some part of the device, which we will represent by a dot, can be in either of two stable states – here, in the left or the right trough, meaning one or zero, say. The curve displays the potential energy of the dot according to its position in the device. The troughs are the minima of this energy and are favored by the dot: they are of equal depth and are hence equally likely to be occupied at the outset. A useful way to think of this operation is to have the dot as a ball, and the curve as an actual shape constraining it. Putting energy into the ball makes it move up and down the sides of its trough; with enough energy the ball will go over the hill and into the next trough – equivalent to our model changing its bit-state. The height of the hill, the amount of energy needed for the transition to occur, is called the barrier potential. In actual operation, we would want the typical thermal fluctuations of whatever it is the dot represents to be much less than this to keep the device stable. Another way of visualizing this is to imagine the dot to be in a box separated into two halves by a partition. The barrier potential will be the energy required to get the dot from one half into the other.

We suppose both model and copier to be modeled by such a potential, and the model to be in some state. This can be random – we need not know what it is, but for sake of illustration let's say it is as shown in Figure 5.12 (where we have used an X for the model's dot).

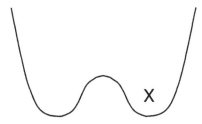

FIGURE 5.12
Initial state of the model.

How does the copier start out? It must be in some standard state. It cannot be in a random state because copying will involve getting it into a definite state, and to do this, we must do work (compressing, if we use the box and partition analogy). Alternatively, you can use phase-space considerations, comparing the number of possible model-copier options before copying (four, if the model is randomly set) and after (just two): this would be a logically irreversible step. Let's say the copier starts out in the state opposite to the model (Figure 5.13).

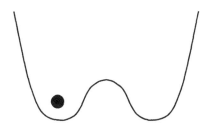

FIGURE 5.13
Initial state of the copier.

Clearly, copying is going to involve somehow getting the dot from one trough to the other. To do this, we need to be able to manipulate the potential curve; we have to make the other trough energetically more favorable to the dot. We shall assume that there are two parameters associated with the copier that we can adjust: the barrier height, and the relative depths of the troughs. Furthermore, we assume that the depths of the troughs can be altered by some force of interaction between the copier and the model. (Don't worry if this is all horribly confusing and abstract! All will become clear.) We'll call this a "tilt" force because it tilts the graph. We will combine these two operations to move the copier dot, but we will combine them in such a way – and this is important – that there will always be a unique minimum accessible to the dot at all times.

What we do is this. We start with the model some way away from the copier. Even at a distance, it will exert a slight tilt force on the copier. We take this force to have the consequence of increasing the depth of whichever trough of the copier corresponds to that occupied in the model. The copier potential will hence be slightly distorted at the outset, as shown in Figure 5.14.

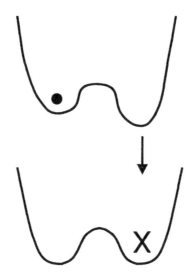

FIGURE 5.14
Initial copier distortion.

The first step in the copy process involves gently lowering the copier's potential barrier. This removes the obstacle to the dot switching positions: it can now wander over to the other bit state. What will make it do this? This is where the "tilt" from the model comes in. In step two, we slowly bring the model up closer to the copier, and in the process, its tilt force increases. This gradually distorts the copier's potential even more, lowering the energy of the appropriate trough as shown in Figure 5.15.

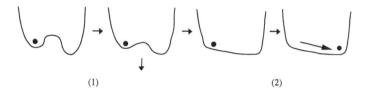

FIGURE 5.15
Lowering the potential barrier and tilting.

The dot now slides smoothly down the potential curve, occupying the new, energetically more favorable trough. In step three, we replace the potential barrier to secure the dot in its new position, and finally, step four, we take the model away, restoring the copier's potential to its normal state (Figure 5.16).

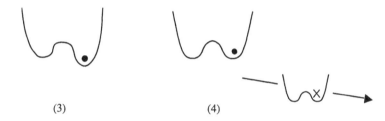

FIGURE 5.16
Final state of the system.

That is the basic idea of this copy machine. It's possible to play around with it further. For example, for appropriate physical systems, we can envisage bringing the model up to the copier in step one in such a way that the tilt force lowers the state the dot is already in so that the dot is held steady while we lower the potential barrier if this is a concern. The model is then moved over to the other side to provide the new tilt. This is one variation, but it does not significantly alter the basic idea.

The crucial thing about this process is that it needs to be carried out slowly and carefully. There are no jumps or sudden changes. The easiest way to get the dot from one trough to the other would be to bring the model up rapidly to bias the troughs in the desired way, then rip away the potential barrier. The dot would then slosh over into its new trough, but the whole process, while nice and quick, would invariably involve dissipation in a real system. However, if the procedure is graceful enough, the lowering of the barrier, the tilting of the trough, and the copying can be done for nothing. This is basically because the physical quantities that contribute to the energy dissipation – such as the kinetic energy of the dot moving to its new state, the work done in raising and lowering the barrier – are negligible under such circumstances. You should be able to see, incidentally, that this procedure will work even if we don't know what state the model is in.

When Bennett discovered all this, no one knew it could be done, although much of the preliminary groundwork had been carried out by his IBM colleague, Landauer, as far back as 1961. There was a lot of prejudice around that had to be argued against. I see nothing wrong with his arguments. I was asked by Carver Mead at Caltech to look into the energy consumption of computers, so I looked at all this stuff and gradually concluded that there was no minimum energy. This was something of a surprise to me! Bennett's result was four years old by then but there were still people fighting over it. Also, it's nice to work this sort of thing out for yourself: as I said in Chapter 1 – OK, you're not the first, but at least you understand it!

5.2.3 A Physical Implementation

Let me return to the preceding example and give you something that is essentially a physical realization of it. It is also fun to think about! We need some kind of bistable physical device, and here it is: two compass needles – just two magnetic dipoles on pivots. One end is North and the other South, and as we all know North attracts South and vice versa; otherwise, we have repulsion. Now suppose that both the model and the copier are made up of such a pair. To make the analysis easier, we insist that each member of a pair is linked to the other, in such a way that both members must point in the same direction. This means that we can analyze each system in terms of just one variable, the angle φ the needles make with the horizontal. So we have the allowed and disallowed situations shown in Figure 5.17a,b).

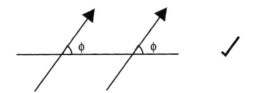

FIGURE 5.17A
Allowed angular configuration.

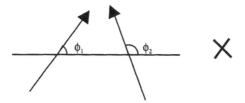

FIGURE 5.17B
Disallowed angular configuration.

The disallowed case would, in any case, clearly be unstable. Now, not all alignments of the needles within a pair have the same potential energy. This is obvious by comparing the states shown in Figure 5.18.

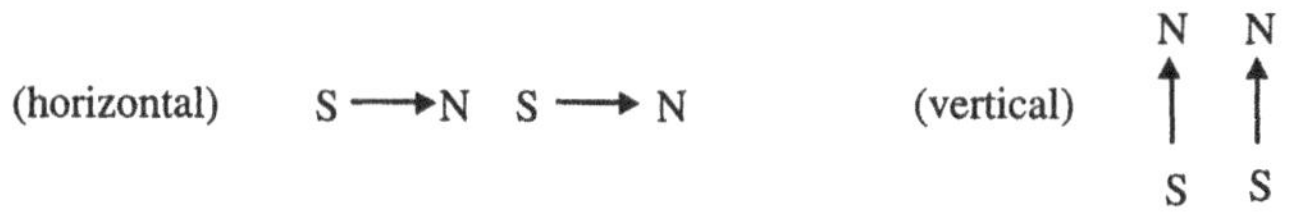

(horizontal)   S —→ N   S —→ N   (vertical)

FIGURE 5.18
Stable and unstable states.

The first is evidently quite stable, with the tip of one needle attracting the base of the other. The second, with both arrows vertical, is quite unstable: the North poles will repel, and the needles will seek to occupy the first state or its mirror image. We can actually calculate the potential energy for a state with angle φ. It is approximately (close enough for us) given by:

$$Potential\ energy \approx \sin^2 \phi \tag{5.17}$$

This potential energy function looks like the graph in Figure 5.19.

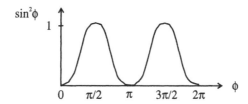

FIGURE 5.19
Potential energy as a function of φ.

Note how similar this is to our abstract potential well. The minima are at $\phi = 0$ and $\phi = \pi$, corresponding to the stable "horizontal" states, while the maxima correspond to the vertical states at $\pi/2$ and $3\pi/2$. (Remember that the graph wraps around at $\pi = 0$ and 2π.) The system is clearly bistable and we can see that once the needles are in one of the two minima, energy would have to be expended to push them to the other.

To manipulate the barrier in this case, we introduce a vertical magnetic field B. It can be shown that this adds a term:

$$- B \sin \phi \tag{5.18}$$

to the potential energy. As we increase B, the effect is to lower the barrier between the 0 and π states as shown in Figure 5.20.

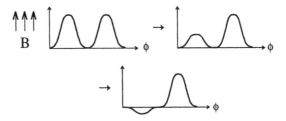

FIGURE 5.20
Barrier manipulation in the dipole copier.

(You can play with numbers to gauge the exact effect of this.) The tilt force, as before, results from bringing the model closer to the copier; this time, we can see what it is about the model that causes this force – it is the magnetic field from the data bit. The force is perpendicular to B, and in the direction of the needles in the model. If we call it b, then it contributes:

$$- b \cos \phi \tag{5.19}$$

to the potential energy. This clearly removes the symmetry about $\pi/2$ and $3\pi/2$ and represents a tilting. We can now see how the copying process works. We start with the copier in a standard state, which we take to be the $\phi = 0$ state (->->). We gently turn up the field B – or alternatively slowly move the copier from a region of weak B to one of high B – until the barrier is removed. At this stage, the dipole is vertical (Figure 5.21).

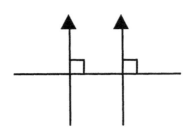

FIGURE 5.21
Initial (unstable) copier state.

Now we bring in the model. This has already been slightly perturbing the copier pair, but not enough to have a noticeable effect so far. Now, as it gets closer, its field biases the copier needles to flip over – but not suddenly! – into a new state. (This is if a new state is appropriate: if the standard state and the model state coincide, the needles will simply return to their original position.) The model is removed, the copier taken out of the field B to restore the barrier, and the copying is finished.

Once again, you can check that this copying method will work if we do not know what the model state is. It is not difficult to see that, if performed slowly, it will cost no energy – no current, no nothing. My previous discussion was to show you the principles; this specific example is probably easier to understand.

5.2.4 A Living Computer

The foregoing example of two dipoles has a certain physical basis but is undeniably artificial. However, here's a copying process that really is found in Nature and is one that involves thermodynamical, rather than mechanical, forces.* It occurs as one of the steps in the synthesis of proteins in a living cell. Now you probably know what proteins are – long, twisted molecular chains of amino acids (such as tryptophan or alanine) – and you may know how central they are both to the structure and functioning of living things. However, a proper understanding of the complex business that is their manufacture would require an understanding of biochemistry lying way beyond this course! I can't make up for that here, so I'll just try to give you enough background to let you see how the copying "machine" I have in mind behaves.

A living creature typically contains a huge number of different types of protein, each uniquely defined by some combination of specific amino acids. If the cell is to manufacture these molecules, then clearly a set of "design rules" for each protein-type must be available somewhere. This information actually resides in the DNA (deoxyribonucleic acid) molecule, the famous "double-helix" structure that sits in the cell nucleus. DNA comprises a double chain, each strand of which is made up of alternating phosphate and pentose sugar groups. To each sugar group is attached one of four bases, A (adenine), T (thymine), C (cytosine), and G (guanine) – a base-sugar-phosphate group is called a nucleotide. It is a certain sequence of bases that provides the code for protein synthesis.

We can break down the synthesis of proteins into two stages. The first stage, and it is only this that interests us, requires the formation of another, linear, strand of sugar phosphates with bases attached, called messenger RNA (or mRNA). The code on the DNA is copied onto the RNA strand base

---

* For a discussion of this topic in the literature, see C. Bennett, Int. J. Theor. Phys 21, pp. 905–940 [1982]. [RPF]

by base (according to a strict matching rule, which I shall come on to), and the mRNA, once completed, leaves the nucleus and travels elsewhere to assist in the making of the protein. The machine that does the copying is an enzyme called RNA polymerase. What happens is this. The DNA and enzyme are floating around in a crazy biological soup that contains, among other things, lots of triphosphates – such as ATP (adenosine triphosphate), CTP, GTP, and UTP (U is another base, uracil). These are essentially nucleotides with two extra phosphates attached. The polymerase attaches itself to whichever part of one of the DNA strands it is to duplicate and then moves along it, building its RNA copy base by base by reacting the growing RNA strand with one of the four nucleotides present in the soup. (A crucial proviso here is that RNA is built out of the four bases A, G, C, and U (not T) and that the RNA strand must be complementary to that on the DNA; the complementarity relationships are that As on the DNA must match with Us on the RNA, Ts with As, Cs with Gs, and Gs with Cs). The nucleotides are provided in the triphosphate form, and during the addition process two of the phosphates are released back into the soup, still bound together (as a pyrophosphate). The nucleotide chosen must be the correct one, that is, complementary to the base on the DNA strand that is being copied. For example: suppose the enzyme, traveling along the DNA strand, hits a C base. Now, at this particular stage of its operation, a bonding between the polymerase and a GTP molecule from the soup happens to be more energetically favorable than one between it and UTP or ATP: complementarity is actually enforced by energy considerations. Preferentially, then, it will pick up a GTP molecule. It releases a pyrophosphate back out into the soup, moves along the DNA and looks for the next complementary nucleotide.

Schematically, we have the picture shown in Figure 5.22.

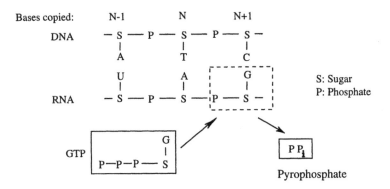

$$\text{RNA (N bases)} + \text{(G-S-P-P-P)} \rightarrow \text{RNA (N+1 bases)} + P_iP \text{ (pyrophosphate)}$$

FIGURE 5.22
Formation of mRNA.

Now the role of enzymes in biochemical processes is as catalysts: they influence the rate at which reactions occur, but not the direction in which they proceed. Chemical reactions are reversible, and it would be just as possible for the polymerase reaction to go the other way – that is, for the enzyme to undo the mRNA chain it is working on. In such an event, it would extract a pyrophosphate from the surrounding soup, attach it to a base on the mRNA, and then release the whole lot back into the environment as one of our triphosphates. The enzyme could just move along the wrong way, eat a G, move along, eat a C, move along, … undoing everything.* Which way the reaction goes depends on the relative concentrations of pyrophosphates and triphosphates in the soup. If there is a lot of ATP, GTP, and so on, but not much free pyrophosphate, then the rate at which the enzyme can run the reaction backward is lowered, because it can't find much pyrophosphate with which to pull off the mRNA nucleotides. On the other hand, if there is an excess of free pyrophosphates over triphosphates, the reaction will tend to run the wrong way, and we'll be uncopying and ruining our copy.

We can actually interpret these relative concentrations in terms of the number of possible states available to our system at any given computational point. If there are plenty of triphosphates around, then there are plenty of forward-moving, and comparatively few backward-moving, states available: the RNA polymerase will tend to enter the former state, in the process lowering its entropy. The difference in free energies, measured by the differing concentrations, determines the way it goes. If we get the concentrations just right, the copier will oscillate forever, and we will never get around to making copies. In an actual cell, the pyrophosphate concentration is kept low by hydrolysis, ensuring that only the copying process occurs, not its inverse. The whole RNA polymerase system is not particularly efficient as far as energy use goes – it dissipates about $100kT$ per bit. Less could be wasted if the enzyme moved a little more slowly (and of course, the reaction rate does vary with concentration gradient), but there has to be a certain speed for the sake of life! Still, $100kT$ per bit is considerably more efficient than the 10^8kT thrown away by a typical transistor!

To reiterate: the lesson of this section is that there is no absolute minimum amount of energy required to copy. There *is* a limit, however, if you want to copy at a certain speed.

---

* Bennett has nicely christened machines like this "Brownian computers" to capture the manner in which their behavior is essentially random but in which they nevertheless progress due to some weak direction of drift imposed on their operation. [RPF]

5.3 Computation: Energy Cost versus Speed

The question of speed is important and I would like to write down a formula for the amount of free energy it takes to run a computation in a finite time. This at least makes our discussion a bit more practical. There is little room for reversible computing in the computer world at the moment, although one can foresee applications that are a little more immediately useful than getting from $10^8 kT$ to under kT. (You can actually get to 2 or 3 kT irreversibly, but you can't get under this.) For example, we can look at the problem of errors arising in parallel processing architectures where we might have thousands of processors working together. The question of error correction through coding in such a situation has arisen and is unsolved. It occurs to me that maybe the devices in the machine could all be made reversible, and then we could notice the errors as we go. What would be the cost of such reversible devices? Maybe these things will find application soon. That would make this discussion more practical to you and since computing is engineering you might value this! In any case, I shouldn't make any more apologies for my wild academic interest in the far future.

An example we gave of reversible computing was that of the chemical process of copying DNA. This involved a machine (if you like) that progressed in fits and starts, going forward a bit, then backward, but more one than the other because of some driving force, and so ended up doing some computation (in this case, copying). We can take this as a model for more general considerations and will use this "Brownian" concept to derive a formula for the energy dissipation in such processes. This will not be a general formula for energy dissipation during computation but it should show you how we go about calculating these things. However, we will precede this discussion by first giving the general formula,* and then what follows can be viewed as illustration.

Let us suppose we have a reversible computer. Ordinarily, the free energy expended in running it reversibly will be zero when the process is infinitesimally slow, but let us suppose that we are actually driving it forward at a rate r. In other words, at any given stage, it is r times more likely to make a forward calculational step than a backward one. Then, the general result is that minimum energy that must be expended per computational step in the process is:

$$kT \log r. \qquad (5.20)$$

---

* This rule is pretty general, but there will be exceptions, requiring slight corrections. We will discuss one such, a "ballistic" computer, in Section 5.5. [RPF]

Note that the smaller r is, the lower the energy.

Let us illustrate this rule by looking at a Brownian-type computer. Imagine we have a system, or device, in a particular state which has a particular energy associated with it. It can go forward or backward into a new state, each transition corresponding either to doing a computation (forward) or undoing it (backward). We can model this situation using the energy level diagram of Figure 5.23.

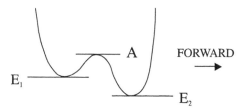

FIGURE 5.23
The general transition.

We assume our computer to be sitting in one of the two states, with energy E_1 or E_2. These energies will not generally be equal. Now our device can go from E_1 to E_2, a forward step – the idea is that the energies are lower in the direction of computation – or from E_2 to E_1, a backward step. The energies of the two states might be equal, but one of them could be effectively lowered by the imposition of an external driving force. We have introduced into this diagram the "activation energy" A, which is the energy that must be supplied to the system to cause a transition of any kind. We will focus on the effects of thermal fluctuations, which will, quite randomly, cause the computer to move between states whenever the energy of these fluctuations exceeds A. Such fluctuations can make the device go either way, and we can calculate the rate at which it goes in either direction. These will not be equal. Roughly, the chance of the system going into the state with energy E_i is the chance that by accident it acquires enough energy to get past the barrier (that is, A) and into E_i. Clearly, the energy needed to get from E_1 to E_2, a forward step, is $(A-E_1)$, while to get from E_2 to E_1 it is $(A-E_2)$. It is a standard result in statistical mechanics that the probability of a transition from one state to another differing in (positive) energy δE is:

$$C \exp(-\delta E / kT, \tag{5.21}$$

where C is a factor that carries information about the thermal fluctuations in the environment. This can be calculated through a phase-space (entropy-type) analysis, examining the probabilities of ensemble transitions between states. However, we are interested in the transition rates between states and this is describable by a similar formula. We simply have to insert another factor, say X, giving us:

$$forward\ rate = CX\ \exp\left[-\left(A-E_1\right)/kT\right] \tag{5.22}$$

and

$$backward\ rate = CX\ \exp\left[-\left(A-E_2\right)/kT\right]. \tag{5.23}$$

The factor X depends on a variety of molecular properties of the particular substance (the mean free path, the speed, and so on), but the property that interests us is that it does not depend on E (consider the transition rates for the case $E_1 = E_2$). We can therefore write for the ratio of the forward to backward rates:

$$\exp\left[\left(E_1-E_2\right)/kT\right]. \tag{5.24}$$

This depends only on the energy difference between successive states. This gives us some insight into the rate at which our computation (= reaction) proceeds, and the energy difference between each step required to drive it. The bigger the energy difference $E_1 - E_2$, the quicker the machine hops from E_1 to E_2, and the faster the computation.

We can tie this result into our earlier general formula by setting the above rate equal to r. We then have, for the energy expended per step:

$$kT\ \log r = E_1 - E_2 \tag{5.25}$$

which makes sense.

Let me give you one more illustration of driving a computer in a particular direction. This time we will look at computational states that do not differ in their energy, but in their availability. That is, our computer selects which state of a certain kind to go into next, not on the basis of the energy of the state, but on the number of equivalent states of that kind available for it to go into. We have an example of such a situation in our DNA copier. A calculational step there involved the RNA enzyme attaching bases to the RNA chain and liberating pyrophosphates into the surroundings. The inverse step involved taking up phosphates from the solution and breaking off bases. Each step is energetically equivalent but one can be favored over the other, depending on the relative concentrations of chemicals in the soup. Suppose there is a dearth of phosphates but a wealth of bases available. Then, the number of (forward) states of the system in which a base is attached to the RNA strand and a phosphate is released – and we consider all such states equivalent – exceeds the number of states in which a phosphate is grabbed and a base released (again, all such states we take to be the same). So we can envisage a computer designed so that it proceeds by diffusion, in the sense that it is more likely to move into a state with greater, rather than lower, availability. Schematically, we have the situation shown in Figure 5.24.

FIGURE 5.24
The availability of states.

where n_i is the number of available states. It is possible to show in this situation (although it takes a little thought) that the ratio of the forward rate to the backward rate is:

$$r = n_2 / n_1. \tag{5.26}$$

If you recall, we defined the entropy of a configuration of a system to be:

$$S \approx k \log W \tag{5.27}$$

where W is the probability of finding the system in that configuration. Hence, we may write:

$$kT \log r = kT \left(\log n_2 - \log n_1 \right) = \left(S_2 - S_1 \right) T \tag{5.28}$$

(with various constant factors canceling to leave the equality). In other words, for this process, the energy loss per step is equal to the entropy generated in that step, up to the usual temperature factor. Again, this makes sense.

So we can see that our general formula reduces to the specific formulae we have obtained in these instances. An interesting question that arises is: in a real-world situation, can we minimize the energy taken per computational step? We know that if we have an effectively reversible computer, the chances of forward and backward movement are equal, and we have no energy loss. The price we pay for this is that a computation will take an infinite time. We will never know when we're finished. So as we've said, to get it going we want to give things a tug, lower the energies of successive steps, make them more available, or whatever. Let us suppose that we have the forward rate, f, just a little bigger than the backward rate, b, so the computation just goes. We write:

$$f = b + \Theta \tag{5.29}$$

where Θ is small. Our general formula now gives:

$$energy \ per \ step = kT \ \log\left[1+\left(\Theta/b\right)\right] \approx kT\Theta/b = kT\left(f-b\right)/b \tag{5.30}$$

for small Θ. We can provide a nice physical interpretation of this expression, although at the cost of mathematical inaccuracy. We replace the formula above with one that is nearly equal to it:

$$energy \ per \ step = kT \frac{\left(f-b\right)}{\left(f+b\right)/2}. \tag{5.31}$$

This differs from the original formula by terms of order Θ^2. Now the numerator of this fraction is the speed at which we go forward and do the calculation. It is a bit like a velocity, in that it represents the rate at which the computer drifts through its calculation, measured in steps per second. The denominator is the average rate of transition – it is a measure of the degree to which our computer is oscillating back and forth. We can interpret this roughly as the fastest speed at which you could possibly go, backward or forward, which would be the speed found if the computer underwent a series of steps in one direction with no reverses: it is the greatest possible drift. So we can write, approximately:

$$energy \ loss \ per \ step = kT \frac{v_{drift}}{v_{max}}. \tag{5.32}$$

Alternatively, we can emphasize time as our variable and write:

$$energy \ loss / step = kT \frac{minimum \ time \ taken / step}{time / step \ actually \ taken} \tag{5.33}$$

Let us now take a look at more general issues in reversible computing.

5.4 The General Reversible Computer

We have repeatedly stated that, if a computation is to be reversible, then we have to store a lot of information that we would ordinarily lose or throw away in order that we can subsequently use it to undo something.

The logic gates of such a machine give us not just the answer to the logical calculation we want, but a whole lot of extra bits. A simple illustration of this for a realistic gate is a simple adder built out of reversible gates. In Section 2.3, I set a problem for you, the construction of a full three-bit adder from reversible gates – specifically, using CN and CCN gates (or alternatively, just CCN gates, out of which all others can be built). An easier example, the simple two-bit adder, is built as follows (Figure 5.25).

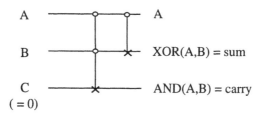

FIGURE 5.25
A two-bit adder from reversible gates.

The C-input is kept set at zero (the full three-bit adder requires the addition of a fourth input line, kept set at zero). As well as the sum and carry of $A+B$, we find this gate feeds the A-line through. We can see that this bit is necessary if we are to be able to reconstruct the input (A,B) from the output. If you look at the three-bit adder, you will find two spare bits at the output. Generally, then, we will always need a certain amount of junk to remind us of the history of the logical operation. We can summarize the main constraint on reversible gates as follows: it is obvious that, when running a computer forward, there must be no ambiguity in the forward step – if you have a "go to", you have to know where to go to. With a reversible machine, there cannot be any ambiguity in *backward* steps either. You should never have a situation where you do not know where to go back to. It is this latter feature that makes reversible computing radically different from ordinary, irreversible computing.

We can, following Bennett, consider the most general computational process, and also answer a criticism leveled at advocates of reversible computing. Let us suppose we have a system of (reversible) logic units tied together, and we put into it some input data. We also have to feed in a set of "standard" zeroes, the bits that are kept set at particular values to control the reversible gates. (If we want a "standard" one instead of a zero, we can just NOT one of the zeroes: this is reversible, of course!) The logic unit will do its business – dup, dup, dup – and at the end, we will find an output – the answer we want plus a pile of garbage bits, forming the history tape. This is shown in Figure 5.26.

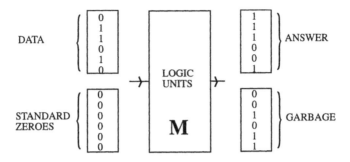

FIGURE 5.26
The general reversible computation.

Now this picture makes it look like you start up with a blank tape (or a pre-set one) and end up with a lot of chaos. Not surprisingly, everyone said that was where the entropy was going: "This randomization of zeroes is (in Bennett's picture) fueling the running of your machine. How can keeping this data make your computation practically reversible? It's rather like claiming that you can make an irreversible heat engine reversible by keeping the water that all the heat has gone into, rather than throwing it away. If you don't throw the water away, sure you have all the information you need about the history of the system, but that hardly means the engine is going to be able to run backward, reversing the motions of water molecules!" In the thermodynamic case, that would indeed be silly. But it isn't so for computing. By adding one more tape to the system, and feeding the results through another machine, we can bypass this difficulty (Figure 5.27):

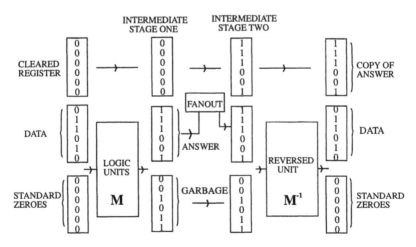

FIGURE 5.27
A zero entropy loss reversible computer.

Let us try to make sense of this! The new logic unit that we have added is the reverse of the original (hence we have labeled it $\mathbf{M}^{-1}$) and is also reversible. $\mathbf{M}^{-1}$ is such that if we feed the output of $\mathbf{M}$ through this, it undoes all the work on it and feeds us back the original inputs to $\mathbf{M}$. The new tape is a cleared register that we will use to copy the answer to our computation. We begin as before, feeding into $\mathbf{M}$ the input data and the standard bits for control. $\mathbf{M}$ gives us an output and a history tape (marked garbage in the diagram). The history tape we feed directly into $\mathbf{M}^{-1}$. We also feed the data output tape in. However, before we do this, we make a copy of it onto the cleared register. We have shown this schematically as a fanout, but this actually represents a copy process (which is, of course, a reversible operation).

The reverse machine $\mathbf{M}^{-1}$ now undoes all the work done by $\mathbf{M}$, producing as its output the standard bits and the input data. At the end of the whole process, we are left with the answer to the computation, plus an exact copy of the inputs we started with. So our grand machine has done a calculation for no entropy loss (ideally – in practice, we would have to drive the system a little as discussed) and reversible computing really can save us work. Of course, there will be an energy loss when we wipe our tapes clean to do another calculation.

Reversible computing is quite a strange concept for those used to thinking in classical Boolean terms, so let me suggest a few problems for you to work on to help you become more comfortable with the ideas.

> **Problem 5.3** Suppose a reversible computer is carrying out a calculation and it needs to execute a subroutine. So it gets sent off to some other place to execute a compact set of instructions. Now these instructions must be reversible, as are the basic computing elements, and so there is a chance that once we are into our subroutine, we might find ourselves running backward. It might even happen that we get back to the start of the routine – and then have to re-enter the main body of the program where we left it! The question is: given that this same subroutine might be used several times throughout the computation, how does the machine know where to return to when this reverse happens? You might like to think about this. Somehow you have to have a number of memory stacks to keep track of where you have to go to find the subroutine, but also where to go back to should you reverse. This is your first problem in reversible computing – how to handle subroutines.

> **Problem 5.4** A related problem concerns how to get "if" clauses to work. What if, after having followed an "if... then..." command, the machine starts to reverse? How can the machine get back to the original condition that dictated which way the "if" branched?

Of course, a set of initial conditions can result in a single "if" output ("if x = 2, 3, 4 or 6.159 let $F = d$"), so this condition may not be uniquely specified. Here is a nice way to analyze things. Simply bring in a new variable at each branch and assign a unique value to this variable for each choice at a branch point. You might like to work this through in detail.

Problem 5.5 A simple question to ask about a general reversible computer is: how big a history tape do we need? The gates we have considered so far have had the number of outputs equal to the number of inputs. Is this always necessary for reversibility? As far as I know, this question hasn't even been asked by theorists. See if you can work it out. Certainly, the minimum has something to do with the number of possible inputs that the output could represent, and we'll apparently need a number of bits to keep track of that (on top of the actual outputted answer). So the questions are: first, what is the minimum number of bits needed to keep a gate reversible in principle, and second, could we actually accomplish it?

5.5 The Billiard Ball Computer

To give you a demonstration of a reversible computer that can actually do calculations, I am now going to discuss an ingeniously simple machine invented by Fredkin, Toffoli, and others. In this device, the movement of billiard balls on a plane is used to simulate the movement of electronic signals (bits) through logic gates. We fire balls into the machine to represent the input, and the distribution of balls coming out gives us our output. The balls all move diagonally across a planar grid and obey the laws of idealized classical mechanics (i.e., zero friction and perfectly elastic collisions). To introduce you to the basic idea, examine Figure 5.28.

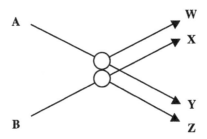

FIGURE 5.28
The basic two-ball collision computation.

This illustrates how a two-ball collision realizes a two-input four-output logic function. The data to this gate is represented by the presence of a ball at a particular position (1) or its absence (0). For example, the gate has two input channels, A and B. If we fire a ball in at A, then the input at A is binary 1. If there is no ball, it is zero. Similarly with B. If we find a ball coming out at point X, this means output X is 1, and so on. There are four possible input states, and for each, we use basic mechanics to calculate the configuration of balls coming out of the device. There are four possible outputs, two corresponding to one input ball being absent and the other going straight through, and two corresponding to a collision.

Let us suppose there is no ball at A. If there is a ball at B, it will continue on through the "machine", coming out at X. We can see that we will only get a ball at X if there is no ball at A *and* one ball at B. In logic terms, X is 1 if and only if B is 1 and A is 0, so we have:

$$X = B \ AND \ NOT \ A \tag{5.34}$$

Similarly, we find that:

$$Y = A \ AND \ NOT \ B \tag{5.35}$$

Output W is a little trickier. We will find a ball there only if there is a ball at both A and B. Likewise for output Z. Hence, both W and Z realize the same AND function:

$$W, Z = A \ AND \ B \tag{5.36}$$

Let us summarize this with some fancy notation (Figure 5.29).

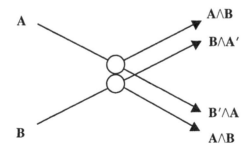

FIGURE 5.29
Logical structure of the basic collision computation.

This is the fundamental collision of this billiard ball computer and you can see how neatly the logic element drops out of it. We can build other logic functions besides AND with this gate. For example, we can use it to make a FANOUT. If we set $A = 1$ (the billiard equivalent of a control line set to "on") and take our output from W and Z, then clearly this has the effect of branching our B input: a ball at B produces one at each of W and Z; no ball at B leaves both outputs blank. You can also make a CN gate with this unit (try it). However, by itself, the basic collision operation will not make enough elements to build a whole computer – we'd be stuck with pairs of balls going along two lines, and we could never change anything! How do we re-route balls? We have to introduce two fundamental mechanical devices. The first, which you would never invent if you were a logician, as it seems a damn silly thing to do, I'll call a collision gate; in this device, two balls go in, but four come out (Figure 5.30).

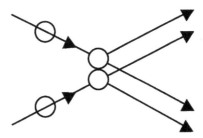

FIGURE 5.30
The collision gate.

This is a sort of all-in "double-FANOUT" process, which we achieve by letting the two incoming balls collide with two stationary ones. (You might find it an interesting exercise to consider the energy and momentum properties of this gate.) The second and more important device is a redirection gate. This is just a mirror to reflect a ball. It can be oriented any way you wish, although we restrict ourselves to four possibilities (Figure 5.31).

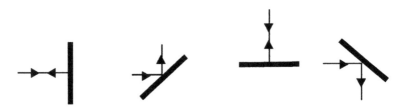

FIGURE 5.31
Four redirection gates.

Mirrors enable us to do a lot of things. For example, we can use mirrors to construct a "crossover" device (Figure 5.32).

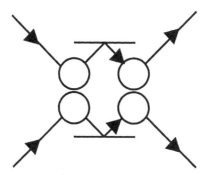

FIGURE 5.32
A crossover device.

Incidentally, this device tells us something important about the balls, namely, that they are indistinguishable. We do not tell them apart and are interested only in their presence or absence. The above crossover device actually switches the incoming balls, but as we can't tell them apart, it looks as if they just pass each other by. Note that if one ball is missing, the other just sails right through.

To show you the sorts of things you can build with these basic structures, I will first give you a unit that acts as a switch (Figure 5.33).

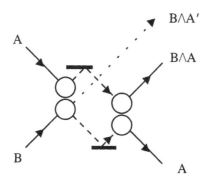

FIGURE 5.33
A switching device.

This is a sort of offset crossover. Note that, irrespective of whether or not there is a B input, the lower right output is always the same as A. This is a "debris" bit, corresponding to the control line fed through the gate. Of course, we are used to such outputs by now.

A question that arises in the context of this chapter is obviously: "OK, show me how to make reversible gates with all these mirrors and balls"; specifically, can we build, say, a CN gate? The answer is that we can, and a CCN gate too if we like. However, it is more enlightening to build a Fredkin, or controlled exchange gate. This is because it is possible to build everything we could want, just out of Fredkin gates! I'll remind you of what such a gate is (Figure 5.34).

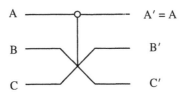

FIGURE 5.34
The Fredkin gate.

Line A goes through unchanged. This is true of B and C also, if $A=0$; but if $A=1$, B and C switch. I won't leave building a Fredkin gate as an exercise. It is constructed from four switching devices of the kind depicted in Figure 5.33, put together with considerable ingenuity as shown in Figure 5.35.

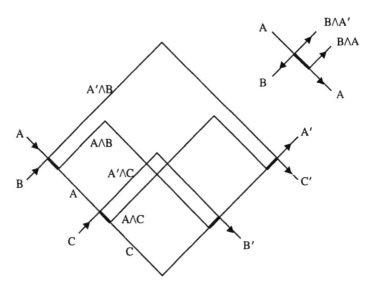

FIGURE 5.35
The Fredkin gate realized by billiard ball gates.

Obviously, there is no point in making a computer like this except for fun. However, it does show how profoundly simple the basic structure of a machine can be.

Now anybody who is familiar with bouncing balls knows that if there's a slight error, it is rapidly magnified. Suppose you have a ball on a table and you drop another onto it from above, right in the middle. You might think: "Oh, it'll go straight down, then straight up, and so on" – everybody has an intuition about this but if you played with balls as a baby, you know that you can't bounce one ball on another. It doesn't work! What happens is that as soon as the ball bounces ever so slightly wrongly, the next bounce is further out and the ball comes down slightly more cock-eyed. When it comes down next time, it is further out still and hits the lower ball in an even more glancing fashion. Next time, the balls will probably miss altogether.

The reasons for this are not hard to fathom. Although at the macroscopic level, balls seem stable and solid, at the microscopic level, they are a seething mass of jiggling molecules. Thermal oscillations, statistical mechanical fluctuations, and whatnot, all contribute corrections to the naive collision of ideal balls. In fact, even the tiniest effects of quantum mechanics get in the way. According to the Uncertainty Principle, we cannot know both the precise location and momentum of a ball, so we cannot drop one perfectly straight. Suppose we have two ideal 1 cm balls, and we drop one onto the other from a height of 10 cm. How many bounces can we get away with before, according to quantum mechanics, things *have* to go awry? We can actually calculate this and the answer is about 17 bounces. Of course, in reality, the disturbances from more classical phenomena are far more significant and we would never get anywhere near this quantum limit. Don't forget, even your hand will be shaking from Brownian motion!

So surely the billiard ball machine idea is nonsense? We may not be dropping balls from a height, but we are colliding them and we would therefore expect errors to accrue just as inevitably. So how can we claim to have a physically implementable reversible computer? After all, all you have to do is give me an error per collision, and I will tell you how long you have before the machine falls apart: 10^{-3} minutes? Five minutes; 10^{-6} minutes? OK, ten minutes. It looks completely hopeless. In order to get this system to work, we have to find some way to keep straightening out the balls. Perhaps we could put them in troughs, guiding them in some way. But if you put a ball in a trough, it'll slosh back and forth, getting worse and worse, unless there are losses – absorption, resistance, dissipation. Even if we design our troughs to cope with these difficulties, inevitably energy will be lost because of friction in the trough. We would have to pull the balls through to drive the machine. Now if you drive it just a little, you can find that the energy required to drive it is a minimum of the ratio:

$$kT \frac{time\ to\ make\ collision}{speed\ at\ which\ it\ happens}. \qquad (5.37)$$

This expression has not been analyzed in any great detail for the billiard ball machine.

5.6 Quantum Computation

The billiard ball computer operates chiefly according to the laws of classical mechanics. However, inspired by the questions it brings up, people have asked me (and others have thought about this too[*]): "What would the situation be if our computer is operating according to the laws of *quantum* mechanics?" Suppose we wanted to make extremely small computers, say the size of a few atoms. Then we would have to use the laws of quantum mechanics, not classical mechanics. Wouldn't the Uncertainty Principle screw things up? Not necessarily. I will wind up this chapter by briefly considering what may become the computers of the future – quantum computers.

We are asking yet another question about absolute limitations! This time, it is: "How *small* can you make a computer?" This is one area where, I think, I've made a contribution. Unlike an airplane, it turns out that we can make it pretty much as small as we want. There will be engineering details about wires,[†] and we will have to find a way of magnifying outputs and whatnot, but we are here discussing questions of principle, not practicality. We cannot get any smaller than atoms[‡] because we will always need something to write on, but all we actually need are bits that communicate. An atom, or a nucleus will do since they are natural "spin systems", that is, they have measurable physical attributes that we can put numbers to and we can consider each different number to represent a state. We can make magnets the size of atoms. (It'll put some chemists out of a job, but that's progress). But the point is that there are no further limitations on size imposed by quantum mechanics, over and above those due to statistical and classical mechanics.

I won't go into too much detail here: I will return to the subject, and all its lovely math – in Chapter 6. For now, I'll just give you the gist of the ideas. Let us begin with some idealized quantum mechanical system (anything very small) and suppose that it can be in one of two states – say "up", which might correspond to an excited state, and "down", corresponding to a de-excited state. Alternatively, the two states might refer to the spin of the quantum system (spin is a crude classical analogy). We can actually allow it to be in other states as well, but for our purposes, it just

---

[*] Notably the physicist Paul Benioff (see, for example, "Quantum Mechanical Models of Turing Machines that Dissipate No Energy", Phys. Rev. Lett. 48, pp. 1581–1585 [1982]). [Editor]

[†] It is interesting to note that most computer theorists treat wire as idealized thin string that doesn't take up any room. However, real computer engineers frequently discover that they just can't get enough wires in! (We'll return to this in Chapter 8.) [RPF]

[‡] I am not allowing for the possibility that some smart soul will build a computer out of more fundamental particles! [RPF]

needs at least two states to represent a binary number: up is one, down is zero. I'll call this quantum mechanical system an atom, so that you can get a grip on its basic nature but bear in mind that it could be something more complex, or even something simpler, like an electron (which has two spin states). Now, the idea is that we build our computing device out of such atoms by stringing them together in a particular way. We start with part or all of the system – a string of atoms in one or other of their two states – representing a number, our input. We then let the whole system evolve over time according to the laws of quantum theory, interacting with itself – the atoms change states, the ones and zeroes move around – until at some point we have a bunch of atoms somewhere that will be in certain states, and these will represent our answer.

We could set the machine running with a single input bit – say firing an atom into the system – and design things such that the machine itself tells us when the calculation is complete, say by firing an atom out of the system. Nothing would be trustable until the output bit was one. You would measure this bit, then change it to zero and freeze the answer for examination. Putting the information in and out is not, incidentally, a particularly quantum mechanical process – it is a matter of amplification. Interestingly, as a rule, one cannot predict the time the computer will take to complete its calculation. It turns out to be ballistic, like Fredkin's, but at the end, you only get a wave packet for the arrival of the answer. We test to see whether or not the answer is in the machine or not. For the simple machine I have designed (see Chapter 6), there exist several quantum mechanical "amplitudes" (certain physical properties of the system) which, upon measurement, tell us how far through the calculation we have gone, but ultimately, we have to wait for the machine to let us know it's finished.

So, in 2050, or before, we may have computers that we can't even see! I will return to these strange beasts in Chapter 6.

6

Quantum Mechanical Computers

CONTENTS

6.1 Introduction

In this chapter,[*] we discuss our part in an effort to analyze the physical limitations of computers due to the laws of physics.[†] For example, Bennett [1] has made a careful study of the free energy dissipation that must accompany computation. He found it to be virtually zero. He suggested to me the question of the limitations due to quantum mechanics and the Uncertainty Principle. I have found that, aside from the obvious limitation to size if the working parts are to be made of atoms, there is no fundamental limit from these sources either. We are here considering ideal machines; the effects of small imperfections will be considered later. This study is one of principle; our aim is to exhibit some Hamiltonian for a system which could serve as a computer. We are not concerned with whether we have the most efficient system, nor how we could best implement it.

Since the laws of quantum physics are reversible in time, we shall have to consider computing engines which obey such reversible laws. This problem already occurred to Bennett [1], and to Fredkin and Toffoli [2], and a great deal of thought has been given to it. Since it may not be familiar to you here, I shall review this and, in doing so, take the opportunity

---

[*] Reprinted with permission from Richard P. Feynman, "Quantum Mechanical Computers," *Optics News*, pp. 11–20 (February 1985). © Optica Publishing Group. Readers will require some understanding of elementary quantum mechanics to fully comprehend this chapter. [Editor]

[†] Although some of the notational and stylistic conventions of this chapter differ from those adopted elsewhere in the book, we have decided to retain them to preserve the flavor of Feynman's original published discourse. There is also some repetition of material discussed in previous chapters. [Editor]

DOI: 10.1201/9781003358817-6   *169*

to review, very briefly, the conclusions of Bennett [3], for we shall confirm them all when we analyze our quantum system.

It is a result of computer science that a universal computer can be made by a suitably complex network of interconnected primitive elements. Following the usual classical analysis, we can imagine the interconnections to be ideal wires carrying one or two standard voltages representing the local 1 and 0. We can take the primitive elements to be just two, NOT and AND. (Actually just the one element NAND = NOT AND suffices, for if one input is set at 1 the output is the NOT of the other input.) They are symbolized in Figure 6.1, with the logical values resulting from different combinations of input wires. From a logical point of view, we must consider the wire in detail, for in other systems, and our quantum system in particular, we may not have wires as such. We see we really have two more logical primitives, FANOUT when two wires are connected to one, and EXCHANGE, when wires are crossed. In the usual computer the NOT and NAND primitives are implemented by transistors, possibly as in Figure 6.2.

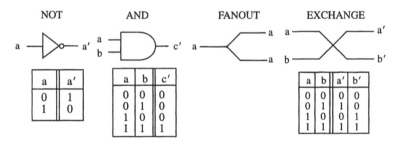

FIGURE 6.1
Primitive elements.

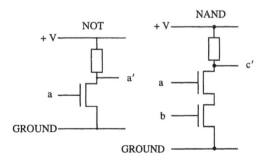

FIGURE 6.2
Transistor circuits for NOT and NAND.

What is the minimum free energy that must be expended to operate an ideal computer made of such primitives? Since, for example, when the AND operates the output line, c' is being determined by one of two values no matter what was before, the entropy change is log2 units. This represents a heat generation of $kT \log 2$ at temperature T. For many years it was thought that this represented an absolute minimum to the quantity of heat per primitive step that had to be dissipated in making a calculation.

The question is academic at this time. In actual machines we are quite concerned with the heat dissipation question, but the transistor system used actually dissipates about $10^8 kT$! As Bennett [3] has pointed out, this arises because to change a wire's voltage we dump it to ground through a resistance; and to build it up again we feed charge, again through a resistance, to the wire. It could be greatly reduced if energy could be stored in an inductance, or other reactive element. However, it is apparently very difficult to make inductive elements on silicon wafers with present techniques. Even nature, in her DNA copying machine, dissipates about 100 kT per bit copied. Being, at present, so very far from this $kT \log 2$ figure, it seems ridiculous to argue that even this is too high and the minimum is really essentially zero. But, we are going to be even more ridiculous later and consider bits written on one atom instead of the present 10^{11} atoms. Such nonsense is very entertaining to professors like me. I hope you will find it interesting and entertaining also.

What Bennett pointed out was that this former limit was wrong because it is not necessary to use irreversible primitives. Calculations can be done with reversible machines containing only reversible primitives. If this is done, the minimum free energy required is independent of the complexity or number of logical steps in the calculation. If anything, it is kT per bit of the output answer. But even this, which might be considered the free energy needed to clear the computer for further use, might also be considered as part of what you are going to do with the answer – the information in the result if you transmit it to another point. This is a limit only achieved ideally if you compute with a reversible computer at infinitesimal speed.

6.2 Computation with a Reversible Machine

We will now describe three reversible primitives that could be used to make a universal machine (Toffoli [4]). The first is the NOT which evidently loses no information, and is reversible, being reversed by acting again with NOT. Because the conventional symbol is not symmetrical we shall use an X on the wire instead (see Figure 6.3a).

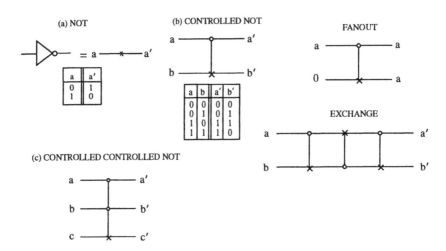

FIGURE 6.3
Reversible primitives.

Next is what we shall call the CONTROLLED NOT (see Figure 6.3b). There are two entering lines, *a* and *b* and two exiting lines *a'* and *b'*. The *a'* is always the same as *a*, which is the control line. If the control is activated *a* = 1 then the output *b'* is the NOT of *b*. Otherwise *b* is unchanged, *b* = *b'*. The table of values for input and output is given in Figure 6.3. The action is reversed by simply repeating it. The quantity *b'* is really a symmetric function of *a* and *b* called XOR, the exclusive or; *a* or *b* but not both. It is likewise the sum modulo two of *a* and *b*, and can be used to compare *a* and *b*, giving a 1 as a signal that they are different. Please notice that this function XOR is itself not reversible. For example, if the value is zero we cannot tell whether it came from (*a*,*b*) = (0,0) or from (1,1) but we keep the other line *a* = *a* to resolve the ambiguity. We will represent the CONTROLLED NOT by putting a 0 on the control wire, connected with a vertical line to an *X* on the wire which is controlled. This element can also supply us with FANOUT, for if *b* = 0 we see that *a* is copied onto line *b'*. This COPY function will be important later on. It also supplies us with EXCHANGE, for three of them used successively on a pair of lines, but with alternate choice for control line, accomplishes an exchange of the information on the lines (Figure 6.3b).

It turns out that combinations of just these two elements alone are insufficient to accomplish arbitrary logical functions. Some element involving three lines is necessary. We have chosen what we can call the CONTROLLED CONTROLLED NOT. Here (see Figure 6.3c) we have two control lines *a*,*b* which appear unchanged in the output and which change the third line *c* to NOT *c* only if both lines are activated (*a* = 1 and *b* =

1). Otherwise $c' = c$. If the third line input c is set to 0, then evidently it becomes 1 ($c' = 1$) only if both a and b are 1 and therefore supplies us with the AND function (see Table 6.1).

TABLE 6.1

The Non-Zero Matrix Elements of G

A	B	C	A'	B'	C'
0	0	0	0	0	0
0	0	1	0	0	1
0	1	0	0	1	0
0	1	1	0	1	1
1	0	0	1	0	0
1	0	1	1	0	1
1	1	0	1	1	1
1	1	1	1	1	0

Three combinations for (a,b), namely (0,0), (0,1), and (1,0) all give the same value, 0, to the AND (a,b) function so the ambiguity requires two bits to resolve it. These are kept in the lines a,b in the output so the function can be reversed (by itself, in fact). The AND function is the carry bit for the sum of a and b.

From these elements it is known that any logical circuit can be put together by using them in combination, and in fact, computer science shows that a universal computer can be made. We will illustrate this by a little example. First, of course, as you see in Figure 6.4, we can make an adder by first using the CONTROLLED CONTROLLED NOT and then the CONTROLLED NOT in succession to produce from a and b and 0, as input lines, the original a on one line, the sum on the second line and the carry on the third:

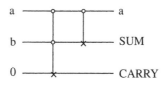

FIGURE 6.4
An adder.

A more elaborate circuit is a full adder (see Figure 6.5) which takes a carry, c (from some previous addition), and adds it to the two lines a and b and has an additional line, d, with a 0 input.

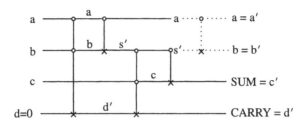

FIGURE 6.5
A full adder.

It requires four primitive elements to be put together. Besides this total sum, the total of the three, a, b, and c and the carry, we obtain on the other two lines two pieces of information. One is the a that we started with, and the other some intermediary quantity that we calculated en route:

This is typical of these reversible systems; they produce not only what you want in output, but also a certain amount of garbage. In this particular case, and as it turns out in all cases, the garbage can be arranged to be, in fact, just the input. If we would just add the extra CONTROLLED NOT on the first two lines, as indicated by the dotted lines in Figure 6.5, we see that the garbage would be a and b, which were the inputs of at least two of the lines. (We know this circuit can be simplified but we do it this way for illustrative purposes.)

In this way, we can by various combinations, produce a general logic unit that transforms n bits to n bits in a reversible manner. If the problem you are trying to do is reversible, then there might be no extra garbage, but in general, there are some extra lines needed to store up the information which you would need to reverse the operation. In other words, we can make any function that the conventional system can, plus garbage. The garbage contains the information you need to reverse the process. And how much garbage? It turns out, in general, that if the output data that you're looking for has k bits, then starting with an input and k bits containing 0, we can produce, as a result, just the input and the output and no further garbage. This is reversible because knowing the output and the input permits you, of course, to undo everything. This proposition is always reversible. The argument for this is illustrated in Figure 6.6.

Suppose we begin with a machine M, which, starting with an input and some large number of 0s, produces the desired output plus a certain amount of extra data which we call garbage. Now we've seen that the copy operation can be done by a sequence of CONTROLLED NOTs, so if we have originally an empty register, with the k bits ready for the output, we can, after the processor M has operated, copy the output from the M onto this new register (Figure 6.6).

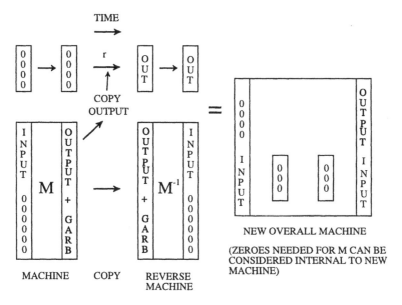

FIGURE 6.6
Clearing garbage.

After that, we can build the opposite machine, the M in reverse, the reverse machine, which would take this output of M and garbage and turn it into the input and 0s. Thus, seen as an overall machine, we would have started with the k 0s of the register for the output, and the input, and ended up with those k 0s occupied by the output data, and repeat the input data as a final product. The number of 0s that was originally needed in the M machine in order to hold the garbage is restored again to 0 and can be considered as internal wires inside the new complete machine (M, M^{-1}, and copy). Overall, then, we have accomplished what we set out to do, and therefore garbage need never be any greater than a repetition of the input data.

6.3 A Quantum Mechanical Computer

We now go on to consider how such a computer can also be built using the laws of quantum mechanics. We are going to write a Hamiltonian, for a system of interacting parts, which will behave in the same way as a large system in serving as a universal computer. Of course the large system obeys quantum mechanics, but it is in interaction with the heat baths and other things that could make it effectively irreversible. What we would like to do is make the computer as small and as simple as possible. Our Hamiltonian will describe in detail all the internal computing actions but not, of course, those interactions with the exterior involved in entering the input (preparing the initial state) and reading the output.

How small can such a computer be? How small, for instance, can a number be? Of course a number can be represented by bits of 1s and 0s. What we're going to do is imagine that we have two-state systems, which we will call "atoms". An n-bit number is then represented by a state of a "register", a set of n two-state systems. Depending upon whether or not each atom is in one or another of its two states, which we call $|1>$ and $|0>$, we can of course represent any number. And the number can be read out of such a register by determining, or measuring, in which state each of the atoms is at a given moment. Therefore one bit will be represented by a single atom being in one of two states, the states we will call $|1>$ and $|0>$.

What we will have to do then can be understood by considering an example: the example of a CONTROLLED CONTROLLED NOT. Let G be some sort of an operation on three atoms a, b, and c, which converts the original state of $a,b,$ and c into a new appropriate state, a',b',c', so that the connection between a', b' and c' and a,b,c, are just what we would have expected if $a,b,$ and c represented wires, and the $a',b',$ and c' were the output wires of a CONTROLLED CONTROLLED NOT. It must be appreciated here that, at the moment, we are not trying to move the data from one position to another, we are just going to change it. Unlike the situation in the actual wired computer in which the voltages on one wire then go over to voltages on another, what we're specifically making is something simpler, that the three atoms are in some particular state, and that an operation is performed which changes the state to new values a',b',c'. What we would have then is that the state, in the mathematical form $|a',b',c'>$ is simply some operation G operating on $|a,b,c>$. In quantum mechanics, state changing operators are linear operators, and so we'll suppose that G is linear. Therefore, G is a matrix, and the matrix elements of G, $G^{a'b'c'a,b,c}$ are all 0 except those in the following table which are, of course, 1 (Table 6.1).

This table is the same table that represents the truth value table for the CONTROLLED CONTROLLED NOT. It is apparent that the operation is reversible, and that can be represented by saying that $G^*G = 1$, where the * means Hermitian adjoint. That is to say G is a unitary matrix. (In fact G is also a real matrix $G^* = G$, but that's only a special case.) To be more specific, we're going to write $A_{ab,c}$ for this special G. We shall use the same matrix A with different numbers of subscripts to represent the other primitive elements.

To take a simple example, the NOT, which would be represented by A_a is the simple matrix:

$$\begin{bmatrix} 0 & 1 \\ 1 & 0 \end{bmatrix}$$

(6.1)

This is a 2 × 2 matrix and can be represented in many ways, in different notations, but the particular one we will use to define this is by the method of creation and annihilation operators. Consider operating in this case, on a single line a. In order to save alphabets, let us call the matrix:

$$\underline{a} = \begin{bmatrix} 0 & 0 \\ 1 & 0 \end{bmatrix}$$

(6.2)

which annihilates the 1 on atom a and converts it to 0; $\underline{a}$ is an operator which converts the state of |1> to |0>. But, if the state of the atom were originally |0>, the operator a produces the number 0. That is, it doesn't change the state, it simply produces the numerical value zero when operating on that state. The conjugate of this thing, of course, is:

$$\underline{a}* = \begin{bmatrix} 0 & 1 \\ 0 & 0 \end{bmatrix}$$

(6.3)

which creates, in the sense that operating on the 0 state, it turns it to the 1 state. In other words, it moves from |0> to |1>. When operating on the |1> state, there is no further state above that you can create, and therefore it gives it the number zero. Every other operator 2 × 2 matrix can be represented in terms of these $\underline{a}$ and $\underline{a}*$. For example, the product of $\underline{a}*\underline{a}$ is equal to the matrix:

$$\underline{a}*\underline{a} = \begin{bmatrix} 1 & 0 \\ 0 & 0 \end{bmatrix}$$

(6.4)

which you might call N_a. It is 1 when the state is |1> and 0 when the state is |0>. It gives the number that the state of the atom represents. Likewise the product:

$$\underline{a}\underline{a}* = \begin{bmatrix} 0 & 0 \\ 0 & 1 \end{bmatrix}$$

(6.5)

is 1 - N_a, and gives 0 for the up-state and 1 for the down-state. We'll use 1 to represent the diagonal matrix:

$$\begin{bmatrix} 1 & 0 \\ 0 & 1 \end{bmatrix}$$

(6.6)

As a consequence of all this, $\underline{a}\underline{a}* + \underline{a}*\underline{a} = 1$.

It is evident then that our matrix for NOT, the operator that produces NOT, is $A_a = \underline{a} + \underline{a}^*$. And further, of course, it is reversible, $A_a{}^*A_a = 1$, and A_a is unitary. In the same way, the matrix $A_{a,b}$ for the CONTROLLED NOT can be worked out. If you look at the table of values for CONTROLLED NOT (Figure 6.3), you see that it can be written this way:

$$a*a\left(\underline{b}+\underline{b}*\right)+\underline{aa}* \tag{6.7}$$

In the first term, the $\underline{a}^*\underline{a}$ selects the condition that the line $a = 1$ in which case we want $\underline{b} + \underline{b}^*$, the NOT, to apply to b. The second term selects the condition that the line a is 0, in which case we want nothing to happen to b and the unit matrix on the operators of b is implied. This can also be written as $1 + \underline{a}^*\underline{a}(\underline{b} + \underline{b}^* - 1)$, the 1 representing all the lines coming through directly, but, in the case that a is 1, we would like to correct that by putting in a NOT instead of leaving the line b unchanged. The matrix for the CONTROLLED CONTROLLED NOT is:

$$A_{ab,c} = 1 + \underline{a}*\underline{ab}*\underline{b}\left(\underline{c} + \underline{c}*-1\right) \tag{6.8}$$

as perhaps you may be able to see.

The next question is what the matrix is for a general logic unit which consists of a sequence of these. As an example, we'll study the case of the full adder which we described before (see Figure 6.5). Now we'll have, in the general case, four wires represented by $a,b,c,$ and d; we don't necessarily have d as 0 in all cases, and we would like to describe how the object operates in general (if d is changed to 1, d' is changed to its NOT). It produces new numbers a', b', c', and d', and we could imagine with our system that there are four atoms labeled a,b,c,d in a state labeled $|a,b,c,d,>$ and that a matrix M operates which changes these same four atoms so that they appear to be in the state $|a',b',c',d'>$ which is appropriate for this logic unit. That is, if $|\psi_{IN}>$ represents the incoming state of the four bits, M is a matrix which generates an outgoing state $|\psi_{out}> = M|\psi_{IN}>$ for the four bits. For example, if the input state were the state $|1,0,1,0>$ then, as we know, the output state should be $|1,0,0,1>$; the first two a',b' should be 1,0 for those two first lines come straight through, and the last two c',d' should be 0,1 because that represents the sum and carry of the first three, a,b,c bits in the first input, as $d = 0$. Now the matrix M for the adder can easily be seen as the result of five successive primitive operations, and therefore becomes the matrix product of the five successive matrices representing these primitive objects:

$$M = A_{a,b}A_{b,c}A_{bc,d}A_{a,b}A_{ab,d} \tag{6.9}$$

The first, which is the one written farthest to the right, is $A_{ab,d}$ for that represents the CONTROLLED CONTROLLED NOT in which a and b are the CONTROL lines, and the NOT appears on line d. By looking at the diagram in Figure 6.5 we can immediately see what the remaining factors in the sequence represent. The last factor, for example, $A_{a,b}$ means that there's a CONTROLLED NOT with a CONTROL on line a and NOT on line b. This matrix will have the unitary property $M^*M = 1$ since all of the As out of which it is a product are unitary. That is to say M is a reversible operation, and M^* is its inverse.

Our general problem, then, is this. Let $A_1, A_2, A_3, \ldots A_k$ be the succession of operations wanted, in some logical unit, to operate on n lines. The $2^n \times 2^n$ matrix M needed to accomplish the same goal is a product $A_k \ldots A_3 A_2 A_1$, where each A is a simple matrix. How can we generate this M in a physical way if we know how to make the simpler elements?

In general, in quantum mechanics, the outgoing state at time t is $e^{iHt}\psi_{IN}$ where ψ_{IN} is the input state, for a system with Hamiltonian H. To try to find, for a given special time t, the Hamiltonian which will produce $M = e^{iHt}$ when M is such a product of non-commuting matrices, from some simple property of the matrices themselves appears to be very difficult.

We realize, however, that at any particular time, if we expand the e^{iHt} out (as $1 + iHt - H^2t^2/2\ldots$) we'll find the operator H operating an innumerable arbitrary number of times, once, twice, three times, and so forth, and the total state is generated by a superposition of these possibilities. This suggests that we can solve this problem of the composition of these As in the following way. We add to the n atoms, which are in our register, an entirely new set of $k + 1$ atoms, which we'll call "program counter sites". Let us call q_i and q_i^* the annihilation and creation operators for the program site i for i = 0 to k. A good thing to think of, as an example, is an electron moving from one empty site to another. If the site is occupied by the electron, its state is $|1\rangle$, while if the site is empty, its state is $|0\rangle$.

We write, as our Hamiltonian:

$$H = \sum_{i=0}^{k-1} q_{i+1}^* \, q_i A_{i+1} + complex \; conjugate$$

$$= q_1^* q_0 A_1 + q_2^* q_1 A_2 + q_3^* q_2 A_3 + \ldots + q_0^* q_1 A_1^* + q_1^* q_2 A_2^* + q_2^* q_3 A_3^* + \ldots \quad (6.10)$$

The first thing to notice is that, if all the program sites are unoccupied so that all the program atoms are initially in the state 0, nothing happens because every term in the Hamiltonian starts with an annihilation operator and it gives 0 therefore.

The second thing we notice is that, if only one or another of the program sites is occupied (in state $|1\rangle$), and the rest are not (state $|0\rangle$), then this is always true. In fact the number of program sites that are in state $|1\rangle$ is a conserved quantity. We will suppose that, in the operation of this computer, either no sites are occupied (in which case nothing happens) or just one site is occupied. Two or more program sites are never both occupied during normal operation.

Let us start with an initial state where site 0 is occupied, is in the $|1>$ state, and all the others are empty, in the $|0>$ state. If later, at some time, the final site k is found to be in the $|1>$ state (and therefore all the others in $|0>$) then, we claim, the n register has been multiplied by the matrix M, which is $A_k...A_2A_1$ as desired.

Let me explain how this works. Suppose that the register starts in any initial state, ψ_{IN}, and that the site, 0, of the program counter is occupied. Then the only term in the entire Hamiltonian that can first operate, as the Hamiltonian operates in successive times, is the first term, $q_1^*q_0A_1$. The q_0 will change site number 0 to an unoccupied site, while q_1^* will change the site number 1 to an occupied site. Thus the term $q_1^*q_0$ is a term which simply moves the occupied site from the location 0 to the location 1. But this is multiplied by the matrix A_1 which operates only on the n register atoms, and therefore multiplies the initial state of the n register atoms by A_1. Now, if the Hamiltonian happens to operate a second time, this first term will produce nothing because q_0 produces 0 on the number 0 site because it is now unoccupied. The term which can operate now is the second term, $q_2^*q_1A_2$ for that can move the occupied point, which I shall call a "cursor". The cursor can move from site 1 to site 2 but the matrix A_2 now operates on the register, therefore the register has now got the matrix A_2A_1 operating on it. So, looking at the first line of the Hamiltonian, if that is all there was to it, as the Hamiltonian operates in successive orders, the cursor would move successively from 0 to k, and you would acquire, one after the other, operating on the n register atoms, the matrices, A, in the order that we would like to construct the total M.

However, a Hamiltonian must be Hermitian, and therefore the complex conjugate of all these operators must be present. Suppose that, at a given stage, we have gotten the cursor on site number 2, and we have the matrix A_2A_1 operating on the register. Now the q_2 which intends to move that occupation to a new position needn't come from the first line, but may have come from the second line. It may have come, in fact, from $q_1^*q_2A_2^*$ which would move the cursor back from the position 2 to the position 1. But note that, when this happens, the operator A_2^* operates on the register, and therefore the total operator on the register is $A_2^*A_2A_1$ in this case. But $A_2^*A_2$ is 1 and therefore the operator is just A_1. Thus we see that, when the cursor is returned to the position 1, the net result is that only the operator A, has really operated on the register. Thus it is that, as the various terms of the Hamiltonian move the cursor forwards and backwards, the As accumulate, or are reduced out again. At any stage, for example, if the cursor were up to the j site, the matrices from A_1, to A_j have operated in succession on the n register. It does not matter whether or not the cursor on the j site has arrived there by going directly from 0 to j, or going further and returning, or going back and forth in any pattern whatsoever, as long as it finally arrived at the state j. Therefore it is true that, if the cursor is found at the site k, we have the net result for the n register atoms that the matrix M has operated on their initial state as we desired.

How then could we operate this computer? We begin by putting the input bits onto the register, and by putting the cursor to occupy the site 0. We then check at the site k, say, by scattering electrons, that the site k is empty, or that the site k has a cursor. The moment we find the cursor at site k, we remove the cursor so that it cannot return down the program line, and then we know that the register contains the output data. We can then measure it at our leisure. Of course, there are external things involved in making the measurements, and determining all of this, which are not part of our computer. Surely a computer has eventually to be in interaction with the external world, both for putting data in and for taking it out.

Mathematically it turns out that the propagation of the cursor up and down this program line is exactly the same as it would be if the operators A were not in the Hamiltonian. In other words, it represents just the waves which are familiar from the propagation of the tight binding electrons or spin waves in one dimension, and are very well known. There are waves that travel up and down the line, and you can have packets of waves and so forth. We could improve the action of this computer and make it into a ballistic action in the following way: by making a line of sites in addition to the ones inside, that we are actually using for computing, a line of, say, many sites both before and after. It's just as though we had values of the index i for q_i, which are less than 0 and greater than k, each of which has no matrix A, just a 1 multiplying there. Then we'd have a longer spin chain, and we could have started, instead of putting a cursor exactly at the beginning site 0, by putting the cursor with different amplitudes on different sites representing an initial incoming spin wave, a wide packet of nearly definite momentum. This spin wave would then go through the entire computer in a ballistic fashion and out the other end into the outside tail that we have added to the line of program sites, and there it would be easier to determine if it is present and to steer it away to some other place, and to capture the cursor. Thus the logical unit can act in a ballistic way.

This is the essential point and indicates, at least to a computer scientist, that we could make a universal computer, because he knows if we can make any logical unit we can make a universal computer. That this could represent a universal computer for which composition of elements and branching can be done, is not entirely obvious unless you have some experience, but I will discuss that to some further extent later.

6.4 Imperfections and Irreversible Free Energy Loss

There are, however, a number of questions that we would like to discuss in more detail such as the question of imperfections. There are many sources of imperfections in this machine, but the first one we would like to consider is the possibility that the coefficients in the couplings, along the program line, are not exactly equal. The line is so long that in a real calculation little irregularities would produce a small probability of scattering, and the waves would not travel exactly ballistically but would go back and forth. If the system, for example, is built so that these sites are built on a substrate of ordinary physical atoms, then the thermal vibrations of these atoms would change the couplings a little bit and generate imperfections. (We should even need such noise for with small imperfections there are shallow trapping regions where the cursor may get caught.) Suppose then, that there is a certain probability, say p per step of calculation (that is, per step of cursor motion $i \rightarrow i + 1$) for scattering the cursor momentum until it is randomized ($1/p$ is the transport mean free path). We will suppose that the p is fairly small. Then in a very long calculation, it might take a very long time for the wave to make its way out the other end, once started at the beginning – because it has to go back and forth so many times due to the scattering. What one then could do, would be to pull the cursor along the program line with an external force. If the cursor is, for example, an electron moving from one vacant site to another, this would be just like an electric field trying to pull the electron along a wire, the resistance of which is generated by the imperfection or the probability of scattering. Under these circumstances we can calculate how much energy will be expended by this external force.

This analysis can be made very simply; it is an almost classical analysis of an electron with a mean free path. Every time the cursor is scattered, I'm going to suppose it is randomly scattered forward and backward. In order for the machine to operate, of course, it must be moving forward at a higher probability than it is moving backward. When a scattering occurs, therefore, the loss in entropy is the logarithm of the probability that the cursor is moving forward, divided by the probability that the cursor was moving backward. This can be approximated by (the probability forward – the probability backward)/(the probability forward + the probability backward). That was the entropy lost per scattering. More interesting is the entropy lost per net calculational step which is, of course, simply p times that number. We can rewrite the entropy cost per calculational step as:

$$p \quad v_D / v_R \tag{6.11}$$

where v_D is the drift velocity of the cursor, and v_R is its random velocity. Or, if you like, it is p times the minimum time that the calculation could be done in, (that is, if all the steps were always in the forward direction), divided by the actual time allowed. The free energy loss per step, then, is $kT \times p \times$ the minimum time that the calculation could be done, divided by the actual time that you allow yourself to do it. This is a formula that was first derived by Bennett. The factor p is a coasting factor, to represent situations in which not every site scatters the cursor randomly, but it has only a small probability to be thus scattered. It will be appreciated that the energy loss per step is not kT but is that divided by two factors. One $(1/p)$ measures how perfectly you can build the machine, and the other is proportional to the length of time that you take to do the calculation. It is very much like a Carnot engine in which, in order to obtain reversibility, one must operate very slowly. For the ideal machine where p is 0, or where you allow an infinite time, the mean energy loss can be 0.

The Uncertainty Principle, which usually relates some energy and time uncertainty, is not directly a limitation. What we have in our computer is a device for making a computation, but the time of arrival of the cursor and the measurement of the output register at the other end (in other words, the time it takes in which to complete the calculation), is not a definite time. It's a question of probabilities, and so there is a considerable uncertainty in the time at which a calculation will be done. There is no loss associated with the uncertainty of cursor energy; at least no loss depending on the number of calculational steps. Of course, if you want to do a ballistic calculation on a perfect machine, some energy would have to be put into the original waves, but that energy can, of course, be removed from the final waves when it comes out of the tail of the program line. All questions associated with the uncertainty of operators and the irreversibility of measurements are associated with the input and output functions. No further limitations are generated by the quantum nature of the computer per se; nothing that is proportional to the number of computational steps.

In a machine such as this there are very many other problems due to imperfections. For example, in the registers for holding the data, there will be problems of cross-talk, interactions between one atom and another in that register, or interaction of the atoms in that register directly with things that are happening along the program line that we didn't exactly bargain for. In other words, there may be small terms in the Hamiltonian besides the ones we've written. Until we propose a complete implementation of this, it is very difficult to analyze. At least some of these problems can be remedied in the usual way by techniques such as error correcting codes and so forth, that have been studied in normal computers. But until we find a specific implementation for this computer, I do not know how to proceed to analyze these effects. However, it appears that they would be very important in practice. This computer seems to be very delicate and these imperfections may produce considerable havoc.

The time needed to make a step of calculation depends on the strength or the energy of the interactions in the terms of the Hamiltonian. If each of the terms in the Hamiltonian is supposed to be of the order of 0.1 electron volts, then it appears that the time for the cursor to make each step, if done in a ballistic fashion, is of the order 6×10^{-15} sec. This does not represent an enormous improvement, perhaps only about four orders of magnitude over the present values of the time delays in transistors, and is not much shorter than the very short times possible to achieve in many optical systems.

6.5 Simplifying the Implementation

We have completed the job we set out to do – to find some quantum mechanical Hamiltonian of a system that could compute, and that is all that we need to say. But it is of some interest to deal with some questions about simplifying the implementation. The Hamiltonian that we've written involves terms which can involve a special kind of interaction between five atoms. For example, three of them in the register for a CONTROLLED CONTROLLED NOT and two of them as the two adjacent sites in the program counter. This may be rather complicated to arrange. The question is, can we do it with simpler parts? It turns out, we can indeed. We can do it so that in each interaction there are only three atoms. We're going to start with new primitive elements instead of the ones we began with. We'll have the NOT all right, but we have in addition to that simply a "switch" (see also Priese [5]).

Supposing that we have a term, $q^*cp + r^*c^*p$ + its complex conjugate in the Hamiltonian (in all cases we'll use letters in the earlier part of the alphabet for register atoms, and in the latter part of the alphabet for program sites (see Figure 6.7)):

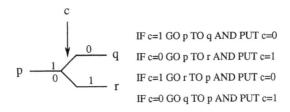

FIGURE 6.7
Switch.

This is a switch in the sense that, if c is originally in the $|1>$ state, a cursor at p will move to q, whereas if c is in the $|0>$ state, the cursor at p will move to r. During this operation the controlling atom c changes its state. (It is possible also to write an expression in which the control atom does not change its state, such as $q^*c^*cp + r^*cc^*p$ and its complex conjugate, but there is no particular advantage or disadvantage to this, and we will take the simpler form.) The complex conjugate reverses this. If, however, the cursor is at q and c is in the state $|1>$ (or cursor at r, c in $|0>$) the H gives 0, and the cursor gets reflected back. We shall build all our circuits and choose initial states so that this circumstance will not arise in normal operation, and the ideal ballistic mode will work.

With this switch we can do a number of things. For example, we could produce a CONTROLLED NOT as in Figure 6.8.

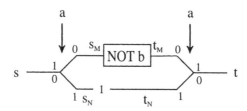

FIGURE 6.8
CONTROLLED NOT realized by switches.

The switch a controls the operation. Assume the cursor starts at s. If $a = 1$ the program cursor is carried along the top line, whereas if $a = 0$ it is carried along the bottom line, in either case terminating finally in the program site t. In these diagrams, horizontal or vertical lines will represent program atoms. The switches are represented by diagonal lines, and in boxes we'll put the other matrices that operate on registers such as the NOT b. To be specific, the Hamiltonian for this little section of a CONTROLLED NOT, thinking of it as starting at s and ending at t, is given below:

$$H_c(s,t) = s_m^* a s + t^* a^* t_M + t_m^* \left(b + b^*\right) s_M + s_N^* a^* s + t^* a t_N + t_N^* s_N + c.c. \quad (6.12)$$

(The c.c. means to add the complex conjugate of all the previous terms.) Although there seem to be two routes here which would possibly produce all kinds of complications characteristic of quantum mechanics, this is not so. If the entire computer system is started in a definite state for a, by the time the cursor reaches s the atom a is still in some definite state (although possibly different from its initial state due to previous computer operations on it). Thus only one of the two routes is taken. The expression may be simplified by omitting the $s_N^* t_N$ term and putting $t_N = s_N$. One need not be concerned in that case that one route is longer (two cursor sites) than the other (one cursor site), for again there is no interference. No scattering is produced in any case by the insertion into a chain of coupled sites, an extra piece of chain of any number of sites with the same mutual coupling between sites (analogous to matching impedances in transmission lines).

To study these things further, we think of putting pieces together. A piece (see Figure 6.9) M might be represented as a logical unit of interacting parts in which we only represent the first input cursor site as s_M and the final one at the other end as t_M. All the rest of the program sites that are between s_M and t_M are considered internal parts of M, and M contains its registers. Only s_M and t_M are sites that may be coupled externally (Figure 6.9).

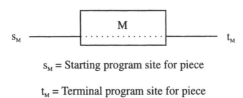

s_M = Starting program site for piece

t_M = Terminal program site for piece

FIGURE 6.9
One "piece".

The Hamiltonian for this sub-section we'll call H_M, and we'll identify s_M and t_M as the names of the input and output program sites by writing $H_M(s_M, t_M)$. So therefore H_M is the part of the Hamiltonian representing all the atoms in the box and their external start and terminator sites.

An especially important and interesting ease to consider is when the input data (in the regular atoms) comes from one logical unit, and we would like to transfer it to another (see Figure 6.10).

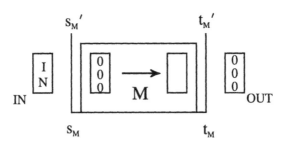

FIGURE 6.10
Piece with external input and output.

Suppose that we imagine that the box M starts with its input register with 0 and its output (which may be the same register) also with 0. We could now use it in the following way. We could make a program line, let's say starting with s_M' whose first job is to exchange the data in an external register which contains the input, with M's input register which at the present time contains 0s. Then the first step in our calculation starting, say, at s_M', would be to make an exchange with the register inside of M. That puts 0s into the original input register and puts the input where it belongs inside the box M. The cursor is now at s_M. (We have already explained how exchange can be made of CONTROLLED NOTs.) Then, as the program goes from s_M to t_M, we find the output now in the box M. The output register of M is now cleared as we write the results into some new external register provided for that purpose, originally containing 0s. This we do from t_M to t_M' by exchanging data in the empty external register with the M's output register.

We can now consider connecting such units in different ways. For example, the most obvious way is succession. If we want to do first M and then N we can connect the terminal side of one to the starting side of the other as in Figure 6.11 to produce a new effective operator K:

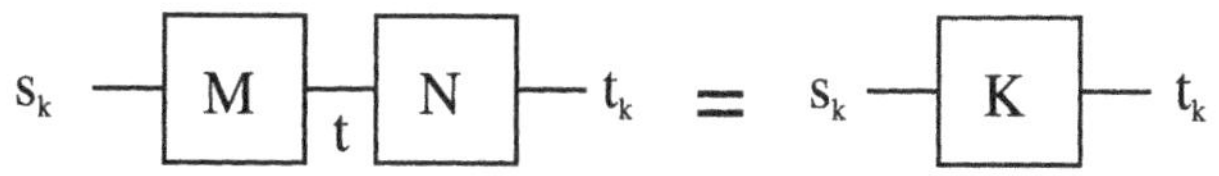

$$s_k \;-\!\boxed{\;M\;}\!-\!\boxed{\;N\;}\!-\; t_k \;=\; s_k \;-\!\boxed{\;K\;}\!-\; t_k$$

FIGURE 6.11
Operations performed in succession.

The Hamiltonian for H_K is then:

$$H\big(s_K, t_K\big) = H_M\big(s_K, t\big) + H_N\big(t, t_K\big) \tag{6.13}$$

The general conditional, if $a = 1$ do M, but if $a = 0$ do N, can be made, as in Figure 6.12.

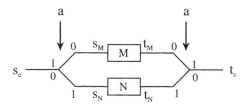

FIGURE 6.12
Conditional: if a = 1 then M, else N.

For this:

$$H_{cond}\left(s_c,t_c\right) = (s_m^* a s_c + t_c^* a^* t_m + s_N^* a^* s_c + t_c^* a t_N + c.c.) + H_M\left(s_M,t_M\right) + H_N\left(s_N,t_N\right)$$

(6.14)

The CONTROLLED NOT is the special case of this with M = NOT b for which H is:

$$H_{NOT\,b}\left(s,t\right) = s^*\left(b + b^*\right)t + c.c.$$

(6.15)

and N is no operation: s^*t.

As another example, we can deal with a garbage clearer (previously described in Figure 6.6) not by making two machines, a machine and its inverse, but by using the same machine and then sending the data back to the machine in the opposite direction, using our switch (Figure 6.13).

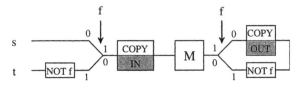

FIGURE 6.13
Garbage clearer.

Suppose in this system we have a special flag f which is originally always set to 0. We also suppose we have the input data in an external register, an empty external register available to hold the output, and the machine registers all empty (containing 0s). We come on the starting line s. The first thing we do is to copy (using CONTROLLED NOTs) our external input into M. Then M operates, and the cursor goes on the top line in our drawing. It copies the output out of M into the external output register. M now contains garbage. Next it changes f to NOT f, comes down on the other line of the switch, backs out through M clearing the garbage and uncopies the input again. When you copy data and do it again, you reduce one of the registers to 0, the register into which you copied the first time. After the copying, it goes out (since f is now changed) on the other line where we restore f to 0 and come out at t. So between s and t we have a new piece of equipment which has the following properties. When it starts we have, in a register called IN, the input data. In an external register, which we call OUT, we have 0s. There is an internal flag set at 0, and the box, M, is empty of all data. At the termination of this, at t, the input register still contains the input data, the output register contains the output of the effort of the operator M. M, however, is still empty, and the flag f is reset to 0.

Also important in computer programs is the ability to use the same sub-routine several times. Of course, from a logical point of view, that can be done by writing that bit of program over and over again each time it is to be used, but in a practical computer it is much better if we could build that section of the computer which does a particular operation just once, and use that section again and again. To show the possibilities here, first just suppose we have an operation we simply wish to repeat twice in succession (Figure 6.14).

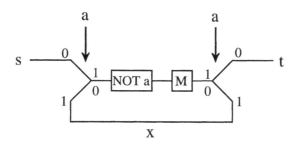

FIGURE 6.14
The operation "do M twice".

We start at s with the flag a in the condition 0, and thus we come along the line and the first thing that happens is we change the value of a. Next we do the operation M. Now, because we changed a, instead of coming out at the top line where we went in, we come out at the bottom line which recirculates the program back into changing a again, and it restores it. This time as we go through M, we come out and we have the a to follow on the upper line and thus come out at the terminal t. The Hamiltonian for this is:

$$H_{MM}(s,t) = \left(s_N^* a^* s + s_M^* \left(a^* + a \right) s_N + x^* a^* t_M + s_N^* a x + t^* a t_M + c.c \right) + H_M \left(s_M, t_M \right)$$

$$(6.16)$$

Using this switching circuit a number of times, of course, we can repeat an operation several times. For example, using the same idea three times in succession, a nested succession, we can do an operation eight times by the apparatus indicated in Figure 6.15:

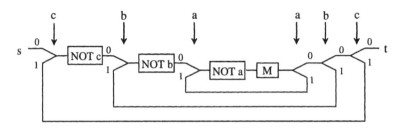

FIGURE 6.15
The operation "do M eight times".

In order to do so, we have three flags, *a*, *b*, and *c*. It is necessary to have flags when operations are done again for the reason that we must keep track of how many times it's done and where we are in the program or we'll never be able to reverse things. A subroutine in a normal computer can be used and emptied and used again without any record being kept of what happened. But here we have to keep a record – and we do that with flags – of exactly where we are in the cycle of the use of the subroutine. If the subroutine is called from a certain place and has to go back to some other place, and is called another time, its origin and final destination are different. We have to know and keep track of where it came from and where it's supposed to go individually in each case, so more data has to be kept. Using a subroutine over and over in a reversible machine is only slightly harder than in a general machine. All these considerations appear in papers by Fredkin, Toffoli, and Bennett.

It is clear by the use of this switch, and successive uses of such switches in trees, that we would be able to steer data to any point in a memory. A memory would simply be a place where there are registers into which you could copy data and then return to the program. The cursor will have to follow the data along and I suppose there must be another set of tree switches set the opposite direction to carry the cursor out again after copying the data so that the system remains reversible.

In Figure 6.16 we show an incremental binary counter (of three bits *a,b,c* with *c* the most significant bit) which keeps track of how many net times the cursor has passed from *s* to *t*.

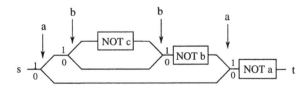

FIGURE 6.16
Increment counter (three-bit).

These few examples should be enough to show that indeed we can construct all computer functions with our SWITCH and NOT. We need not follow this in more detail.

6.6 Conclusions

It is clear from these examples that this quantum machine has not really used many of the specific qualities of the differential equations of quantum mechanics. What we have done is only to try to imitate as closely as possible the digital machine of conventional sequential architecture. It is analogous to the use of transistors in conventional machines where we don't properly use all the analog continuum of the behavior of transistors, but just try to run them as saturated on or off digital devices so the logical analysis of the system behavior is easier. Furthermore, the system is absolutely sequential – for example, even in the comparison (exclusive OR) of two k bit numbers, we must do each bit successively. What can be done, in these reversible quantum systems, to gain the speed available by concurrent operation has not been studied here.

Although, for theoretical and academic reasons, I have studied complete and reversible systems, if such tiny machines could become practical there is no reason why irreversible and entropy creating interactions cannot be made frequently during the course of operations of the machine. For example, it might prove wise, in a long calculation, to ensure that the cursor has surely reached some point and cannot be allowed to reverse again from there. Or, it may be found practical to connect irreversible memory storage (for items less frequently used) to reversible logic or short term reversible storage registers, etc. Again, there is no reason we need to stick to chains of coupled sites for more distant communication where wires or light may be easier and faster. At any rate, it seems that the laws of physics present no barrier to reducing the size of computers until bits are the size of atoms, and quantum behavior holds dominant sway.

References[*]

1. C.H. Bennett, "Logical Reversibility of Computation," *IBM Journal of Research and Development*, 6 (1979), pp. 525–532.

---

[*] I would like to thank T. Toffoli for his help with the references. [RPF]

2. E. Fredkin and T. Toffoli, "Conservative Logic," *International Journal of Theoretical Physics*, 21 (1982), pp. 219–253.
3. C.H. Bennett, "Thermodynamics of Computation – A Review," *International Journal of Theoretical Physics*, 21 (1982), pp. 905–940.
4. T. Toffoli, "Bicontinuous Extensions of Invertible Combinatorial Functions," *Mathematical Systems Theory*, 14 (1981), pp. 13–23.
5. L. Priese, "On a Simple Combinatorial Structure Sufficient for Sublying Non-trivial Self Reproduction," *Journal of Cybernetics*, 6 (1976), pp. 101–137.

7

Quantum Computing 40 Years Later

John Preskill

CONTENTS

DOI: 10.1201/9781003358817-7

Forty years ago, Richard Feynman proposed harnessing quantum physics to build a more powerful kind of computer. Realizing Feynman's vision is one of the grand challenges facing 21st-century science and technology. In this article, we'll recall Feynman's contribution that launched the quest for a quantum computer and assess where the field stands 40 years later.

After a brisk account in Sections 7.1 and 7.2 of how quantum computing has developed over the past few decades, I sketch the foundations of the subject, discussing salient features of quantum information in Section 7.3, then formulating a mathematical model of quantum computation in Section 7.4 and highlighting some implications of the model. In Sections 7.5 and 7.6, I review two particularly promising applications of quantum computing foreseen by Feynman, simulating the dynamics of complex quantum systems, and computing their static properties. In Section 7.7, I explain the concept of quantum error correction, the basis of our belief that quantum computers can be scaled up to large systems that solve very hard problems. I offer some concluding thoughts in Section 7.8 and reminisce about some of my interactions with Feynman in the Reminiscences section of the book.

7.1 Feynman and Quantum Computation

7.1.1 Feynman's 1981 Talk

Richard Feynman was renowned for his wide-ranging intellect, and computation was one of the many topics that fascinated him. He was interested in the scientific applications of computing, but also deeply curious about how hardware and software really work, and about fundamental issues in the theory of computation. These interests are amply reflected in *The Feynman Lectures on Computation*.

In May 1981, Feynman spoke at a conference on the topic "Simulating Physics with Computers". There he proposed the idea of using quantum computers to simulate quantum systems that are too hard to simulate using conventional classical digital computers. Feynman's talk, later published as a lightly edited transcript [1], is justly remembered for its role in launching quantum computing as a field of study.

In the talk, Feynman clearly states his goal, which is to simulate quantum systems using resources that scale well with the size of the system.

> The rule of simulation that I would like to have is that the number of computer elements required to simulate a large physical system is only to be proportional to the spacetime volume of the physical system.

He points out why digital computers are not adequate for the task, because there is no succinct way to classically describe a quantum state of many particles:

> Now I explicitly go to the question of how we can simulate with a computer ... the quantum mechanical effects ... But the full description of quantum mechanics for a large system with R particles is given by a function which we call the amplitude to find the particles at $x_1, x_2, \ldots x_R$, and therefore because it has too many variables, *it cannot be simulated with a normal computer*. [Italics mine.]

In a crucial passage, he speculates that a different kind of computer might be up to the job:

> Can you do it with a new kind of computer – a quantum computer? Now it turns out, as far as I can tell, that you can simulate this with a quantum system, with quantum computer elements. *It's not a Turing machine, but a machine of a different kind.* [Italics mine.]

And Feynman challenges computer scientists to study the power of this new model of computation:

> I present that as another interesting problem: To work out the classes of different kinds of quantum mechanical systems which are really intersimulatable – which are equivalent – as has been done in the case of classical computers.

About half of the talk is devoted to elaborating the argument that digital computers will be inadequate for efficiently simulating quantum systems. He emphasizes that quantum theory will not admit a local hidden variable description, and there follows a lucid discussion of Bell inequalities and the experimental evidence that these are violated (without any references and without ever mentioning Bell).

> If you take the computer to be the classical kind I've described so far (not the quantum kind described in the last section) and there're no changes in any laws, and there's no hocus-pocus, the answer is certainly, "No!" This is called the hidden variable problem. It is impossible to represent the results of quantum mechanics with a classical universal device.

Perhaps most famously, Feynman brought the talk to a stirring conclusion.

> Nature isn't classical, dammit, and if you want to make a simulation of Nature, you'd better make it quantum mechanical, and by golly it's a wonderful problem because it doesn't look so easy.

Feynman, then nearly 63, was calling for a fundamentally new type of computing machine, and he foresaw its natural application: making "a simulation of Nature". Bringing that vision to fruition is indeed "a wonderful problem" and 40 years later it still "doesn't look so easy".

7.1.2 Manin and Benioff

Around the same time, a few others were considering quantum models of computation, including the distinguished mathematician Yuri Manin. In his 1980 book *Computable and Uncomputable* [2], Manin, like Feynman, cogently emphasized the exponential cost of simulating a many-particle particle system with a classical computer. Manin wrote [translated from the Russian by Victor Albert]:

> These objects [quantum automata] may show us mathematical models of deterministic processes with highly unusual features. One of the reasons for this is because *the quantum phase space is much bigger than classical*: where classical space has N discrete levels, a quantum system allowing their superposition will have c^N Planck cells. In a union of two classical systems, their sizes N_1 and N_2 multiply, but in the quantum case we have $c^{N_1+N_2}$. These heuristic calculations point to a much larger potential complexity of the behavior of a quantum system when compared to its classical imitator. [Italics mine.]

Also in 1980, Paul Benioff explained how to describe computation from a fundamentally quantum point of view [3]. He wrote:

> These considerations suggest that it may be impossible even in principle to construct a quantum mechanical Hamiltonian model of the computation process. The reason is that any such model evolves as an isolated system with a constant total energy. The point of this paper is to suggest, by construction of such models, that this may not be the case.

Unlike Manin and Feynman, Benioff was not concerned with quantum *complexity*. Rather, he mainly focused on the question of whether a quantum computer can operate without dissipation. Feynman was also deeply interested in this issue and addressed it in detail in his talk at the 1984 CLEO/IQEC conference ("Quantum Mechanical Computers"), which is reprinted in this volume.

Strangely, though Feynman broached the topic of quantum computing in the lecture course that is captured by this book, he never mentioned in his class lectures the deep idea, so clearly articulated in his 1981 talk, that quantum computers can solve certain problems far more efficiently than classical computers. Yet what could be a better fit for a course on "The Potentialities and Limitations of Computing Machines"? I find this omission baffling.

7.1.3 From Feynman to Shor and Beyond

It took a while, but gradually the influence of Feynman's ideas grew. In 1985, David Deutsch formalized the notion of a quantum computer [4], an important advance, and raised the question of whether quantum computers might have an advantage over classical computers at solving problems that have nothing to do with quantum physics. In 1993, Umesh Vazirani and his student Ethan Bernstein formulated a contrived problem that a quantum computer could solve with a superpolynomial speedup over a classical computer [5]. Soon after, Daniel Simon showed that a quantum computer could achieve an exponential speedup in solving an idealized version of the problem of finding the period of function [6]. Though Simon's problem had no obvious applications, it inspired Peter Shor [7], who worked out a very efficient way of performing a Fourier transform using a quantum computer and applied it to formulate an efficient quantum algorithm for computing discrete logarithms. Just days later, Shor used similar ideas to find an efficient quantum algorithm for factoring large numbers.

Shor's discovery, and its obvious implications for cryptanalysis, caused interest in quantum computing to skyrocket. But very good physicists like Rolf Landauer [8], Bill Unruh [9], and Serge Haroche [10] voiced strong skepticism about whether quantum computers could ever work effectively. Those physicists were deeply familiar with the debilitating effects of *decoherence*, which under typical conditions prevent complex many-particle quantum systems from exhibiting quantum behavior, and viewed quantum computing (in the derisive words of Haroche and Raimond) as "the computer scientist's dream [but] the experimenter's nightmare". Remarkably, it was again Shor who led the next pivotal advances – the discovery of quantum error-correcting codes [11, 12] and of fault-tolerant methods for executing a quantum computation reliably using noisy hardware [13]. By the end of 1996, it was understood at least in principle that quantum computing could be scaled up to large devices that solve very hard problems, assuming that errors afflicting the hardware are not too common or too strongly correlated [14–18]. This "accuracy threshold theorem" for quantum computing was already in place two and a half years after the discovery of Shor's algorithm. Meanwhile, Cirac and Zoller had proposed that tools in atomic physics and quantum optics can be exploited to perform quantum logical operations [19], amplifying interest in the experimental physics community [20]. Those were exciting times [21].

7.1.4 Imagining the Future

We don't know exactly how Feynman arrived at the idea of a quantum computer, but we do know that by 1981 he had been thinking about the future of computing for decades. In his remarkable 1959 talk called "There's

Plenty of Room at the Bottom" [22], which foresaw the field of nanotechnology, he mused about the miniaturization of computing circuitry:

> If they had millions of times as many elements, they could make judgements … In many ways, they would have new qualitative features … There is nothing that I can see in the physical laws that says computer elements cannot be made enormously smaller than they are now. In fact, there may be certain advantages.

And he imagined devices assembled by precise placement of single atoms:

> All our devices can be mass produced so that they are absolutely perfect copies of one another … The principles of physics, as far as I can see, do not speak against the possibility of maneuvering things atom by atom.

That willingness to look beyond the limitations of current technology and envision what might be possible in the future is equally apparent in his speculations about "simulating physics with computers". To computer scientists of the early 1980s, quantum physics was viewed as an obstacle that would ultimately block further miniaturization of electrical circuitry. But to Feynman, quantum physics offered a dazzling opportunity. As his lectures on computing attest, Feynman knew enough about the theory of computation to understand and appreciate the "extended Church-Turing thesis", which asserts that any physically realizable universal computer can efficiently simulate any other (under a loose notion of "efficiently"). The great insight trumpeted in his 1981 talk is that this story needs revision because "Nature isn't classical, dammit". That idea could change the world.

7.2 Where We're Going and Where We Are

7.2.1 How Will Quantum Computers Be Used?

Because of Shor's algorithm, the public key cryptographic protocols we use to protect our privacy when we communicate over the Internet will become vulnerable to attacks by quantum computers in the future. To protect against that threat, "quantum-resistant" protocols are now being developed, based on computational problems that we think are too hard to solve even with a quantum computer [23]. An alternative path is quantum cryptography, where quantum states are exchanged over a quantum communication network, and security rests on the principle that eavesdropping on quantum communication produces an unavoidable and detectable disturbance. (That is another fascinating story of quantum technology [24, 25], which unfortunately is beyond the scope of this chapter.) Perhaps both of these approaches will be deployed, depending on the needs of the user [26].

Quantum computing is such a big departure from previous methods for processing information that no one can be expected to foresee its long-term ramifications. But based on what we currently understand about the power of quantum computing, Feynman's proposal to use quantum computers to simulate quantum systems is still the application that seems most likely to have a broad impact on the world. More powerful methods in computational quantum chemistry, for example, may eventually yield significant improvements in human health (pharmaceuticals), agriculture (nitrogen fixation), and the sustainability of our planet (energy storage and production, carbon sequestration) [27]. In contrast, while Shor's algorithm for factoring will have a disruptive effect on electronic commerce in the relatively near term, its long-term implications are not likely to be comparably profound.

It is important to emphasize that quantum computers have limitations. We don't expect, in particular, that quantum computers can find exact solutions to NP-hard optimization problems efficiently [28]. There is a general scheme for speeding up exhaustive search for a solution using quantum computers (Grover's algorithm), but the speedup is quadratic in that case [29]; that is, the quantum computer finds the solution in a time that scales as the square root of the time needed by the classical computer. Under the highly idealized assumption that the classical and quantum computers have the same clock speed (can perform the same number of elementary operations per second), that means that the quantum computer can find a solution that is $2n$ bits long in the time it would take the classical computer to find a solution that is n bits long, for asymptotically large n. That may be important someday. But for factoring large numbers or simulating quantum systems, the speedup is far more spectacular. The runtime for simulating an n-qubit quantum system using a classical computer, in the hardest instances, rises exponentially with n, while the runtime for simulating the system on a quantum computer scales like a power of n. That is a game-changing difference, as Feynman appreciated.

7.2.2 The NISQ Era Unfolds

It is also important to emphasize that quantum computers are not very useful yet. In the 40 years since Feynman's talk, a variety of approaches to building quantum hardware have emerged and progressed, but both the number of qubits and the accuracy of our quantum processors are still quite modest. An indicator of the current status is a milestone declaimed in 2019 by the Google AI Quantum group [30], known as "quantum computational supremacy" [31, 32].

As Feynman emphasized, a remarkable claim about quantum physics is that classical systems cannot in general simulate quantum systems efficiently. Arguably, that is one of the deepest statements known about the difference between quantum and classical, and we have a strong incentive to validate it in the laboratory to the extent that we can. Can we exhibit some task performed by a quantum computer that would require a much longer runtime on any existing classical computer?

Using superconducting quantum technology, the Google group constructed a programmable quantum computer called Sycamore with 53 working qubits arranged in a two-dimensional array, such that entangling two-qubit quantum gates can be performed on neighboring qubits in the array. (We will get to an explanation of what "entangling" means in Section 7.3.) They executed up to 20 layers of two-qubit gates and then measured all the qubits at the end. Because the hardware sometimes makes errors, the final measurement yields the correct output only once in 500 runs of the computation. But by repeating the same computation millions of times in just a few minutes, they extracted a statistically useful signal.

Using the best currently known methods, simulating what Sycamore does in a few minutes would take at least a few days for the most powerful existing classical supercomputer [33]. Furthermore, the cost of the classical simulation rises exponentially with the number of qubits, and so would be very far beyond the classical computer's reach if only a few more qubits could be added. While the classical computer occupies the equivalent of two tennis courts and consumes megawatts of power, Sycamore is just a single chip nested inside a dilution refrigerator. Quantum David overpowers Classical Goliath.

Admittedly, this task performed by Sycamore is of no particular interest for any purpose besides demonstrating quantum computational supremacy. But it signals that quantum hardware is now working well enough to produce meaningful results in a regime where classical simulation is very difficult and motivates us to seek more meaningful applications.

It is convenient to have a word for the new quantum era that is now opening, and the word NISQ has caught on [34]. It stands for *Noisy Intermediate-Scale Quantum*. "Intermediate scale" conveys that today's quantum devices with more than 50 well-controlled qubits cannot be simulated by brute force using the most powerful currently existing classical supercomputers; "noisy" reminds us that these devices are not error corrected, and that the noise limits their computational power. For physicists, NISQ technology is exciting – it provides us with new tools for exploring the properties of highly complex many-particle quantum systems in a regime that has never been experimentally accessible before. It might also have other applications of interest to a broader community of users, but we're not yet sure about that. NISQ will not change the world by itself, at least not right away; instead, we should regard it as a step toward more powerful quantum technologies we hope to develop in the future.

In the most advanced multi-qubit quantum processors that are currently available, the probability that a two-qubit quantum gate makes a sizable error is slightly less than 1%. That was why the 53-qubit Sycamore device was unable to execute circuits with more than 20 time steps. We have no convincing argument that a quantum computation with (say) of order 100 qubits and fewer than 100 time steps can solve practical problems.

One heuristic proposal is to search for approximate solutions to optimization problems using a hybrid quantum/classical approach [35, 36]. It makes sense to rely heavily on our powerful classical processors, and then attempt to boost that power with a NISQ co-processor. But we just don't know yet whether this hybrid method can outperform the best purely classical hardware running the best classical algorithms for solving the same problems. Frankly, it's a lot to ask, considering that the classical methods are well honed after decades of development, and the NISQ processors are becoming available for the first time now. But we'll have to try it and see how well it works. Vibrant discussions are already underway among potential application users, hardware providers, and quantum algorithm experts. As we experiment with NISQ technology, we'll learn more about how it performs and perhaps that will point us toward promising ideas for applications.

7.2.3 Quantum Simulation in the NISQ Era

Though it's not clear how we'll fare in our quest for NISQ applications to practical problems of potential commercial interest, I do feel optimistic about using quantum platforms to advance physics over the next five to ten years. Despite their notable limitations, NISQ processors should be able to prepare and study exotic quantum states that have not been accessible in the physics lab until now.

Classical computers are especially bad at simulating *quantum dynamics* – that is, predicting how a highly entangled quantum state will change with time. Quantum computers should have a big advantage in that task. It is instructive to recall that the theory of classical chaos (the extreme sensitivity to initial conditions in classical dynamical systems, which accounts for our inability to predict the weather more than two weeks out) advanced rapidly in the 1960s and 1970s after it became possible to simulate chaotic dynamical systems using classical computers. We may anticipate that the emerging ability to simulate chaotic quantum systems (those in which entanglement spreads very rapidly) will promote advances in our understanding of quantum chaos. Valuable insights might already be gleaned using noisy devices with about 100 qubits.

I should comment on the distinction between analog and digital quantum simulation. When we speak of an *analog quantum simulator*, we mean a system with many qubits whose dynamics resemble the dynamics of a model system we are trying to study and understand. In contrast, a *digital quantum simulator* is a gate-based universal quantum computer that can be used to simulate any physical system of interest when suitably programmed and can also be used for other purposes.

Analog quantum simulation has been a vibrant area of research for the past 20 years [37, 38], while digital quantum simulation with general-purpose circuit-based quantum computers is just now getting started. Some of the same experimental platforms, for example, trapped ions and superconducting circuits, can be used for both purposes, while other systems, like trapped neutral atoms and molecules, are particularly well suited for use as analog simulators. Analog quantum simulators have been getting notably more sophisticated and are already being employed to study quantum dynamics in regimes that may be beyond the reach of classical simulators [39, 40]. They can also be used to create highly entangled equilibrium states of quantum matter, and to study their static properties [41–43].

Though analog quantum simulators are becoming increasingly programmable, they are still hampered by imperfect control – the actual quantum system in the lab only crudely approximates the target system of interest. For that reason, analog simulators are best suited for studying features that physicists call *universal*, properties that are relatively robust with respect to introducing small sources of error. A major challenge for research using analog quantum simulators is identifying accessible properties of quantum systems that are robust with respect to errors yet are also hard to simulate classically.

We can anticipate that analog quantum simulators will eventually become obsolete. Because they are hard to control, they will be surpassed someday by digital quantum simulators, which can be firmly controlled using quantum error correction. But because of the hefty overhead cost of quantum error correction, the reign of the analog quantum simulator may persist for many years. Therefore, when seeking near-term applications of quantum technology, we should not overlook the potential power of analog quantum simulators.

In the near term, circuit-based simulations of quantum matter may be prohibitively expensive, as realistic simulations of many-particle systems will require many gates. But circuit-based methods have advantages, too, such as greater flexibility in the Hamiltonians we can study and the initial states we can prepare. Therefore, it is important to pursue both digital and analog simulation methods, keeping in mind that experience with near-term digital simulators will help to lay foundations for the more ambitious simulations we will be able to carry out in the future. The same remark applies to applications of NISQ technology more broadly.

7.2.4 From NISQ to FTQC

As I have emphasized, NISQ-era devices will not be protected by quantum error correction, and noise severely limits the scale of computations that can be executed accurately using NISQ technology. In the long run, we expect to overcome these limitations imposed by noise using quantum error correction (QEC) and fault-tolerant quantum computing (FTQC), but QEC carries a high overhead cost in the number of qubits and number of logic gates needed [44, 45]. This cost depends on which algorithms we run and on the quality of our hardware. But if the error rate per entangling two-qubit gate is, say, 0.1% (which is better than today's hardware), we may need more than 100,000 physical qubits to run high-impact applications to quantum chemistry or materials science [46, 47].

It's a huge leap from where we expect to be in the next few years, with a few hundred physical qubits, to hundreds of thousands or millions of physical qubits, and that's likely to take a while. Though I'm confident that quantum computing will have a transformative impact on society eventually, that impact might still be a few decades away. No one knows for sure how long it will take to get there. Keep in mind, though, that the technology is still at an early stage, with many competing approaches, so an unanticipated breakthrough could change the outlook suddenly.

7.3 Quantum Information

Let's now probe a little deeper into what makes a quantum computer so different from an ordinary digital computer. But first, we need to understand that what a quantum processor manipulates is something different from the information processed by today's computers.

7.3.1 Quantum vs. Classical

What is the difference between quantum information and classical information? To a physicist, information is something we can encode and store and process in some physical system using some physical process. Since fundamentally physics is quantum mechanical, information may be viewed as something we store and process in a quantum state.

For practical purposes, we often get away with ignoring the nuances of quantumness. A typical macroscopic system that we might encounter in everyday life is not well isolated from its environment, and interactions with the environment continually "measure" the system, a phenomenon called *decoherence*. A quantum system that is continually observed by its surroundings is well described by classical physics. But information carried by a quantum system (typically a microscopic one) that is sufficiently well isolated from its surroundings has intrinsic features that are not shared by classical information. Here are a few such features.

Randomness. Suppose a radioactive nucleus is about to emit an alpha particle. We cannot predict with certainty whether that nucleus is going to decay in the next second; we can only assign some probability, stating how likely the emission is to occur in the next second. This process is *intrinsically random* in the sense that I am unable to say whether the nucleus will decay or not even if I have the most complete description of the nucleus that the laws of physics will allow. I say *intrinsic* randomness to distinguish from the randomness we often encounter in everyday life, which arises from *ignorance*. If I flip a coin, I know it must have come up either heads or tails, but I don't know which until I look at the coin. So, before I look, I assign probabilities to the possible outcomes, reflecting my ignorance about the true outcome. The intrinsic quantum randomness of alpha decay is something different. It applies even when I have the most complete possible description.

Uncertainty. When I speak of uncertainty, I usually mean that the observables arising in quantum theory, the things that we can measure, don't necessarily commute. That means different possible observations can interfere with one another, not because I'm sloppy but for a fundamental reason. If two operators A and B don't commute, then if I measure A, that will unavoidably influence a measurement of B that I perform afterward. In the classical world, we can in principle measure the properties of a system as accurately as we please without in any way disturbing the system. That's not true in the quantum world.

Entanglement. Quantum entanglement is the principle that even if we know everything about the whole system, we don't necessarily know everything about its parts. The composite quantum system AB could be in what we call a pure state, meaning that we have all of the information that characterizes that state as completely as the laws of physics will allow. But if we observe just part A by itself, or part B by itself, its state is not pure – rather some of the information needed to completely characterize A is missing, and likewise for B. Classical systems are not like that. In the classical world, if I know everything that can be known about the full system, I know everything about each of its parts as well.

7.3.2 The Qubit

The indivisible unit of classical information is the bit. We can encode a bit in any physical system that can be in either one of two possible distinguishable states. It is often convenient to think about bits abstractly, denoting the two possible states 0 and 1 without worrying about how the bit is physically instantiated.

The corresponding indivisible unit of quantum information is the *qubit*. It too can be realized physically in many possible ways, but here we will mostly think about qubits abstractly, without any concern about the physical quantum system that encodes the qubit; it could be an atom, an electron, a photon, an electrical circuit, or something else. While a bit is a two-state system that can be either 0 or 1, a qubit can be described mathematically as a vector in a complex Hilbert space, with two mutually orthogonal basis states that we can label $|0\rangle$ and $|1\rangle$:

$$|\psi\rangle = a|0\rangle + b|1\rangle, \quad a,b \in \mathbb{C}, \quad |a|^2 + |b|^2 = 1, \quad |\psi\rangle \sim e^{i\alpha}|\psi\rangle. \quad (7.1)$$

It may seem like two complex numbers a and b are needed to describe the state of a qubit, but, in fact, just two real parameters suffice. That's because we don't really care about the overall normalization of the vector (by convention we usually set it equal to one). And we also don't care about its overall phase – we can multiply the vector by a complex number of modulus one without changing anything that's physically relevant. Once we have fixed the normalization, and removed the freedom to multiply by an overall phase, the general state of a qubit can be written as

$$|\psi(\theta,\varphi)\rangle = e^{-i\varphi/2}\cos(\theta/2)|0\rangle + e^{i\varphi/2}\sin(\theta/2)|1\rangle, \quad 0 \le \theta \le \pi, \quad 0 \le \phi < 2\pi. \quad (7.2)$$

A bit is just the special case of a qubit where we know for sure the vector is always either $|0\rangle$ or $|1\rangle$. Imagine a game where Alice prepares a state of a qubit and sends the qubit to Bob; then Bob measures the qubit and tries to guess what state Alice sent. If Alice promises to always send either $|0\rangle$ or $|1\rangle$ then Bob can win the game every time. Orthogonal basis states can be perfectly distinguished by Bob if he does the appropriate measurement.

The game becomes more challenging if the state sent by Alice is not necessarily chosen from among a set of mutually orthogonal states. In that case, there is no possible strategy for Bob that enables him to win the game for sure. Suppose for example that Alice promises to send to Bob one of the two states $|0\rangle$ or $|+\rangle$, where

$$|+\rangle = \frac{1}{\sqrt{2}}(|0\rangle + |1\rangle) \quad \Rightarrow \quad \langle 0|+\rangle = \frac{1}{\sqrt{2}}. \quad (7.3)$$

Now the possible states are not orthogonal, and no measurement by Bob can distinguish them perfectly. If the two states are equally likely, it turns out that if Bob plays the best possible strategy, he wins the game with probability $\cos^2(\pi/8) \approx .853$. Even though Bob knows for sure that Alice sent either $|0\rangle$ or $|+\rangle$, Bob has no way to find out for sure which of the two states he received. That's one important way a qubit is different from a bit.

More general measurements are possible, but we'll mostly be content to consider the simplest case of a measurement of a qubit, in which we force the qubit to become classical by declaring itself to be either $|0\rangle$ or $|1\rangle$. Then the rules of quantum theory say if the qubit state (7.2) is measured, the outcome will be $|0\rangle$ with probability $p_0 = \cos^2(\theta/2)$ and will be $|1\rangle$ with probability $p_1 = \sin^2(\theta/2)$. Even if Bob knows for sure the qubit is in the state $|\psi(\theta,\varphi)\rangle$, he can't predict the measurement outcome with certainty unless $\cos^2(\theta/2)$ is 0 or 1. That's another important way a qubit is different from a bit.

7.3.3 The Tensor Product

To understand quantum computing, we need to understand how we describe composite systems in quantum mechanics. Consider two individual quantum systems A and B. How should I mathematically describe the composite system AB?

Let's say that system A has dimension d_A. That means its Hilbert space $\mathcal{H}_A$ is spanned by a set of orthonormal basis states $\{|i\rangle_A, i = 1, 2, \ldots, d_A\}$. System B has dimension d_B; it's Hilbert space $\mathcal{H}_B$ is spanned by the orthonormal basis states $\{|a\rangle, a = 1, 2, \ldots, d_B\}$. Our rule for building the composite system should be compatible with the notion that if states are orthogonal then they can be perfectly distinguished. That means we should define our inner product on the composite system this way:

$$\left(\langle j| \otimes \langle b|\right)\left(|i\rangle \otimes |a\rangle\right) = \delta_{ij}\delta_{ab} \tag{7.4}$$

If I combine a basis state from system A with a basis state from system B, then if the system A basis states are mutually orthogonal, that means it's possible to perfectly distinguish the composite states. I can perfectly distinguish them just by looking at system A. Likewise, if I consider basis states for the composite system that are mutually orthogonal on system B, then those can be distinguished just by looking at system B. Reflecting this observation, the basis states in (7.4) are orthogonal if either $i \neq j$ or $a \neq b$. What we have constructed is called the *tensor product* $\mathcal{H}_A \otimes \mathcal{H}_B$ of the two Hilbert spaces.

For example, in the case of two qubits, we can choose our mutually orthogonal basis states to be

$$\{|00\rangle, |01\rangle, |10\rangle, |11\rangle\}, \tag{7.5}$$

just as you would label the possible states of two bits. (To save writing, we sometimes use the notation $|00\rangle$ for the tensor product of vectors $|0\rangle \otimes |0\rangle$, etc.) These four vectors can be perfectly distinguished, and so should be regarded as mutually orthogonal vectors in the Hilbert space of the composite system.

We can generalize this idea to many qubits. The Hilbert space for n qubits is

$$\mathbb{C}^{2^n} = \underbrace{\mathbb{C}^2 \otimes \mathbb{C}^2 \otimes \cdots \otimes \cdots \otimes \mathbb{C}^2 \otimes \mathbb{C}^2}_{n \text{ times}}. \tag{7.6}$$

This is the 2^n-dimensional complex Hilbert space spanned by 2^n mutually orthogonal vectors labeled by bit strings:

$$|x\rangle = |x_{n-1}\rangle \otimes |x_{n-2}\rangle \otimes \cdots \otimes |x_1\rangle \otimes |x_0\rangle, \quad x \in \{0,1\}^n, \tag{7.7}$$

such that $\langle x | y \rangle = \delta_{x,y}$. If $x_i \neq y_i$ for the ith qubit, then I can distinguish the basis states just by observing the ith qubit, and correspondingly the basis states are orthogonal, no matter what the value of the other bits.

The possible *pure* quantum states of this n-qubit system (those for which we have the most complete possible description) are vectors in that 2^n-dimensional space with complex coefficients,

$$|\psi\rangle = \sum_{x=0}^{2^n-1} a_x |x\rangle, \quad a_x \in \mathbb{C}. \tag{7.8}$$

For a typical pure quantum state, specified as completely as the laws of physics will allow, a full description of that state in classical language requires a vast amount of classical data. For 300 qubits, say, the state is a vector in a space of dimension $2^{300} \approx 10^{90}$. We could not possibly write down a complete classical description of that state, even if every atom in the visible universe were at our disposal. This seems to be a fundamental principle about the extravagance of the quantum world: there is no succinct way to describe classically a typical quantum state of n qubits.

In the classical world, suppose I have a memory that can store n bits. I might choose to store one of the possible strings of n bits, and not tell you which string I stored. If you made a list of all the possible strings I might have chosen, that list would be unmanageably long. But in that classical setting, for any one of those 2^n possibilities, I can easily describe to you what I stored by sending to you just n bits. The quantum situation is radically different. For just one typical completely specified quantum state, there is no possible succinct classical description.

There is an important caveat, though. While in the sense I just described, a quantum state seems to be vastly more complex than a classical bit string, that very extravagant description of the quantum state is not directly accessible. If I send to you an n-qubit quantum state, and don't tell you anything about the state, you can try to learn something about the state by measuring it. Measuring forces the state to yield one particular n-bit string x from among the 2^n possible strings. That is, the measurement yields only n bits of classical data. You could try to extract classical information from my quantum state using some more clever strategy, but a rigorous result (*Holevo's theorem* [48]) says there is no way to acquire more than n bits of classical information from a single copy of an n-qubit quantum state, no matter what you do. The art of designing useful quantum algorithms hinges on somehow taking advantage of the hidden extravagance of the quantum world, overcoming our inability to glimpse any more than a meager shadow of the underlying quantum reality whenever we read out our quantum device.

When we speak of building a quantum system out of many qubits, one could ask: given a Hilbert space of exponentially large dimension, how should I decompose it into qubits (or other lower-dimensional systems)? From a mathematical viewpoint, the choice of a decomposition is completely arbitrary. But from a physical viewpoint, there are preferred ways to do the decomposition. Typically, the way we choose to decompose a large quantum system as a tensor product of small systems is dictated by spatial locality. That is, we consider the different qubits to be at different locations. They might be in different cities, in Pasadena and New York, say. When we consider a quantum system shared between Pasadena and New York, it is very natural to decompose it in terms of qubits in Pasadena and qubits in New York. If I have n atoms, which are spatially separated from one another, my preference would be to describe the system in terms of qubits residing in the individual atoms.

The reason for that preference is that interactions between qubits are typically local in space. We expect that only systems in close proximity to one another interact significantly. The structure of these interactions provides a natural way to divide the big quantum systems into small subsystems. If qubits are located in n different cities A_1, A_2, ... A_n, it is quite easy to prepare a so-called product state

$$|\psi\rangle = |\psi_1\rangle_{A_1} \otimes |\psi_2\rangle_{A_2} \otimes \cdots \otimes |\psi_n\rangle_{A_n}. \tag{7.9}$$

I can just call my friends in the n cities and ask each one to prepare a single-qubit state. As we have seen, each of the single-qubit states is described by 2 real parameters, so any product state can be described succinctly by $2n$ real parameters.

States that are not product states are called *entangled* states. These are the states that cannot be created locally, no matter what I ask my n friends to do in each city. Entangled states can be created only by quantum communication (sending qubits from one city to another) or by interactions among the qubits. My n friends can talk on the (classical) telephone all they want, but if they don't have entanglement to begin with they cannot create entanglement – they are stuck with a product state. To create entanglement shared among n cities, we must either allow the qubits to come together to interact, or we could create an entangled state in one city, and then send each of those entangled qubits to different cities. These days, technological limitations make it hard to send a qubit a long way (say from Pasadena to New York) without the state of the qubit being damaged as it travels. But eventually we should be able to do that, allowing us to share a many-qubit entangled quantum state among the nodes of a quantum network distributed around the world.

As we've seen, a typical entangled state of n qubits is described by a number of classical parameters that are exponential in n, not linear in n as for product states. But it turns out that if I want to make an entangled state of n qubits, in principle I can create any such state, starting with a product state, if the qubits come together to interact two at a time. Mathematically, this is not a very deep fact. But physically it is quite significant. It means that by allowing particles to interact just two at a time, we can in principle control the quantum world.

There is a catch, though; we can't in general make any quantum state we desire efficiently, because the space of possible n-qubit states is so immense. The n-qubit Hilbert space has a dimension that is exponential in n, and a volume that is exponential in the dimension, hence *doubly exponential* in n. If we consider all the ways I can bring together qubits to interact pairwise T times in succession, roughly speaking the number of quantum states I can closely approximate is exponential in T. That means we need T to be exponential in n to create a state that closely approaches some specified typical n-qubit state. In this operational sense, the vast Hilbert space of n qubits is to some degree illusory. Future quantum engineers will know the joy of exploring Hilbert space far more extensively than we can today, but no matter how powerful their technology, most quantum states will remain far beyond their grasp. It's humbling to contemplate.

7.4 What Is a Quantum Computer?

Now we know enough about quantum information to formulate a mathematical model of a quantum computer.

7.4.1 The Quantum Circuit Model

Hilbert space. To begin, we specify the arena in which a quantum computation takes place, namely the Hilbert space $\mathcal{H} = \mathbb{C}^{2^n}$. And it is important that the Hilbert space is not just a very large vector space but is also equipped with a natural decomposition of the big Hilbert space as a tensor product of small subsystems, namely n qubits. As already emphasized, this natural decomposition is in practice dictated by spatial locality.

The reason it is so important to have a preferred decomposition is that we would like to be able to speak of the complexity of quantum computations or of quantum states. For a quantum state, I may define the complexity of the state as the number of steps need to prepare the state, but that number of steps depends on what the starting point is, and on what kind of operations we are allowed to perform in each step. Each elementary step should be something that's relatively easy to do in the lab, and we'll be guided by the idea that, while operations that act collectively on many qubits may be "hard", operations that act on a small number of qubits (like one or two) are "easy".

Initial state. The natural starting point, the initial state in a computation, is a product state. When we say "product state" we already have in mind a preferred decomposition into qubits. We may by convention say that each one of n qubits starts out in the state $|0\rangle$. We consider starting with a product state because it is easy to prepare; we don't want to hide complexity in the initial state preparation itself. The preparation is easy because I can set each qubit to $|0\rangle$ separately, using operations that act on only one qubit at a time. For example, I might measure each qubit in the basis $\{|0\rangle, |1\rangle\}$ (see below), and then flip the qubit if necessary to obtain the state $|0\rangle$.

Universal quantum gates. Now, we consider building up a quantum state, or performing a computation, starting with the state $|0\rangle^{\otimes n}$. For that purpose, we'll need a specified set of elementary operations, which we can compose together in a circuit. Here, too, we need to distinguish between operations that are "easy" and operations that are "hard", and again we make use of the preferred decomposition into qubits to make that distinction. The operations that act on a small number of qubits (a constant number, independent of n) are regarded as "easy", while generic operations that act on many qubits (a number that increases with n) are considered to be "hard". We don't want to hide complexity in our choice of elementary operations. That way, it makes sense to quantify the complexity of a computation according to the number of elementary operations needed to execute the computation.

Specifically, we suppose that there is a finite alphabet

$$\mathcal{G} = \{U_1, U_2, \ldots, U_{n_G}\} \tag{7.10}$$

Of unitary transformations, each acting on a constant number of qubits, which are hardwired in our quantum processor. A complex $m \times m$ matrix U is unitary if $U^\dagger U = I$, where $U^\dagger$ denotes the adjoint of U. We consider unitary transformations because these are the ones that are allowed under the rules of how a finite-dimensional quantum system can evolve. (More general transformations are allowed if we pad our set of qubits with extra qubits, perform a joint unitary transformation on our qubits plus the extra ones, and then discard the extra ones. But in that case, we could just consider those extra qubits to be included in the quantum computer's Hilbert space, so there is no loss of generality if we stick with only unitary operations.) We call these hardwired elementary unitaries our *quantum gates*, as these are the quantum computer's counterpart of the elementary Boolean gates in the classical circuit model of computation. (Just to save syllables, we'll sometimes say "unitaries" as a shorthand for "unitary transformations" when the context makes the meaning clear.)

As previously mentioned, quantum gates acting on just two qubits are already universal – with a circuit of two-qubit gates we can approximate any n-qubit unitary transformation as accurately as we please. Since two-qubit gates are usually easier to do in the physics lab than k-qubit gates for $k > 2$, we normally include in our alphabet only gates acting on one or two qubits at a time.

Now, unitary transformations, unlike Boolean classical logic gates, form a continuum of possible operations. An experimentalist who executes the gates has some classical parameters she can adjust to perform the gates; these parameters are real numbers, and as they vary continuously, so does the executed unitary transformation. So it may seem perverse to insist that our alphabet of quantum gates is a finite set, but there is actually a very good reason to do so. Since quantum hardware is not very reliable, we need to use quantum error-correcting codes to make our quantum gates more robust. And once we choose our code, only a finite set of one-qubit and two-qubit gates that are compatible with the structure of our code can be done efficiently and accurately. Only the robust gates from this finite set are good candidates to include in our gate set, and we should be sure these are chosen so as to maintain universality.

If we want to accurately approximate any unitary transformation acting on a single qubit, it suffices to build circuits (sequences of gates) from just two noncommuting elementary gates. One popular choice that works (popular because these gates arise naturally when we use quantum error-correcting codes with nice properties) is

$$H = \frac{1}{\sqrt{2}} \begin{pmatrix} 1 & 1 \\ 1 & -1 \end{pmatrix}, \quad T = \begin{pmatrix} 1 & 0 \\ 0 & e^{i\pi/4} \end{pmatrix}; \tag{7.11}$$

H is often called the "Hadamard gate", and T (for lack of a better name) is simply called the "T gate". Since these two gates don't commute, sequences of gates chosen from this pair can reach a number of single-qubit unitaries that grows exponentially with the length of the sequence, and these densely fill the unitary group as the length increases. What is less obvious, but true and important, is that there is an efficient classical algorithm for finding a gate sequence that approximates a given desired one-qubit unitary within a specified error [16, 49].

With single-qubit gates and an initial product state, we can reach only product states. But augmenting these single-qubit gates with just one entangling two-qubit gate is enough to achieve universality. A standard choice for this two-qubit gate (again, because it is convenient to perform on quantum information protected by a quantum error-correcting code) is the controlled-NOT (CNOT) gate with action

$$\text{CNOT} = |0\rangle\langle 0| \otimes I + |1\rangle\langle 1| \otimes X, \quad X = \begin{pmatrix} 0 & 1 \\ 1 & 0 \end{pmatrix}. \tag{7.12}$$

That is, if the first (control) qubit is $|0\rangle$, the CNOT gate acts trivially on the second (target) qubit, but if the control qubit is $|1\rangle$, the gate applies a bit flip (X) operation to the target qubit. The CNOT is a classical operation, in the sense that it maps our standard basis states to other standard basis states, but it can create entanglement when it acts on a control qubit that is in a superposition of basis states, for example:

$$\text{CNOT} : \frac{1}{\sqrt{2}}\left(|0\rangle + |1\rangle\right) \otimes |0\rangle \rightarrow \frac{1}{\sqrt{2}}\left(|00\rangle + |11\rangle\right). \tag{7.13}$$

Once we have fixed our universal gate set, we have a notion of how hard it is to reach a particular n-qubit unitary transformation, just as the circuit model of classical computation provides a notion of how hard it is to compute a particular Boolean function. We ask: what is the size of the minimal circuit that generates the desired unitary? A difference from the classical case is that, since the unitary transformations form a continuum, we in general have to accept some small error – it may be that no circuit reaches the desired unitary exactly, in which case we settle for constructing a unitary that is only distance ε away from the desired unitary according to some appropriate notion of distance.

So far, we have considered just one possible choice of the gates in our universal set $\mathcal{G} = \{H, T, \text{CNOT}\}$. We had good motivation for this choice, but for reasons of your own, you might prefer a different universal gate set. Maybe you are using a different kind of hardware than I am, and some operations are easy for you but not as easy for me, and vice versa. Even though we have different gate sets, we can agree on which unitaries are easy to construct and which ones are hard. That's because I can efficiently simulate your universal gates using mine, and you can efficiently simulate my universal gates using yours, with only modest overhead. Since my gates are universal, and each of your universal gates acts on a constant number of qubits, I can approximate any one of your universal gates, with an error ε, using polylog $(1/\varepsilon)$ of my gates. (That is, the number of gates I use scales like a polynomial in $\log(1/\varepsilon)$ for asymptotically small ε.) This fundamental result is called the *Solovay–Kitaev theorem* [16, 49].

Let's say you can reach some n-qubit unitary U, with error δ, using T of your gates. Suppose I simulate each of your gates with error δ/T, which I can do with $O(\text{polylog}(T/\delta))$ of my gates. Since the error accumulates at worst linearly with the number of gates, I can simulate your circuit using $O(T(\text{polylog}(T/\delta)))$ of my gates, making an error which is at worst 2δ. If your circuit is "efficient" (which we usually take to mean that T is bounded above by some polynomial in the number of qubits n), then so is mine.

Classical control. We do not want to hide complexity in the *design* of our quantum circuit. The same issue arises in the circuit model of classical computation as well. In the classical case, we have to augment the circuit model with an additional computer, such as a Turing machine, which designs the circuit when we choose the problem we want to solve and specify the size of the input to the problem. In that case, we demand that the runtime of this additional classical computer is also polynomial in the size of the input. Since a polynomial-size quantum circuit has a succinct classical description (even though the way the circuit acts on an input quantum state does not), we can use the same idea to augment the quantum circuit model. As in the classical circuit model, we insist that the family of quantum circuits that solve a problem with variable input size has the property of being *uniform*. Informally, this just means that once we have found the circuit of size poly(n) that works for the problem instance of specified size, the problem of finding the appropriate circuit for a larger input size is not much harder.

Readout. So far, we have been talking mostly about the complexity of a unitary transformation acting on a specified initial state. But when we use a quantum computer to solve a problem, we want the output of the computer to be classical information, which we can write down and share with our friends. To obtain a classical result, we need to do a measurement at the end of our quantum computation, and we don't want to hide complexity in our choice of readout procedure. Therefore, let's assume that we do the final readout by measuring the qubits (or some subset of the qubits) in the standard basis, obtaining the outcome 0 or 1 for each measured qubit. That completes our description of the quantum circuit model of computation. The initial state preparation and the final measurement are easy to do; what determines whether a quantum computation is easy or hard is the number of gate operations we need to do between the initial preparation and the final readout.

I should emphasize that quantum computing is a randomized model of computation because the measurement of a quantum state is not deterministic. If, for example, we are trying to solve a decision problem, for which the answer is either YES or NO, we might not get the correct answer every time we run the quantum computation. That's not really an issue, as long as we get the right answer with a sufficiently high success probability. The standard convention is to demand that our final readout yields the correct answer with probability at least two-thirds. Then, if we run the computation a modest number of times, and take a majority vote (concluding that the answer is really YES if most of the computations gave the answer YES, and that the answer is NO if most of the computations gave the answer NO), then we will solve the problem correctly with probability close to one.

7.4.2 Computability and Efficiency

Now that we have formulated our model of quantum computation, we want to understand the power of the model. What computations can it run? What problems can it solve? In particular, what quantum algorithms can we design that achieve speedups relative to the best classical algorithms we know that solve the same problems?

To summarize, the features of the quantum model are these.

1. Scalable number of qubits.
2. Preparation of standard initial state.
3. Universal set of quantum gates.
4. Classical computer to design uniform quantum circuit families.
5. Readout in the standard basis.

We should emphasize that every feature of this model can be simulated by an ordinary classical computer if equipped with a random number generator to capture the nondeterministic nature of the final quantum measurement. All the classical computer needs to do is keep track of a vector in a Hilbert space as we act on the vector with a sequence of matrices. For the final readout, we project the vector onto our standard set of axes and assign probabilities to the different measurement outcomes accordingly. Since a (randomized) classical computer can do whatever a quantum computer does, there is no difference in *computability* – whatever is computable by a quantum computer is also computable by a classical computer.

The important distinction between the quantum and classical models is all about *efficiency*. In general, for the classical computer to simulate the quantum computer, it has to deal with vectors in a space whose dimension is exponential in the number of qubits. For the hardest problem instances, we just don't know any way to do that simulation on the classical computer without using resources that scale exponentially with the number of qubits.

From the viewpoint of physics (or the foundations of computer science), we should ask whether our abstract model of quantum computation is a good one for capturing the information processing that can really be done efficiently in the natural physical world. We don't know for sure whether that is the case or not. It's a hypothesis, what we might call the *extended quantum Church–Turing thesis*. Whether this thesis is correct is not entirely obvious. To describe elementary particles, physicists use what they call local quantum field theory. Formally, quantum field theory involves a number of degrees of freedom per unit spatial volume which is infinite. We can't expect to simulate exactly an infinite number of degrees of freedom using any finite machine. It is an article of faith among physicists that we never really need all those degrees of freedom, that a good approximation can be attained by retaining only a finite number of degrees of freedom per unit volume. The number we need is controlled by input parameters of the simulation, such as the total energy of the process we wish to study. With a limited amount of energy, we cannot probe physics at arbitrarily short distance scales, so we only need a limited number of degrees of freedom to describe things accurately. In that case, we can argue persuasively that a physical process described by quantum field theory can be accurately and efficiently simulated using the quantum circuit model [50, 51].

Physicists believe that most phenomena in the physical universe can be accurately described by local quantum field theory. Possible exceptions arise, though, in situations where both quantum physics and gravitational physics play an important role. For example, we have only an incomplete understanding at present of how black holes process quantum information. Whether the quantum circuit model suffices for efficiently describing how black holes behave is something we don't yet know for sure. If the quantum circuit model really does capture everything that happens in physics, then we'll be able to use quantum computers in the future to explore fundamental physics at a very deep level. But if not, that's even more exciting. It means that Nature will ultimately allow even more powerful information processors than the quantum computers we currently foresee.

7.4.3 Quantum Hardware

The details of quantum hardware are beyond the scope of this chapter, but let's pause for a few words about how physical systems that we can really build and operate today align with the abstract model of quantum computing we have formulated. Though actual qubits are never perfect, we desire qubits that closely resemble the qubits described by the ideal model. Let's revisit the criteria our quantum hardware should satisfy, this time with an eye on whether actual devices are up to the task [52].

1. Scalable system with controllable qubits.
2. Sufficiently accurate qubit preparation in the state $|0\rangle$.
3. Qubit coherence time sufficiently long compared to gate execution times.
4. Sufficiently accurate universal set of quantum gates.
5. Sufficiently accurate qubit measurement in the standard basis.

Various quantum systems have the potential to meet these desiderata; I'll just mention two. I apologize for the paucity of references in this discussion, but you can find more details in two recent reviews [53, 54].

When Shor's algorithm precipitated a surge in interest in quantum computing in the mid-1990s, it was a happy coincidence that experimental tools relevant to quantum computing were already being developed for other reasons. For one, advances in the technology for manipulating individual atomic ions trapped by electromagnetic fields had been motivated by the desire for more accurate atomic clocks.

In an ion trap, a single electrically charged atom (i.e., an ion), which can be in either its ground (lowest energy) state or some long-lived excited state, may serve as a qubit, and tens of such qubits may be loaded into a trap while maintaining precise control of each qubit. If we choose the right ion and the right excited state, then the other criteria can be met; in particular, idle qubits have very low error rates, that is, the coherence time is very long, longer than a second for some ions. Furthermore, information processing, state preparation, and measurement can all be achieved by addressing the ions with pulses of light from a very stable laser.

For readout, one illuminates an atom with light of an appropriate frequency so that atoms in the ground state strongly scatter the light, while atoms in the excited state are transparent. Just by observing whether the illuminated ion glows or not, we can determine with high confidence whether the state of the qubit is $|0\rangle$ or $|1\rangle$. Measurement error rates below 10^{-4} can be achieved by interrogating an ion for a few hundred microseconds. Initial state preparation can also be achieved efficiently and accurately via laser manipulation of the ion's internal atomic state.

Single-qubit quantum gates in ion traps are also easy and quite accurate. A laser pulse induces a coherent coupling between the two basis states of the qubit for a prescribed time to apply a desired unitary transformation to the qubit. Single-qubit gates with error rates below 10^{-4} can be executed in a few microseconds.

For ion traps, as for most other quantum platforms, the most challenging task is performing entangling two-qubit gates, which require that two atoms interact sufficiently strongly. The electrostatic repulsion of the ions provides the needed interaction. Because of the repulsion, the ions have shared normal modes of vibration in the trap. A laser pulse couples a normal mode shared by two ions to the internal state of the pair of ions, guiding that mode on an excursion during which the two-qubit state acquires a phase that depends on the internal states of the two atoms; the result is an entangling two-qubit gate [55, 56]. The speed of the gate depends on the optical power of the laser and the vibrational frequencies of the ions in the trap; it typically takes at least tens of microseconds. Gates are usually executed sequentially rather than in parallel to avoid unwanted couplings between qubits that might compromise gate fidelity. In the best current multi-qubit devices, the error rate per entangling two-qubit gate is typically around 1%, though error rates below 10^{-3} have been achieved under highly favorable conditions.

As an alternative to actual atoms, engineered "artificial atoms" may serve as qubits. In particular, reasonably high-quality qubits can be realized using superconducting electrical circuits, which conduct electricity with negligible resistance at sufficiently low temperatures. These circuits have an energy-level structure reminiscent of an atom's if the circuit includes nonlinear elements (Josephson junctions), and a qubit can be encoded using the circuit's lowest energy state $|0\rangle$ and its first excited state $|1\rangle$. The energy splitting between these levels is typically around 5 GHz, and the device is kept at a sufficiently low temperature (10–20 mK $\approx$ 200–400 MHz) that the thermal state of the qubit is very close to its quantum-mechanical ground state. Coherence times of tens to hundreds of microseconds can be routinely achieved.

The scheme for executing single-qubit gates is conceptually comparable to the scheme used in ion traps, except that the qubit's evolution is driven by a microwave pulse rather than a laser. If the pulses are carefully shaped to avoid unwanted transitions to higher energy levels, single-qubit gate error rates well below 1% can be achieved in a few tens of nanoseconds.

There are several different schemes for performing entangling two-qubit gates. For example, one can tune the frequency of a qubit by applying a magnetic flux, and the desired gate can be obtained by bringing two quantum states of a pair of qubits to nearly coincident frequencies for a specified time. In multi-qubit devices, two-qubit gate error rates comparable to 1% can be achieved in tens of nanoseconds. (As is the case for ions, two-qubit gate error rates below 10^{-3} have now been achieved under highly favorable conditions.)

To read out a qubit, one couples it to a microwave resonator, and the resonator's frequency shifts by an amount that depends on whether the qubit's state is $|0\rangle$ or $|1\rangle$. The frequency shift can then be detected by observing the resonator's response to a microwave drive. A measurement error rate of about 1% can be achieved in hundreds of nanoseconds.

Ion traps and superconducting circuits are currently the two leading qubit technologies, and each has characteristic advantages and disadvantages. For example, atoms are all alike, and have exceptionally long coherence times. In addition, a two-qubit gate can be performed acting on any pair of ions in a trap, with a fidelity that is not very sensitive to the spatial separation between the ions. In contrast, the coherence times of superconducting qubits are limited by imperfections in how they are fabricated; furthermore, their properties vary from qubit to qubit and can evolve over time as well. Therefore, the qubits need to be carefully and frequently calibrated. Also, though schemes for long-range coupling have been proposed, in today's state-of-the-art quantum processors, high-quality two-qubit gates are performed only between neighboring qubits laid out in a one-dimensional or two-dimensional array.

On the other hand, quantum gates are much faster in superconducting devices, and many gates can be executed in parallel without the gate fidelity being seriously diminished. That may be a big advantage in the future when we judge the performance of a quantum computation according to the total time needed to find a solution.

Scaling from the tens of qubits we have now to the millions of physical qubits we expect to need in the future will be a daunting challenge for both ion traps and superconducting circuits, just as for all the other currently known quantum technologies. With more than about 100 ions in a trap, it becomes too difficult to control all of the coupled vibrational modes. To scale up further will probably require some sort of modular design, with many relatively small traps networked together into a large system. To share quantum information among the modules, several ideas have been suggested. We might engineer optical interconnects, which allow a photon to travel coherently from one trap to another. Or we might shuttle ions between traps while maintaining the coherence of the ions' internal states. Both these approaches are under development, but still have far to go before a large-scale modular quantum computer becomes practical.

For superconducting circuits as well, control of the system becomes increasingly challenging as the number of qubits increases, in part because of the proliferation of microwave control lines that exacerbates problems like crosstalk. Some of these issues can be mitigated through advances in engineering, but there are big opportunities in basic research, too. Superconducting circuits can support a rich variety of device designs, and there are many possibilities yet to be explored.

Many other promising approaches to quantum hardware are being pursued, but we won't delve into the subject any further here. Above all, I want to emphasize that we are still in the early stages of developing quantum computing systems, and no doubt big surprises lie ahead. The brief synopsis above is likely to be badly out of date soon!

7.5 Simulating Quantum Dynamics

Next, we'll look in more depth at how quantum computers can be used to solve problems in quantum physics, as foreseen by Feynman. An especially important application is solving the time-dependent Schrödinger equation, i.e. finding out how an *n*-quantum system evolves in time, as governed by some many-body Hamiltonian. In special cases we know how to solve this problem efficiently with a classical computer; sometimes we can even find an analytic solution. But, in general, the best classical algorithms have a runtime that scales exponentially with *n*. Simulating time evolution with a quantum computer, in contrast, scales polynomially with *n* if the Hamiltonian *H* is *local* [57]. Here we'll explain why this exponential quantum speedup is possible, without attempting to exhibit the best state-of-the-art quantum algorithms.

For a system of *n* qubits, we say that *H* is *k-local* if

$$H = \sum_a H_a, \tag{7.14}$$

where each term H_a acts nontrivially on at most *k* qubits – i.e., $H_a = \tilde{H}_a \otimes I^{n-k}$, and $\tilde{H}_a$ acts on some set of at most *k* qubits. (Of course, we may use a similar definition for a system of *d*-dimensional subsystems for constant $d > 2$, rather than qubits.) We say that *H* is local if it is *k*-local for some constant *k*.

There is a stronger notion of locality we sometimes use, which can be called *geometrical locality* or *spatial locality*. A *k*-local Hamiltonian is geometrically local in *D* dimensions if the qubits can be arranged in (flat) *D*-dimensional space with a bounded number of qubits per unit volume, and the *k* qubits upon which H_a acts nontrivially are all contained in a ball of constant radius. In this sense, there are no *long-range* interactions among the qubits. *H* is geometrically local if it is geometrically *k*-local in *D* dimensions for some constant *D* and *k*.

If we write $H = \sum_a H_a$ where there is a unique H_a for each set of *k* qubits, then the expansion of a *k*-local *H* contains at most $\binom{n}{k} = O(n^k)$ terms, and the expansion of a geometrically local *H* contains $O(n)$ terms (each of the *n* qubits is contained in a constant number of interacting sets). Let us also assume that each H_a is bounded:

$$\|H_a\|_\infty \leq h \text{ for all } a, \text{ where } h \text{ is a constant.} \tag{7.15}$$

Physicists are interested in geometrically local Hamiltonians because they seem to provide an accurate description of Nature. Therefore, it is noteworthy that quantum circuits can simulate quantum evolution governed by a local Hamiltonian efficiently: evolution of n qubits for time t can be simulated to constant accuracy using a circuit whose size is polynomial in n and t.

We can formulate the problem this way: suppose we are given an initial quantum state $|\psi(0)\rangle$, or a classical description of a quantum circuit that prepares the state. Our goal is to construct

$$|\psi(t)\rangle = U(t)|\psi(0)\rangle \tag{7.16}$$

where $U(t)$ satisfies $\dfrac{d}{dt}U(t) = -iH(t)U(t)$ and the boundary condition $U(0) = I$. (Thus, $U(t) = e^{-iHt}$ in the case where H is time independent.) We will settle for computing $|\psi(t)\rangle$ to accuracy δ, i.e., constructing $\tilde{\psi}(t)\rangle$ where

$$\left\| \tilde{\psi}(t)\rangle - |\psi(t)\rangle \right\| < \delta. \tag{7.17}$$

Depending on the situation, we might be satisfied if δ is a sufficiently small constant, or we might impose the stricter requirement that the error is smaller than some specified power of the size n of the system. To relate this simulation task to a task that can be described classically, suppose the goal is to sample from the probability distribution

$$\langle \psi(t)|\Pi_a|\psi(t)\rangle \tag{7.18}$$

where Π_a projects onto an eigenstate with eigenvalue a of an observable A that can be measured efficiently by a quantum computer. Classically, this task is believed to be hard at least in some cases, because the unitary matrix $U(t)$ is exponentially large ($2^n \times 2^n$). But quantumly we can do the simulation efficiently if H is a local Hamiltonian.

To simulate continuous time evolution on a classical or quantum computer, we choose a small step size Δ, and approximate evolution for time t by a sequence of t/Δ steps. (If H is actually time dependent, assume Δ is small enough that the change of H during a time interval of width Δ can be neglected.) We wish to attain accuracy

$$\left\| \tilde{U}(t) - U(t) \right\|_\infty < \delta, \tag{7.19}$$

where $\tilde{U}$ is the simulated unitary and U is the ideal unitary. Hence the error per time step should be less than $\delta\Delta/t$.

Suppose $H = \sum_a H_a$ is a sum of M k-local terms, and let's consider the geometrically local case, where $M = O(n)$. We will show below that a single time step can be simulated by a product of M local "gates" (unitary transformations that act on a constant number of qubits) where each such "gate" has an error $O(\Delta^2 h^2)$. Therefore, the simulation of evolution for time t uses all together Mt/Δ gates where we require

$$\frac{Mt}{\Delta}\Delta^2 h^2 \approx \delta \implies \Delta = O\left(\frac{\delta}{h^2 Mt}\right). \tag{7.20}$$

Therefore, the total number of gates is

$$L = O\left(\frac{h^2(Mt)^2}{\delta}\right). \tag{7.21}$$

Furthermore, each "gate" can be simulated to accuracy $O(\Delta^2 h^2)$ with a universal gate set using $\text{polylog}\left(\frac{1}{\Delta^2 h^2}\right) = \text{polylog}\left(\frac{h^2(Mt)^2}{\delta^2}\right)$ gates, according to the Solovay–Kitaev theorem. We conclude that the simulation can be done with a quantum circuit of size

$$L = O\left(\frac{h^2(Mt)^2}{\delta}\text{polylog}\left(\frac{h^2(Mt)^2}{\delta^2}\right)\right). \tag{7.22}$$

In the case where H is geometrically local, $M = O(n) = O(V)$, where V is the spatial volume of the system. Since h is a constant, we find that the cost of simulating time evolution with fixed accuracy scales like

$$L = O(\Omega^2 \text{ polylog } \Omega), \tag{7.23}$$

where $\Omega = Vt$ is the simulated volume of spacetime.

Now we need to explain how to simulate a single time step. We'll use the idea that $\exp\left(\sum_a A_a\right)$ can be approximated by $\prod_a \exp(A_a)$ if $||A|| \ll 1$. To check the accuracy, we expand the exponentials:

$$\exp\left(\sum_a A_a\right) - \prod_a \exp(A_a) \tag{7.24}$$

$$= \left(1 + \sum_a A_a + \frac{1}{2}\sum_{a,b} A_a A_b + \dots\right) - \prod_a\left(1 + A_a + \frac{1}{2}A_a^2 + \dots\right)$$

$$= \left(1 + \sum_a A_a + \frac{1}{2}\sum_{a,b} A_a A_b + \dots\right) - \left(1 + \sum_a A_a + \sum_a \frac{1}{2}A_a^2 + \sum_{a<b} A_a A_b + \dots\right)$$

$$= \frac{1}{2}\left(\sum_{a<b} A_a A_b + \sum_{a<b} A_b A_a\right) - \sum_{a<b} A_a A_b + \ldots$$

$$= -\frac{1}{2}\sum_{a<b}[A_a, A_b] + \ldots$$

(where $+\ldots$ denotes terms higher order in A_a). Writing $H = \sum_a H_a$, then, we find that

$$e^{-iH\Delta} - \prod_a e^{-iH_a\Delta} = \frac{1}{2}\Delta^2 \sum_{a<b}[H_a, H_b] + \text{higher order.} \qquad (7.25)$$

Now, how many non-vanishing commutators $\{[H_a, H_b]\}$ can occur in this sum? Let's suppose the Hamiltonian is geometrically local, in which case there are $O(n)$ terms in H, and each term fails to commute with a constant number of terms. So, there are $O(n) = O(M)$ non-vanishing commutators. We conclude that (in the geometrically local case)

$$\left\| e^{-iH\Delta} - \prod_a e^{-iH_a\Delta} \right\| = O(M\Delta^2 h^2). \qquad (7.26)$$

Since $\prod_a e^{-iH_a\Delta}$ is a product of M "gates", we have verified that the accuracy per gate is $O(\Delta^2 h^2)$. (Note that terms arising from the higher-order terms in the expansion of the exponential are of order $M\Delta^3 h^3$, and therefore systematically suppressed by another factor of $\Delta h = O(\delta / hMt) = O((\delta / L)^{1/2})$.)

We have shown that, for a geometrically local H that is a sum of bounded terms, evolution in a spacetime volume Ω can be achieved with a quantum circuit of size

$$L = O(\Omega^2 \text{ polylog } \Omega), \qquad (7.27)$$

The simulation can be achieved with quantum resources that scale like the *square* of the simulated volume (up to a polylog factor). With more sophisticated methods, the scaling with Ω and also the scaling with the error δ can be improved. We will not discuss these improvements here, even though they may be of great practical importance in the future if they can substantially reduce the runtime on a quantum computer for problems of interest to physicists and chemists. We have settled for making the crucial point – that the quantum runtime scales polynomially with the size of the physical system, while the best general-purpose classical algorithms scale exponentially.

7.6 Energy Eigenvalues and Eigenstates

Aside from simulating time evolution, physicists and chemists are also interested in "diagonalizing" many-body Hamiltonians, that is. finding energy eigenvalues and properties of energy eigenstates. Here again there are special cases where we can find analytic solutions or obtain good approximate solutions efficiently using classical computers. But there are many cases of physical interest where the problem seems to be hard classically, simply because the Hamiltonian is an extremely large $2^n \times 2^n$ matrix. With a quantum computer, we can "solve" the problem efficiently, subject to some important caveats which we'll come to.

The algorithm for estimating eigenvalues and preparing eigenstates of a local Hamiltonian H using a quantum computer makes use of the algorithm described in Section 7.5 for simulating time evolution. Once we have constructed an efficient quantum circuit for the time-evolution operator $U(t) = \exp(-iHt)$, we apply a general procedure for estimating eigenvalues of unitary matrices. This general procedure, called *phase estimation* [58], leverages a very efficient procedure for evaluating the Fourier transform on a quantum computer. Phase estimation is an essential primitive used in a variety of quantum algorithms, including Shor's factoring algorithm.

7.6.1 Quantum Fourier Transform

Before explaining phase estimation, let's see how to Fourier transform a function using a quantum computer. We suppose that the function is encoded in the *amplitudes* of an m-qubit quantum state:

$$\sum_{x=0}^{N-1} f(x)|x\rangle; \tag{7.28}$$

here $x = x_{m-1}x_{m-2}\ldots x_1 x_0$ is shorthand for the integer x expanded in binary notation, and $N = 2^m$. The discrete quantum Fourier transform (QFT) acts on this state according to

$$\text{QFT}: \sum_{x=0}^{N-1} f(x)|x\rangle \to \sum_{k=0}^{N-1} \left(\frac{1}{\sqrt{N}} \sum_{x=0}^{N-1} e^{2\pi ikx/N} f(x) \right)|k\rangle; \tag{7.29}$$

it is an $N \times N$ unitary matrix with matrix elements $\{(e^{2\pi i/N})^{kx} / \sqrt{N}\}$. Here N might be exponentially large, but thanks to the simple structure of the QFT, it can be implemented by a quite efficient quantum circuit containing only $O(m^2)$ gates.

If we express x and k as binary expansions

$$x = x_{m-1} \cdot 2^{m-1} + x_{m-2} \cdot 2^{m-2} + \ldots + x_1 \cdot 2 + x_0,$$

$$k = k_{m-1} \cdot 2^{m-1} + k_{m-2} \cdot 2^{m-2} + \ldots + k_1 \cdot 2 + k_0, \tag{7.30}$$

then in the product of x and k, we may discard any terms containing m or more powers of 2, as these make no contribution to $\exp(2\pi i k x / 2^m)$. Hence

$$\frac{kx}{2^m} \equiv k_{m-1}(.x_0) + k_{m-2}(.x_1 x_0) + k_{m-3}(.x_2 x_1 x_0) + \ldots$$

$$+ k_1(.x_{m-2} x_{m-3} \ldots x_0) + k_0(.x_{n-1} x_{m-2} \ldots x_0), \tag{7.31}$$

where the factors in parentheses are binary expansions; for example,

$$.x_2 x_1 x_0 = \frac{x_2}{2} + \frac{x_1}{2^2} + \frac{x_0}{2^3}. \tag{7.32}$$

Using (7.31), we can see that the quantum Fourier transform maps each computational basis state to a product state of m qubits:

$$\text{QFT} : |x\rangle \rightarrow \frac{1}{\sqrt{N}} \sum_{k=0}^{N-1} e^{2\pi i k x / N} |k\rangle$$

$$= \frac{1}{\sqrt{2^m}} \left(\underbrace{|0\rangle + e^{2\pi i (.x_0)} |1\rangle}_{k_{m-1}} \right) \otimes \left(\underbrace{|0\rangle + e^{2\pi i (.x_1 x_0)} |1\rangle}_{k_{m-2}} \right) \otimes$$

$$\ldots \otimes \left(\underbrace{|0\rangle + e^{2\pi i (.x_{m-1} x_{m-2} \cdots x_0)} |1\rangle}_{k_0} \right); \tag{7.33}$$

as a result, it can be efficiently implemented. To be concrete, consider the case $m = 3$. We can readily see that the circuit

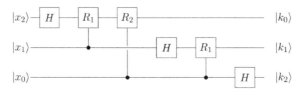

does the job (but note that the order of the bits has been reversed in the output). Each Hadamard gate H acts as

$$H:|x_j\rangle \rightarrow \frac{1}{\sqrt{2}}\Big(|0\rangle + (-1)^{x_j}|1\rangle\Big) = \frac{1}{\sqrt{2}}\Big(|0\rangle + e^{2\pi i(.x_j)}|1\rangle\Big). \qquad (7.34)$$

The other contributions to the relative phase of $|0\rangle$ and $|1\rangle$ in the jth qubit are provided by the two-qubit controlled rotations, where

$$R_d = \begin{pmatrix} 1 & 0 \\ 0 & e^{i\pi/2^d} \end{pmatrix}, \qquad (7.35)$$

and $d = (j - \ell)$ is the "distance" between the qubits. (The controlled R_d shown in the circuit diagram applies the nontrivial phase $e^{i\pi/2^d}$ only if the two-qubit state is $|11\rangle$.)

In the case $m = 3$, the QFT is constructed from three H gates and three controlled-R_d gates. For general m, the obvious generalization of this circuit requires m H gates and $\binom{m}{2} = \frac{1}{2}m(m-1)$ controlled R_ds. A two-qubit gate is applied to each pair of qubits, again with controlled relative phase $\pi / 2^d$, where d is the "distance" between the qubits. Thus, the circuit family that implements the QFT has a size of order $(\log N)^2$. On a quantum computer, the Fourier transform is remarkably easy to implement, even when N is exponentially large. In contrast, the classical "fast Fourier transform" algorithm has a runtime $O(N \log N)$.

7.6.2 Phase Estimation

Phase estimation is a quantum algorithm that estimates eigenvalues of a unitary operator U, using the QFT as a subroutine. The quantum circuit makes use of an auxiliary register that records an integer-valued "time" parameter t; this time register is initialized in a uniform superposition of all values of t running from $t = 0$ to $t = 2^m - 1$. Then the unitary U is executed t times acting on a data register, controlled by the time register. If the initial state of the data register is $|\psi\rangle$, this procedure prepares the state

$$\frac{1}{\sqrt{2^m}}\left(\sum_{t=0}^{2^m-1}|t\rangle \otimes U^t |\psi\rangle\right). \qquad (7.36)$$

To be more concrete, the circuit that prepares this state is shown here for the case $m = 3$.

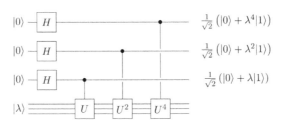

The three Hadamard gates acting on $|0\rangle^{\otimes 3}$ prepare the uniform superposition of 2^3 computational basis states $\{|t_2t_1t_0\rangle\}$. Then U is applied conditioned on the least significant bit $|t_0\rangle$, U^2 conditioned on the next bit $|t_1\rangle$, and so on. If the initial state of the data register happens to be an eigenstate $|\lambda\rangle$ of U with eigenvalue λ, this circuit yields the state

$$\frac{1}{\sqrt{2^3}}\left(\underbrace{|0\rangle + \lambda^4|1\rangle}_{t_2}\right) \otimes \left(\underbrace{|0\rangle + \lambda^2|1\rangle}_{t_1}\right) \otimes \left(\underbrace{|0\rangle + \lambda|1\rangle}_{t_0}\right) = \frac{1}{\sqrt{2^3}}\sum_{t=0}^{7}\lambda^t|t\rangle \otimes |\lambda\rangle. \quad (7.37)$$

To recover the value of λ, we can now apply the QFT to the time register and measure in the computational basis. If $\lambda = e^{-2\pi i k/2^m}$, where $k = k_{m-1}k_{m-2}\ldots k_1k_0$ is an integer less than 2^m, then the measurement outcome will be k with probability one. More generally, if $\lambda = e^{-2\pi i\phi}$ where the binary expansion of ϕ does not necessarily terminate after m bits, the measurement finds ϕ to about m bits of precision with high success probability. In other words, we can estimate ϕ with an accuracy $\delta \approx 2^{-m}$ by conditionally applying U up to $2^m \approx 1/\delta$ times.

If the initial state of the data register is not an eigenstate of U, it can be expanded in terms of U eigenstates. If we apply the phase estimation circuit and obtain the measurement outcome k, then the data is projected onto the U eigenstates with eigenvalues that are close to $e^{-2\pi i k/2^m}$. Once an (approximate) eigenstate of U has been prepared in this fashion, we can perform additional measurements to collect further information about the properties of this state. Since both this preparation of the U eigenstate and the additional measurements we perform on that eigenstate are nondeterministic, we may need to repeat the whole procedure multiple times to acquire statistically useful information. Furthermore, the probability of finding a particular eigenstate will of course depend on the initial state of the data register to which phase estimation is applied.

7.6.3 Hamiltonian Eigenstates

If we can simulate quantum evolution governed by a Hamiltonian H, then we can use the phase estimation algorithm to find eigenvalues and prepare eigenstates of H. To obtain eigenvalues to m bits of accuracy, we choose a convenient unit of time T, and execute the time evolution operator $e^{-iHs} = U^t$, where $U = e^{-iHT}$, conditioned on $t \in \{1,2,4,8,..2^{m-1}\}$. That is, the control parameter t used in phase estimation may now be interpreted as the evolution time s expressed in units of T. Note that if there is an efficient circuit for U^t, then U^t conditioned on a single control qubit is also efficient, with a comparable gate count.

As in Section 7.6.2, phase estimation then suffices to find the fractional part of $\dfrac{ET}{2\pi}$ to m-bit accuracy, where E is an eigenvalue of the Hamiltonian H. We should choose the step size in the simulation of e^{-iHs} so that the accuracy is $\delta \approx 2^{-m}$ for $s = 2^m T$. If the Hamiltonian is geometrically local, we have seen in (7.22) that this approximation can be achieved with a circuit size

$$L = \tilde{O}\left(\frac{h^2(ns)^2}{\delta}\right) = \tilde{O}\left(h^2(nT)^2 \times \frac{2^{2m}}{2^{-m}}\right) = \tilde{O}\left((hT)^2 n^2 2^{3m}\right). \tag{7.38}$$

(Here we use the $\tilde{O}$ notation to indicate that a polylog factor has been neglected.) To compute the energy eigenvalue to accuracy polynomial in the system size n, we choose

$$\delta \approx 2^{-m} \approx 1/n^c \Rightarrow m = c \log_2 n, \tag{7.39}$$

where c is a constant. The algorithm is efficient – the quantum circuit size is

$$\tilde{O}\left((hT)^2 n^2 2^{3m}\right) = \tilde{O}(n^2 n^{3c}), \tag{7.40}$$

which is polynomial in n. The approximations we have used can be improved significantly; my goal here was just to explain as simply as possible why there is an exponential quantum advantage. The phase estimation algorithm for measuring e^{-iHT} is shown schematically in Figure 7.1.

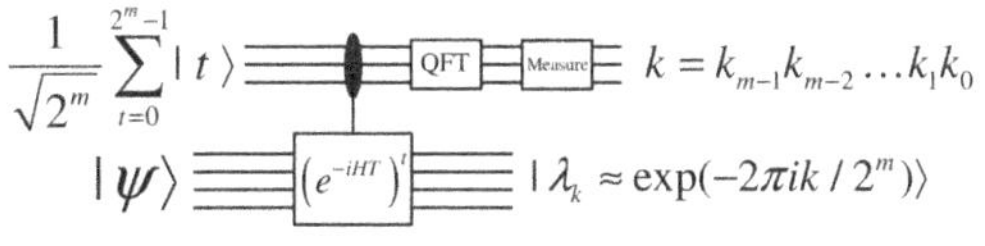

$$\frac{1}{\sqrt{2^m}}\sum_{t=0}^{2^m-1}|t\rangle \quad \boxed{\text{QFT}}\ \boxed{\text{Measure}}\ k = k_{m-1}k_{m-2}\ldots k_1 k_0$$

$$|\psi\rangle \quad \boxed{\left(e^{-iHT}\right)^t}\quad |\lambda_k \approx \exp(-2\pi i k / 2^m)\rangle$$

FIGURE 7.1

Phase estimation algorithm for measuring eigenvalues of e^{-iHT}.

For a particular preparation of the input state $|\psi\rangle$, suppose we repeat the computation many times, and plot a histogram of the results. Then the *location* of each narrow peak estimates an energy eigenvalue E_a, modulo $2\pi/T$. The *height* of the peak estimates $|\langle E_a | \psi \rangle|^2$ – the overlap $|\psi\rangle$ with the corresponding energy eigenstate $|E_a\rangle$.

However, if we want to estimate (say) the ground state energy E_0 to polynomial accuracy in quantum polynomial time, we must be able to prepare a state $|\psi\rangle$ whose overlap with the ground state $|E_0\rangle$ is no worse than polynomially small:

$$|\langle E_0 | \psi \rangle|^2 > 1 / \text{poly}(n). \tag{7.41}$$

If that is the case, we can get a good estimate of E_0 in only polynomially many trials. As a bonus, when we obtain the value E_0 for the measured eigenvalue E_0, then we have projected the state $|\psi\rangle$ onto the ground state $|E_0\rangle$, and therefore, we can compute further properties of $|E_0\rangle$, such as the distribution $\text{Prob}(a) = \langle E_0 | \Pi_a | E_0 \rangle$, where Π_a is a projector onto an eigenspace of an efficiently measurable observable.

7.6.4 Initial State Preparation

However, there is a catch – preparing an initial state that overlaps substantially with the ground state could be very hard in some cases. This is already true classically; finding a good approximation to the ground state of a classical spin glass is NP-hard, as hard as any problem whose solution can be checked efficiently by a classical computer [59]. Finding the ground state for a quantum system with a local Hamiltonian seems to be even harder; it is QMA-hard [49], as hard as any problem whose solution can be checked efficiently by a quantum computer, and we expect that QMA is a larger class than NP. Surprisingly, computing the ground-state energy seems to be a hard problem for a quantum computer even in the case of a geometrically local translationally invariant quantum system in one dimension [60]. That is, it follows from rather weak complexity-theoretic assumptions that there are hard instances of the one-dimensional version of the ground-state-energy problem, even though it is often easy in cases of physical interest.

A general procedure for preparing ground states is adiabatic evolution. We can prepare a state having a sizable overlap with the ground state of H by starting with the easily prepared ground state of a simpler Hamiltonian $H(0)$, then slowly deforming the Hamiltonian along a path $H(s)$ connecting $H(0)$ to $H(1) = H$. This procedure succeeds in polynomial time provided the energy gap $\Delta(s)$ between the ground and first excited states of $H(s)$ is no smaller than inverse polynomial in n for all $s \in [0,1]$

along the path. For problem instances that are quantumly hard, then, the gap becomes superpolynomially small somewhere along the path [61].

Though the general problem is quantumly hard, we may surmise that there are many local quantum systems for which computing the ground-state energy is quantumly easy yet classically hard. For example, the electronic structure of a molecule with atomic nuclei at fixed positions can be accurately described by a local Hamiltonian, and chemists assert (without proof) that it is possible to evolve adiabatically from the *Hartree–Fock* Hamiltonian (which they can solve classically) to the *full configuration interaction* (FCI) Hamiltonian (which they want to solve, but don't know how to solve classically in general), while the gap Δ exceeds a nonzero constant everywhere along the adiabatic path [62]. If that is true, someday fully scalable fault-tolerant quantum computers will be powerful tools for advancing molecular chemistry.

7.7 Quantum Error Correction

Classical digital computers exist and have had a transformative impact on our lives. Large-scale quantum computers do not yet exist. Why not?

Building reliable quantum hardware is challenging because of the difficulty of controlling quantum systems accurately. Small errors in quantum gates accumulate in a large circuit, eventually leading to large errors that foil the computation. Furthermore, qubits in a quantum computer inevitably interact with their surroundings; decoherence arising from unwanted correlations with the environment is harmless in a classical computer (and can even be helpful by introducing friction that impedes accidental bit flips), but decoherence in a quantum computer can irreparably damage the delicate superposition states processed by the machine.

Quantum information can be better protected against noise by using a quantum error-correcting code, in which "logical" information is encoded redundantly in a block of many physical qubits [11, 12]. Quantum error correction is in some ways much like classical error correction, but more difficult, because while a classical code need only protect against bit flips, a quantum code must protect against both bits flips and phase errors.

7.7.1 Conditions for Quantum Error Correction

Suppose for example, that we want to encode a single logical qubit, with orthonormal basis states denoted $|\bar{0}\rangle$ and $|\bar{1}\rangle$, which is protected against all the errors spanned by a set $\{E_a\}$. For the distinguishability of the basis states to be maintained even when errors occur, we require

$$E_a \,|\, \bar{0}\rangle \perp E_b \,|\, \bar{1}\rangle, \tag{7.42}$$

where E_a, E_b are any two elements of the error basis. This condition by itself would suffice for reliable storage of a classical bit.

But for storage of a qubit, we also require protection against phase errors, which occur when information about whether the state is $|\bar{0}\rangle$ or $|\bar{1}\rangle$ leaks to the environment; equivalently, distinguishability should be maintained for the dual basis states $(|\bar{0}\rangle \pm |\bar{1}\rangle)/\sqrt{2}$:

$$E_a \left(|\bar{0}\rangle + |\bar{1}\rangle\right) \perp E_b \left(|\bar{0}\rangle - |\bar{1}\rangle\right), \tag{7.43}$$

where E_a, E_b are any two errors. In fact, the two distinguishability conditions (7.42) and (7.43) suffice to ensure the existence of a recovery map that corrects any error spanned by $\{E_a\}$ acting on any linear combination of $|\bar{0}\rangle$ and $|\bar{1}\rangle$ [63].

Together, (7.42) and (7.43) imply

$$\langle\bar{0}\,|\, E_a^\dagger E_b \,|\, \bar{0}\rangle = \langle\bar{1}\,|\, E_a^\dagger E_b \,|\, \bar{1}\rangle; \tag{7.44}$$

no measurement of any operator in the set $\{E_a^\dagger E_b\}$ can distinguish the two basis states of the logical qubit. Typically, because we expect noise acting collectively on many qubits at once to be highly suppressed, we are satisfied to correct *low-weight* errors, those that act nontrivially on a sufficiently small fraction of all the qubits in the code block. Then (7.44) says that all the states of the logical qubit look the same when we examine a small subsystem of the code block. To be well protected, the logical states should be highly entangled, so that no logical information is accessible locally.

7.7.2 Protected Quantum Memory and Topological Order

It is useful to formulate the distinction between classical and quantum error correction in more physical terms (see Figures 7.2 and 7.3). The prototype for a protected classical memory is a ferromagnet, where a single bit is encoded according to whether most of the spins are up or down. The encoded bit can be read out by performing local measurements on all spins, and then executing a majority vote to protect against errors that flip a minority of the spins. Errors in the memory create domain walls where neighboring spins misalign, and a logical error occurs when a domain wall sweeps across the sample, inducing a global operation acting on

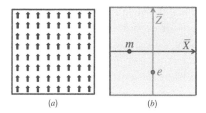

FIGURE 7.2
(a) A prototypical classical memory is a ferromagnet. (b) A prototypical quantum memory is a topologically ordered medium.

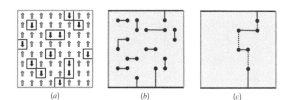

FIGURE 7.3
(a) In a two-dimensional ferromagnet, domain walls surround droplets of flipped bits. (b) In a two-dimensional topological quantum memory, pointlike anyons appear at the ends of chains of flipped qubits. (c) A logical error occurs if the actual errors (dark colored) combine with our error diagnosis (dotted black) to produce a path that traverses the medium.

many spins. The memory is robust at a sufficiently small nonzero temperature because large droplets of flipped spins have a large energy cost and are therefore unlikely to occur due to thermal fluctuations. This memory is a particularly simple physically motivated example of a classical error-correcting code; there are more sophisticated examples.

The prototype for a protected quantum memory is a medium in two dimensions with $\mathbb{Z}_2$ topological order [64]. In contrast to the ferromagnet, errors in the medium create pointlike excitations ("anyons") rather than domain walls. There are two types of anyons, which we denote as e (for "electric") and m (for "magnetic"). We speak of "$\mathbb{Z}_2$ topological order" because when an e anyon travels around an m anyon (or an m anyon travels around an e anyon), the many-body wave function acquires a "topological phase" of −1. This phase is topological in the sense that it does not matter what path the e anyon follows; all that matters is that it winds around the m anyon an odd number of times. $\mathbb{Z}_2$ just means that the topological phase takes two possible values, +1 and −1.

There are two possible types of one-dimensional edge for this two-dimensional medium, shown in Figure 7.2b. An e anyon can appear or disappear at the edge on the top and the bottom of the sample, while an m anyon can appear or disappear at the edge on the left and on the right.

The protected code space is the space of quantum states in which no anyons are present. There are nontrivial physical processes that preserve this code space. Namely, an e anyon can appear at the bottom, propagate across the sample and disappear at the top. This process applies a unitary operator to system, which we call $\bar{Z}$. Or an m anyon can appear at the left, propagate across, and disappear at the right. This process applies a different unitary operator to the system, which we call $\bar{X}$. Because of the topological phase -1 that arises when e winds around m, these operators do not commute; rather

$$\bar{X}^{-1}\bar{Z}^{-1}\bar{X}\bar{Z} = -I, \tag{7.45}$$

Thus, two anticommuting operators both preserve the code space, which means the code space cannot be one dimensional. In fact, it is two dimensional, and we may interpret $\bar{Z}$ and $\bar{X}$ as the Pauli operators acting on the protected qubit:

$$\bar{Z} = \begin{pmatrix} 1 & 0 \\ 0 & -1 \end{pmatrix}, \quad \bar{X} = \begin{pmatrix} 0 & 1 \\ 1 & 0 \end{pmatrix}. \tag{7.46}$$

I have written a bar above $\bar{Z}$ and $\bar{X}$ to distinguish these "logical" Pauli operators, which act on the encoded qubit, from the physical Pauli operators we will discuss in Section 7.7.3.

There is another process we might consider. A pair of e anyons (or m anyons) are created in the bulk of the sample, away from any boundary. These anyons wander around for a while, without ever approaching the boundary, until finally they find one another, annihilate, and disappear. This process also preserves the code space, but in contrast to the $\bar{Z}$ and $\bar{X}$ operators, it acts trivially on the protected qubit; that is, it commutes with both $\bar{Z}$ and $\bar{X}$. For example, the path followed by the m anyon from the left to right edge can be deformed so it stays away from the pair of e anyons wandering in the bulk. This does not change how $\bar{X}$ acts on the code space but makes clear that the diffusing pair of e anyons can't have any effect on $\bar{X}$.

The system is protected by a nonzero energy gap, the energy cost of creating a pair of anyons. Hence, quantum information can be stored for a long time if the temperature is small compared with the gap, but unlike the case of the two-dimensional ferromagnet, the storage time does not improve as the system size increases. In the ferromagnet, the energy cost of a bubble of flipped spins increases as the bubble grows; in contrast, once a pair of anyons is thermally excited, no further energy barrier prevents the anyons from wandering to opposite sides of the sample, producing a logical $\bar{Z}$ or $\bar{X}$ error. However, if we *monitor* the particles as they diffuse through the sample, then a logical error occurs only if particles propagate across the sample without being noticed, an event that *does* become increasingly unlikely as the system size grows [65]. The scheme for performing robust quantum computation described in Section 7.7.3 builds on this observation.

7.7.3 Surface Code Accuracy Threshold

To create a stable quantum memory, we need not synthesize a topologically ordered material; instead, we can *simulate* the material using whatever quantum computing hardware we prefer. Kitaev constructed a simple two-dimensional lattice model (the *surface code*), with a qubit at each lattice site, which exhibits $\mathbb{Z}_2$ topological order just as described in Section 7.7.2 [64, 66, 67]. Though it was first proposed nearly 25 years ago, the surface code still offers a particularly promising route toward scalable fault-tolerant quantum computation. It has two major advantages. First, the quantum processing needed to diagnose and correct errors is remarkably simple. Second, and not unrelatedly, it can tolerate a relatively high gate error rate.

Errors afflicting a quantum memory can be expanded in terms of multiqubit Pauli operators, and each such Pauli operator can be expressed as a product of an X-type error, where either X or the identity acts on each qubit, and a Z-type error, where either Z or the identity acts on each qubit. (A $Y = -iZX$ error is just the case where both X and Z act on the same qubit.) Therefore, our quantum memory will be well protected if we can correct both X-type and Z-type errors with high success probability. In the case of the surface code, there are two separate procedures for correcting X errors and correcting Z errors, and both work in essentially the same way, so it will suffice to discuss only how the Z errors are corrected.

In (one version of) the surface code, the physical qubits reside on the edges of a square lattice, and e anyons may reside on the sites of the lattice. Suppose an unknown quantum state $\alpha|\bar{0}\rangle + \beta|\bar{1}\rangle$ has been stored in the code space, where $|\bar{0}\rangle$ and $|\bar{1}\rangle$ are the encoded $\bar{Z}$ eigenstates. After this state is encoded, Z errors occur on some of the qubits, knocking the state out of the code space by creating e anyons. A snapshot of a typical error configuration is shown in Figure 7.3b; edges on which the qubits have Z errors (dark colored), define a set of connected "error chains", and pairs of anyons appear at the endpoints of each error chain. The positions of the anyons (and hence the endpoints of the error chains) can be identified by a simple quantum computation. After finding their positions, we can remove these anyons two at a time; we select a pair of anyons, and apply Z to all the qubits along a "recovery chain" that connects the pair, in effect bringing the pair of anyons together to annihilate. Alternatively, we can remove a single anyon by choosing a recovery chain that connects that anyon to the top or bottom edge. Our goal is to remove all of the anyons, returning the state to the code space, and (we hope) restoring the initial encoded state.

The anyon positions are said to constitute an error "syndrome" because they help us to diagnose the damage sustained by the physical qubits in the code block. Even though this syndrome locates the boundary points of the error chains, we don't know the configuration of the error chains themselves, so our recovery chains won't necessarily coincide with the error chains, or even connect together the same pairs of anyons. But they don't have to. If each connected path resulting from combining the error chain with the recovery chain forms a closed loop in the bulk, or an open path with both its endpoints lying on the same edge (either top or bottom), then error recovery is successful. This works because of the properties of the topologically ordered medium noted earlier: creation of a pair of anyons followed by pair annihilation, or creation of a single anyon at the bottom (top) edge followed by annihilation at the bottom (top) edge are processes that act trivially on the code space. On the other hand, if the error chain combined with the recovery chain produces a path connecting the bottom and top edges as in Figure 7.3c, then (if there are an odd number of such paths) a logical $\bar{Z}$ error occurs and our recovery procedure fails.

To keep things simple, consider a stochastic independent noise model, in which each qubit in the code block experiences a Z error with probability ε. Suppose we choose our recovery chains to have the minimal possible weight; that is, we return to the code space by applying Z to as few qubits as possible. Given the known positions of the anyons, this minimal chain can be computed efficiently with a classical computer. For this recovery procedure, we can find an upper bound on the probability of a logical error by the following argument [65].

We denote by d the minimal weight of a connected path from the bottom edge to the top edge; that is, d (the *distance* of the code) is the minimal weight of a $\bar{Z}$ logical operator. If our attempt to recover resulted in a logical $\bar{Z}$ error, there must be a path connecting the bottom and top edges of the code block such that each edge of the lattice on this path is in either an error chain or a recovery chain. Let's say this connected path has length $\ell \geq d$, and denote the path by C_ℓ. The number of errors on C_ℓ must be at least $\ell/2$ if ℓ is even, or $(\ell+1)/2$ if ℓ is odd; otherwise we could have found a lower weight recovery chain by applying Z on the error chains contained in C_ℓ, rather than to the qubits on C_ℓ that are complementary to the error chains on C_ℓ. The number of ways that the edges with Z errors could be distributed along C_ℓ is no more than 2^ℓ (each qubit on C_ℓ either has an error or does not). Since, for each physical qubit, Z errors occur with probability ε, the probability that C_ℓ is contained in the union of error chains and recovery chains obeys

$$P(C_\ell) \le 2^\ell \varepsilon^{\ell/2}. \tag{7.47}$$

Let N_ℓ denote the number of paths connecting the bottom and top edges with length ℓ. For a logical error to occur, the combination of error chains and recovery chains must produce at least one path connecting the bottom and top edges. Using the upper bound (7.47) on the probability of each such path, and applying the union bound, we conclude that the probability of a logical $\bar{Z}$ error satisfies

$$P_{\text{logical}} \le \sum_{\ell=d}^{n} N_\ell 2^\ell \varepsilon^{\ell/2}. \tag{7.48}$$

The lower limit on the sum is $\ell = d$, the length of the shortest path connecting the bottom and top edges. The upper limit is n, the total number of qubits in the code block, which is therefore the maximum length of any path.

We can also find a simple upper bound on N_ℓ. Let's say our square lattice is $d \times d$. A path from the bottom to the top edge can begin at any one of d positions along the bottom edge, and in each of the ℓ steps along the path, there are three possible moves: straight ahead, left turn, or right turn. Therefore (even if we don't insist that the path reach the top edge), we have

$$N_\ell \le d\, 3^\ell \implies P_{\text{logical}} \le d \sum_{\ell=d}^{n} \left(36\varepsilon\right)^{\ell/2}. \tag{7.49}$$

Now suppose that $\varepsilon < 1/36$, so that the terms in the sum over ℓ decrease as ℓ increases. For a square lattice, the number of edges (qubits) in the code block is $n = O(d^2)$, so the number of terms in the sum is also $O(d^2)$, and we conclude that

$$P_{\text{logical}} \le O(d^3)\left(\varepsilon / \varepsilon_0\right)^{d/2} \quad \text{for} \quad \varepsilon < \varepsilon_0 = 1/36 \approx .028. \tag{7.50}$$

Thus, this argument establishes that the surface code is a quantum memory with an *accuracy threshold* – for any constant $\varepsilon < \varepsilon_0$, the probability of a logical error decays exponentially as the code distance d increases (apart from a possible polynomial prefactor). If the physical error rate is below the threshold value ε_0, we can make the logical error rate arbitrarily small by choosing a sufficiently large code block. Unsurprisingly, in view of the crudeness of this argument, the actual value of the error threshold ε_0 is larger than we estimated. Monte Carlo simulations find $\varepsilon_0 \approx .103$ [68].

7.7.4 Scalable Quantum Computing

To draw quantitative conclusions about the overhead cost of fault-tolerant quantum computing, refinements of this argument are needed. First, we implicitly assumed that the error syndrome measurements are perfect. In fact, measurement errors occur, which means we need to repeat the measurement $O(d)$ times to acquire sufficiently trustworthy information about where the anyons are located. Second, we did not take into account the structure of the quantum circuit used to make these measurements. To determine whether an anyon is present at a particular site, four entangling two-qubit gates are needed, any one of which could be faulty, and a single fault can cause both an error in the measurement outcome and errors in the data qubits. A more complete analysis shows that the threshold error rate for the two-qubit gates is close to 1% [69, 70]. Numerical simulations find that for each round of syndrome measurement, the probability of a logical error rate scales roughly like [71]

$$P_{\text{logical}} \approx 0.1\big(100p\big)^{(d+1)/2}, \tag{7.51}$$

where now p denotes the two-qubit gate error rate, and we have assumed that d is odd.

So far, we have considered only the probability of a storage error for one protected qubit, but in a scalable fault-tolerant quantum computer, we will need many protected qubits, and we will need to perform highly reliable universal quantum gates that act on these qubits. One can envision an architecture in which the logical qubits are arranged like square tiles on a surface, with buffer qubits filling gaps between the tiles [72–74]. I won't go into the details of how the logical gates are executed, but it is helpful to realize that much of the logical processing can be executed by performing entangling measurements on pairs of logical blocks. For example, we can measure $\bar{X}_1 \otimes \bar{X}_2$, where blocks 1 and 2 reside on adjacent tiles, by fusing the blocks together along their horizontal edges and then cutting the blocks apart again, a process called "lattice surgery" [75]. The fusing and cutting are achieved by measurements that activate the buffer qubits in between the edges of the two blocks, followed by measurements that decouple the buffer qubits.

The good news is that the error rates for logical gates are not much worse than the storage error rates we have already discussed, except we should keep in mind that we need to repeat the syndrome measurement $O(d)$ times in each logical gate cycle. The bad news is that (7.51) indicates that we'll need a rather large code distance if we want to make the logical error rate very small. Suppose, for example, that we would like to run Shor's algorithm to factor a 2048-bit number, which would break the RSA cryptosystem, and suppose that the physical two-qubit gate error rate is 10^{-3}, better than in current multi-qubit devices. The analysis in [76] calls for a logical error probability $\approx 10^{-15}$ per round of syndrome measurement, and hence a code distance of $d = 27$. The number of physical qubits per code block, including ancilla qubits needed for syndrome measurement and lattice surgery, is $2(d+1)^2 = 1568$, and the total number of logical qubits used in this version of the factoring algorithm is about 14,000, pushing the physical qubit count above 20 million [76]. That's a lot!

There are many challenges to making large-scale fault-tolerant quantum computing practical, including serious systems engineering issues. There are also issues of principle to consider – what is required for a fault-tolerant scheme to be scalable, and what conditions must be satisfied by the noise model? One essential requirement is some form of cooling to extract the entropy introduced by noise [77]. In the protocol described above, entropy is extracted by measuring and resetting ancilla qubits in each round of syndrome measurement. Parallel operations are also necessary, so noise can be controlled in different parts of the computer simultaneously.

The analysis leading to (7.51) is based on a simple noise model in which gate errors are stochastic (rather than coherent) and there are no correlations among errors in different gates. The fault-tolerant methods should work for more realistic noise models, as long as the errors are sufficiently weak and not too strongly correlated. By benchmarking logical error rates using relatively small quantum codes during the NISQ era, we will gain valuable insight into how effectively quantum error correction protects computations performed on actual quantum hardware.

7.8 Outlook

I cherish the memory of some very enjoyable conversations with Dick Feynman about physics and about other things, too. But quantum computing was one subject we never discussed. I knew Feynman was interested in quantum computation, but I was not very interested back then, so I never asked him about it. Naturally, I regret that now. Six years after his death I became very interested, but by then it was too late.

The key issues we might have discussed in the early 1980s still loom over the subject today. Can we build powerful large-scale quantum computers? How will we do it? When will it happen? And what will we do with those awesome machines? I'm confident that the answer to the first question is yes. But 40 years later, the answers to the other questions are still far from clear. Feynman was right to conclude his 1981 talk with the observation, "it's a wonderful problem because it doesn't look so easy".

Those who aspire to build quantum computing systems face a daunting engineering challenge, but there's more to it than that. It would transform the prospects for practical quantum computing applications if we could improve physical gate error rates (currently around 1% for entangling two-qubit gates) by several orders of magnitude. The progress achieved so far has been driven by advances in qubit design, control technology, fabrication methods, and materials, and further incremental improvements can surely be expected. But quantum hardware is still at an early stage, and truly disruptive progress might flow from fresh ideas about how to encode and manipulate quantum information robustly. The quantum community should continue to think broadly and imaginatively about new approaches to building and operating quantum hardware.

Feynman was on the right track when he suggested using quantum computers to solve problems in quantum physics and chemistry. It is still the most important application we can clearly foresee, and there is plenty of opportunity to flesh out our ideas about how quantum computers can best be used to advance science. Applications of broader interest are also possible. Quantum computers can speed up the exhaustive search for solutions to optimization problems, but because the speedup is only quadratic in that case, this might not be useful until far in the future. More dramatic quantum speedups for optimization and related problems cannot yet be ruled out and should continue to be a goal for research on quantum algorithms.

Though fully scalable fault-tolerant quantum computers may still be a ways off, the advent of the NISQ era already heralds unprecedented opportunities for exploring the properties of highly entangled many-body quantum systems. With NISQ technologies, we will also assess the performance of heuristic hybrid quantum/classical algorithms, which may steer us toward practical applications, and we'll advance our toolkit for mitigating noise and correcting errors in quantum platforms. Today's quantum computers can help us to build tomorrow's more powerful quantum computers.

This chapter has focused on the prospects for building and using quantum computers. But if Feynman were here today, I would be just as eager to tell him about the myriad of ways that quantum information concepts have opened new vistas across many domains of physics. To mention two prominent examples, we have understood that different quantum phases of matter can be distinguished according to the structure of their long-range quantum entanglement [78] and that the spacetime geometry in a model of quantum gravity can admit an alternative description, in which the geometry is encoded in the quantum entanglement of a quantum system that does not involve gravitation at all [79]. Powerful insights like these signal that quantum information science has become an essential force in humanity's struggle to grasp Nature's hidden secrets. From now on, quantum computer science and quantum physical science will advance together, hand in hand.

Bibliography

1. Richard P Feynman. Simulating physics with computers. *International Journal of Theoretical Physics*, 21(6/7):467–488, 1981.
2. Yuri Manin. *Computable and Uncomputable*. Sovetskoye Radio, Moscow, 128, 1980.
3. Paul Benioff. The computer as a physical system: A microscopic quantum mechanical Hamiltonian model of computers as represented by Turing machines. *Journal of Statistical Physics*, 22(5):563–591, 1980.
4. David Deutsch. Quantum theory, the Church–Turing principle and the universal quantum computer. *Proceedings of the Royal Society of London. Series A: Mathematical and Physical Sciences*, 400(1818):97–117, 1985.
5. Ethan Bernstein, and Umesh Vazirani. Quantum complexity theory. *SIAM Journal on Computing*, 26(5):1411–1473, 1997.
6. Daniel R Simon. On the power of quantum computation. *SIAM Journal on Computing*, 26(5):1474–1483, 1997.
7. Peter W Shor. Polynomial-time algorithms for prime factorization and discrete logarithms on a quantum computer. *SIAM Review*, 41(2):303–332, 1999.
8. Rolf Landauer. Is quantum mechanics useful? *Philosophical Transactions of the Royal Society of London. Series A: Physical and Engineering Sciences*, 353(1703):367–376, 1995.
9. William G Unruh. Maintaining coherence in quantum computers. *Physical Review. Part A*, 51(2):992, 1995.
10. Serge Haroche, and Jean-Michel Raimond. Quantum computing: Dream or nightmare? *Physics Today*, 49(8):51–54, 1996.
11. Peter W Shor. Scheme for reducing decoherence in quantum computer memory. *Physical Review. Part A*, 52(4):R2493, 1995.
12. Andrew M Steane. Error correcting codes in quantum theory. *Physical Review Letters*, 77(5):793, 1996.
13. Peter W Shor. Fault-tolerant quantum computation. In *Proceedings of 37th Conference on Foundations of Computer Science*, pages 56–65. IEEE, 1996.
14. Dorit Aharonov, and Michael Ben-Or. Fault-tolerant quantum computation with constant error rate. *Proceedings of the Twenty-Ninth Annual ACM Symposium on Theory of Computing*, pp. 176–188, 1997.
15. Emanuel Knill, Raymond Laflamme, and Wojciech H Zurek. Resilient quantum computation. *Science*, 279(5349):342–345, 1998.

16. Aleksei Yur'evich Kitaev. Quantum computations: Algorithms and error correction. *Uspekhi Matematicheskikh Nauk*, 52(6):53–112, 1997.

17. John Preskill. Reliable quantum computers. *Proceedings of the Royal Society of London. Series A: Mathematical, Physical and Engineering Sciences*, 454(1969):385–410, 1998.

18. John Preskill. Fault-tolerant quantum computation. In *Introduction to Quantum Computation and Information*, edited by Hoi-Kwong Lo, Tim Spiller, and Sandu Popescu, pages 213–269. Singapore: World Scientific, 1998.

19. Juan I Cirac, and Peter Zoller. Quantum computations with cold trapped ions. *Physical Review Letters*, 74(20):4091, 1995.

20. Chris Monroe, David M Meekhof, Barry E King, Wayne M Itano, and David J Wineland. Demonstration of a fundamental quantum logic gate. *Physical Review Letters*, 75(25):4714, 1995.

21. John Preskill. Quantum computing: Pro and con. *Proceedings of the Royal Society of London. Series A: Mathematical, Physical and Engineering Sciences*, 454(1969):469–486, 1998.

22. Richard P Feynman. There's plenty of room at the bottom. In *California Institute of Technology, Engineering and Science Magazine*, pages 22–36. Pasadena, CA: California Institute of Technology, 1960 February.

23. Daniel J Bernstein, and Tanja Lange. Post-quantum cryptography. *Nature*, 549(7671):188–194, 2017.

24. Stephen Wiesner. Conjugate coding. *ACM Sigact News*, 15(1):78–88, 1983.

25. Charles H Bennett, and Gilles Brassard. Quantum cryptography: Public key distribution and coin tossing. *arXiv preprint arXiv:2003.06557*, 2020.

26. Michele Mosca. Cybersecurity in an era with quantum computers: Will we be ready? *IEEE Security and Privacy*, 16(5):38–41, 2018.

27. Sam McArdle, Suguru Endo, Alan Aspuru-Guzik, Simon C Benjamin, and Xiao Yuan. Quantum computational chemistry. *Reviews of Modern Physics*, 92(1):015003, 2020.

28. Charles H Bennett, Ethan Bernstein, Gilles Brassard, and Umesh Vazirani. Strengths and weaknesses of quantum computing. *SIAM Journal on Computing*, 26(5):1510–1523, 1997.

29. Lov K Grover. Quantum mechanics helps in searching for a needle in a haystack. *Physical Review Letters*, 79(2):325, 1997.

30. Frank Arute, Kunal Arya, Ryan Babbush, Dave Bacon, Joseph C Bardin, Rami Barends, Rupak Biswas, Sergio Boixo, Fernando G S L Brandao, David A Buell, et al. Quantum supremacy using a programmable superconducting processor. *Nature*, 574(7779):505–510, 2019.

31. John Preskill. Quantum computing and the entanglement frontier. *arXiv preprint arXiv:1203.5813*, 2012.

32. Aram W Harrow, and Ashley Montanaro. Quantum computational supremacy. *Nature*, 549(7671):203–209, 2017.

33. Cupjin Huang, Fang Zhang, Michael Newman, Junjie Cai, Xun Gao, Zhengxiong Tian, Junyin Wu, Haihong Xu, Yu Huanjun, Bo Yuan, et al. Classical simulation of quantum supremacy circuits. *arXiv preprint arXiv:2005.06787*, 2020.

34. John Preskill. Quantum computing in the NISQ era and beyond. *Quantum*, 2:79, 2018.

35. Edward Farhi, Jeffrey Goldstone, and Sam Gutmann. A quantum approximate optimization algorithm. *arXiv preprint arXiv:1411.4028*, 2014.

36. Alberto Peruzzo, Jarrod McClean, Peter Shadbolt, Man-Hong Yung, Xiao-Qi Zhou, Peter J Love, Alán Aspuru-Guzik, and Jeremy L O'Brien. A variational eigenvalue solver on a photonic quantum processor. *Nature Communications*, 5(1):1–7, 2014.

37. Dieter Jaksch, Christoph Bruder, Juan Ignacio Cirac, Crispin W Gardiner, and Peter Zoller. Cold bosonic atoms in optical lattices. *Physical Review Letters*, 81(15):3108, 1998.

38. Markus Greiner, Olaf Mandel, Tilman Esslinger, Theodor W Hänsch, and Immanuel Bloch. Quantum phase transition from a superfluid to a Mott insulator in a gas of ultracold atoms. *Nature*, 415(6867):39–44, 2002.

39. Hannes Bernien, Sylvain Schwartz, Alexander Keesling, Harry Levine, Ahmed Omran, Hannes Pichler, Soonwon Choi, Alexander S Zibrov, Manuel Endres, Markus Greiner, et al. Probing many-body dynamics on a 51-atom quantum simulator. *Nature*, 551(7682):579–584, 2017.

40. Jiehang Zhang, Guido Pagano, Paul W Hess, Antonis Kyprianidis, Patrick Becker, Harvey Kaplan, Alexey V Gorshkov, Z-X Gong, and Christopher Monroe. Observation of a many-body dynamical phase transition with a 53-Qubit quantum simulator. *Nature*, 551(7682):601–604, 2017.

41. Christie S Chiu, Geoffrey Ji, Annabelle Bohrdt, Muqing Xu, Michael Knap, Eugene Demler, Fabian Grusdt, Markus Greiner, and Daniel Greif. String patterns in the doped Hubbard model. *Science*, 365(6450):251–256, 2019.

42. Biswaroop Mukherjee, Parth B Patel, Zhenjie Yan, Richard J Fletcher, Julian Struck, and Martin W Zwierlein. Spectral response and contact of the unitary fermi gas. *Physical Review Letters*, 122(20):203402, 2019.

43. Giulia Semeghini, Harry Levine, Alexander Keesling, Sepehr Ebadi, Tout T Wang, Dolev Bluvstein, Ruben Verresen, Hannes Pichler, Marcin Kalinowski, Rhine Samajdar, et al. Probing topological spin liquids on a programmable quantum simulator. *arXiv preprint arXiv:2104.04119*, 2021.

44. Daniel Gottesman. An introduction to quantum error correction and fault-tolerant quantum computation. In *Quantum Information Science and Its Contributions to Mathematics, Proceedings of Symposia in Applied Mathematics*, volume 68, edited by Samuel J. Lomonaco, pages 13–58. Washington, DC: American Mathematical Society, 2010.

45. Earl T Campbell, Barbara M Terhal, and Christophe Vuillot. Roads towards fault-tolerant universal quantum computation. *Nature*, 549(7671):172–179, 2017.

46. Ian D Kivlichan, Craig Gidney, Dominic W Berry, Nathan Wiebe, Jarrod McClean, Wei Sun, Zhang Jiang, Nicholas Rubin, Austin Fowler, Alán Aspuru-Guzik, et al. Improved fault-tolerant quantum simulation of condensed-phase correlated electrons via trotterization. *Quantum*, 4:296, 2020.

47. Earl T Campbell. Early fault-tolerant simulations of the Hubbard model. *arXiv preprint arXiv:2012.09238*, 2021.

48. Alexander Semenovich Holevo. Bounds for the quantity of information transmitted by a quantum communication channel. *Problemy Peredachi Informatsii*, 9(3):3–11, 1973.

49. Alexei Yu Kitaev, Alexander Shen, Mikhail N Vyalyi, and Mikhail N Vyalyi. *Classical and Quantum Computation*. Number 47. Providence, RI: American Mathematical Soc., 2002.

50. Stephen P Jordan, Keith S M Lee, and John Preskill. Quantum algorithms for quantum field theories. *Science*, 336(6085):1130–1133, 2012.

51. John Preskill. Simulating quantum field theory with a quantum computer. *arXiv preprint arXiv:1811.10085*, 2018.

52. David P DiVincenzo. The physical implementation of quantum computation. *Fortschritte der Physik: Progress of in Physics*, 48(9–11):771–783, 2000.

53. Colin D Bruzewicz, John Chiaverini, Robert McConnell, and Jeremy M Sage. Trapped-ion quantum computing: Progress and challenges. *Applied Physics Reviews*, 6(2):021314, 2019.

54. Morten Kjaergaard, Mollie E Schwartz, Jochen Braumüller, Philip Krantz, Joel I-J Wang, Simon Gustavsson, and William D Oliver. Superconducting qubits: Current state of play. *Annual Review of Condensed Matter Physics*, 11:369–395, 2020.

55. Klaus Mølmer, and Anders Sørensen. Multiparticle entanglement of hot trapped ions. *Physical Review Letters*, 82(9):1835, 1999.

56. Anders Sørensen, and Klaus Mølmer. Quantum computation with ions in thermal motion. *Physical Review Letters*, 82(9):1971, 1999.

57. Seth Lloyd. Universal quantum simulators. *Science*, 273:1073–1078, 1996.

58. A Yu Kitaev. Quantum measurements and the abelian stabilizer problem. *arXiv preprint quant-ph/9511026*, 1995.

59. Francisco Barahona. On the computational complexity of Ising spin glass models. *Journal of Physics A: Mathematical and General*, 15(10):3241, 1982.

60. Daniel Gottesman, and Sandy Irani. The quantum and classical complexity of translationally invariant tiling and Hamiltonian problems. In *2009 50th Annual IEEE Symposium on Foundations of Computer Science*, pages 95–104. IEEE, 2009.

61. Edward Farhi, Jeffrey Goldstone, Sam Gutmann, and Michael Sipser. Quantum computation by adiabatic evolution. *arXiv preprint quant-ph/0001106*, 2000.

62. Alán Aspuru-Guzik, Anthony D Dutoi, Peter J Love, and Martin Head-Gordon. Simulated quantum computation of molecular energies. *Science*, 309(5741):1704–1707, 2005.

63. Emanuel Knill, and Raymond Laflamme. Theory of quantum error-correcting codes. *Physical Review A*, 55(2):900, 1997.

64. A Yu Kitaev. Fault-tolerant quantum computation by anyons. *Annals of Physics*, 303(1):2–30, 2003.

65. Eric Dennis, Alexei Kitaev, Andrew Landahl, and John Preskill. Topological quantum memory. *Journal of Mathematical Physics*, 43(9):4452–4505, 2002.

66. Sergey B Bravyi, and A Yu Kitaev. Quantum codes on a lattice with boundary. *arXiv preprint quant-ph/9811052*, 1998.

67. Michael H Freedman, and David A Meyer. Projective plane and planar quantum codes. *Foundations of Computational Mathematics*, 1(3):325–332, 2001.

68. Chenyang Wang, Jim Harrington, and John Preskill. Confinement-Higgs transition in a disordered gauge theory and the accuracy threshold for quantum memory. *Annals of Physics*, 303(1):31–58, 2003.

69. Robert Raussendorf, and Jim Harrington. Fault-tolerant quantum computation with high threshold in two dimensions. *Physical Review Letters*, 98(19):190504, 2007.

70. Robert Raussendorf, Jim Harrington, and Kovid Goyal. Topological fault-tolerance in cluster state quantum computation. *New Journal of Physics*, 9(6):199, 2007.

71. Austin G Fowler, Simon J Devitt, and Cody Jones. Surface code implementation of block code state distillation. *Scientific Reports*, 3(1):1–6, June 2013.

72. Austin G Fowler, and Craig Gidney. Low overhead quantum computation using lattice surgery. *arXiv preprint arXiv:1808.06709*, 2018.

73. Daniel Litinski. A game of surface codes: Large-scale quantum computing with lattice surgery. *Quantum*, 3:128, 2019.

74. Daniel Litinski. Magic state distillation: Not as costly as you think. *Quantum*, 3:205, 2019.

75. Clare Horsman, Austin G Fowler, Simon Devitt, and Rodney Van Meter. Surface code quantum computing by lattice surgery. *New Journal of Physics*, 14(12):123011, 2012.

76. Craig Gidney, and Martin Ekerå. How to factor 2048 bit RSA integers in 8 hours using 20 million noisy qubits. *Quantum*, 5:433, 2021.

77. Dorit Aharonov, Michael Ben-Or, Russell Impagliazzo, and Nisan Noam. Limitations of noisy reversible computation. *arXiv preprint quant-ph/9611028*, 1996.

78. Bei Zeng, Xie Chen, Duan-Lu Zhou, and Xiao-Gang Wen. *Quantum Information Meets Quantum Matter*. New York: Springer, 2019.

79. Shinsei Ryu, and Tadashi Takayanagi. Holographic derivation of entanglement entropy from the anti–de Sitter space/conformal field theory correspondence. *Physical Review Letters*, 96(18):181602, 2006.

8

Physical Aspects of Computation

CONTENTS

Editor's Note

This chapter covers the most time-dependent of all the topics in these lectures because the advances in silicon technology over the past 30 years have been truly dramatic. Nonetheless, Feynman's account of the physics of semiconductor devices is still relevant and demonstrates his deep insights. Similarly, his discussion of energy use and heat loss fits in with his concerns about the amount of energy expended in computation. As Feynman explains, 'In a CMOS inverter, no energy is required to *hold* a state, just to *change* it'. This was one of the drivers of the transition to CMOS technology in the late 1970's and early 1980's. Carver Mead was a long-time colleague of Feynman at Caltech and In 1971 Mead had answered a question from Gordon Moore, one of the founders of Intel, by showing that transistor semiconductor technology could be scaled down in size so that circuits got more complex, ran faster, and took less power.

DOI: 10.1201/9781003358817-8

Feynman was excited by this scaling prediction and by Mead's classic book with Lynn Conway on 'Introduction to VLSI systems', published in 1980. This was the first textbook on Very Large Design Integration (VLSI) design that was explicitly aimed at educating electrical engineering and computer science students. Feynman's discussion of VLSI circuit construction closely follows their discussion of the design rules for the various layers of the planar fabrication process. Today, nearly 40 years on, things are much more complicated. The leading fabrication facilities now advertize a '3 nanometer process technology' that can support up to 14 metal layers rather than the one described by Mead and Conway. However, the '3nm' name is more of a marketing term and no longer relates to any physical feature size. Typical gate and metal separations are actually several 10s of nanometers. In addition, new transistor technologies have emerged to replace planar CMOS transistors. With so-called FinFET technology the industry has moved towards fabrication in 3 dimensions. However, in 2023 it seems that FinFET technology is likely to be displaced by the Gate-All-Around FETs (GAAFETs) using nanowires and nanosheets. Feynman would clearly have loved to analyze these new technologies and their constraints!

The unifying theme of this course has been what we can and cannot do with computers and why. We have considered restrictions arising from the organization of the basic elements within machines, the limitations imposed by fundamental mathematics, and even those resulting from the laws of Nature themselves. In this chapter, we come to address perhaps the most practical of obstacles: the constraints that arise from the technology we employ to actually *build* our machines – both from the materials we use and from the way in which we arrange the elementary component parts.

Presently, the majority of computers are based on *semiconductor* technology, which is used to fashion the basic building blocks of machines – devices such as transistors and diodes. Very-large scale integration (VLSI), the field of microelectronics dealing with the construction and utilization of silicon chips – and hence of central importance to computing – is a vast subject in itself and we can only scratch the surface here. The reader will certainly find what follows easier to understand if he or she has some knowledge of electronics. However, we hope that our presentation will be intelligible to those with only a passing acquaintance with electricity and magnetism, and we provide several references in the section on suggested reading for the curious to take their interest further.

To begin with, we shall take a look at one simple kind of device, the *diode*. This is a cunning device that allows current to flow in one direction only. We shall consider the physical phenomena involved in its operation, and how it works in the engineering context of a field effect transistor,

8.1 The Physics of Semiconductor Devices

Our current understanding of the electrical properties of metals and other materials is based on the so-called "Band Theory" of solids. Loosely speaking, this theory predicts that the possible physical states that can be occupied by electrons within a material are arranged into a series of (effectively continuous) strata called "bands", each characterized by a specific range of energies for the allowed electron energy levels within it. These bands arise from the complex interplay of electrons with their parent atoms located within the atomic lattice of the material and are an intrinsically quantum mechanical effect. Electrons in different atomic states occupy different bands. In a general substance, we can identify two essentially distinct types of band relevant to the conduction of electric current: these are the "filled" or "valence" band, and the "conduction" band. States in the filled band correspond to electrons that are bound to their parent atoms and are effectively confined to a certain region within the material – they are not free to roam around. Electrical conduction occurs when electrons leave their parent atoms and are able to move freely through the conductor. Mobile electrons of this type are said to occupy states within the "conduction band". Typically, there will be a *discrete* energy gap between the filled and conduction bands. The size of this gap largely determines whether our material is to be classified as a conductor or an insulator, as we'll see. Let us examine the energy band structure shown in Figure 8.1.

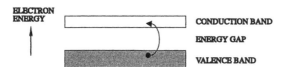

FIGURE 8.1
Band structure.

As you can see, we have valence and conduction bands separated by an energy gap – in the diagram, the energy associated with the bands increases as we move up vertically. When the lower band is full, the material acts like an insulator: there are no available energy states for electrons to gain energy from the applied electric field and form a current. To support an electrical current, we need electrons in the conduction band where there are plenty of empty states available. To produce such electrons, enough energy must be supplied to occupants of the valence band to help them leap above the gap and make the transition into the conduction zone. This minimum energy is called the "band gap energy" and its value largely determines the electrical properties of a substance, as I've said. Good conductors have a plentiful supply of free electrons under normal conditions, the band gap energy being tiny or nonexistent (filled and conduction bands can even overlap). Hence, it will not be difficult to excite a current in such a material. Insulators, however, have prohibitively wide gaps (several eV) and only conduct under pretty extreme conditions. There is, however, a third class of material needing consideration, and that is a sort of hybrid of conductors and insulators – the *semiconductor* – for which the energy gap is relatively small (1 eV or so).

The primary mechanism responsible for getting electrons out of the filled band and into the conduction band is *thermal excitation* (neglecting the application of external electric fields). This is simply the process whereby the energy changes of random thermal fluctuations are themselves enough to supply the energy required to enable electrons to make a transition. Typical thermal energy might be of the order of 25 meV and if this exceeds the band gap energy, it will be sufficient to cause transitions. This is the case for metals but not for insulators – with their large band gap energies of several eV. For any given material, we can calculate how likely it is for a thermal fluctuation to result in a conduction electron. If the temperature of the substance is T, and E is the band gap energy, then the rate at which electrons spontaneously pop up to the higher band is determined by the Boltzmann distribution and is proportional to $exp(-E/kT)$, where k is Boltzmann's constant. At room temperature ($T{\approx}300K$), we have $kT{\approx}1/40$ of an electron-volt or 25 meV. Note that, due to the exponential in the formula, this transition rate rises rapidly with temperature. Nonetheless, for most insulators, this rate remains negligible right up to near the melting point.

Let us take a look at a semiconductor. At zero degrees (and low temperatures generally), the semiconductor silicon (henceforth Si) is effectively an insulator. Its band gap is of the order of 1.1 eV and thermal transitions are rare. However, we can certainly excite a current by supplying energy to the valence electrons and when we do, we find something interesting happening, something which is of central importance in our study of semiconductors. When we excite an electron to the conduction band, not only does it become free to run around and give rise to some conductivity, but it leaves behind, in the lower band, a *hole*. This hole has an effective positive charge and, like the electron in the conduction band, is also able to move about and carry electric current: if a nearby electron fills the space vacated by the thermally excited particle, it will leave a positive charge in its own original location, as if the hole had moved sites. Holes are not "real" free particles – they are just empty spaces in the valence band that behave as if they are particles with positive charge. Holes also appear in insulators but rarely in metals.

There is a special trick that we can perform with Si that modifies its properties so that it is ideal for use in computers. This is the process of *doping.* Doping involves adding atoms of another substance (an "impurity") into the Si lattice.* A common dopant is the element phosphorus (P), which sits next to Si in the periodic table. P has a valency of five rather than the four of Si: this means it has five electrons in its outer shell compared with silicon's four. In an ordinary silicon crystal lattice, all four of these valence electrons play a role in holding the atom in place in the lattice and they are not free to move through the crystal – the valence band is fully occupied. When some impurity P atoms are introduced, each impurity atom bonds to four silicon atoms using up four of the five valence shell electrons of the phosphorus. This leaves an extra electron per P atom free to roam through the material and carry a current (Figure 8.2).

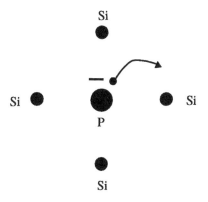

FIGURE 8.2
Liberation of a phosphorus electron during doping.

The resulting material is called an "n-type" semiconductor, as there is an excess of *negative* charge carriers. At modest levels of doping, substances of this sort conduct quite weakly compared with metals; the latter may have one or two free electrons per metal atom whereas an n-type semiconductor has but one electron for each phosphorus atom.

There are very few holes in n-type Si, even when the temperature is high enough to dislodge electrons thermally, because holes in the lower band are filled in by the P electrons preferentially before they fill levels in the conduction band. The venerable "Law of Mass Action", as used for chemical reactions, gives an important relation between electron and hole densities, n_e and n_h, respectively, and one which, interestingly enough, is actually independent of the fraction of dopant in the material:

---

* An undoped semiconductor is usually referred to as an *intrinsic* semiconductor. If it is doped, it is *extrinsic*. [Editor]

$$n_e \, n_k = n_{i^2} \tag{8.1}$$

where n_i is the density of electrons and holes at that temperature for pure, undoped Si. (This relationship is pretty obvious for undoped Si since we must have $n_e = n_h$.) Ideally, we would like to be able to design components that still work when material specifications such as n_i or the temperature are slightly but unpredictably changed.

Another type of doping involves replacing selected Si atoms with atoms from group 3 of the periodic table. Thus, we could add an impurity atom such as boron (B), which has one less electron than Si in its outer shell.* If we do this, then clearly we will find ourselves with an excess of *holes* rather than electrons, and another type of semiconductor. Due to the wonders of the laws of electromagnetism, holes can be viewed rather like bubbles of positive charge in an electric field – just as air bubbles in a liquid go up in a gravitational field (having an effective "negative weight"), so do holes go the "wrong way" in an electric field. Since they act like positive charges, B-doped Si is called "p-type" Si to indicate this. Note that, once again, relation (8.1) still holds.

8.1.1 The np Junction Diode and npn Transistor

We will now look at what it is about semiconductors that makes them useful in the manufacture of parts for computers. We start by examining the particularly interesting situation that occurs when slabs of p-type and n-type silicon are brought into *contact* with each other. This forms the basis of a device called a *diode*. We will give an idealized, qualitative discussion, and not allow ourselves to get bogged down in the murky details. We envisage a situation like that shown in Figure 8.3.

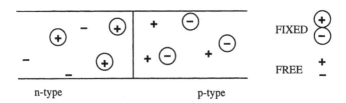

n-type   p-type

FIGURE 8.3
The np junction.

On the left-hand side, we have the n-type material, which we can view as comprising a bunch of fixed positive charges and free negative charges.

---

* Feynman actually used aluminum instead of boron in his lectures. Aluminum immediately precedes Si in the periodic table but has rarely, if ever, been used as a valence-3 dopant. [Editor]

On the right-hand side, we have the opposite situation. We know how many free electrons and free holes there are since they match up with the extra fixed B and P atoms – one electron per P, and one hole per B atom. At room temperature, there will also be extra carriers due to thermal fluctuations.

In a moment we'll stick this device into a circuit and put a voltage across it. First, let's see what's going on in the absence of any such field. The charge carriers will not only move about within their respective halves but will billow out, like steam escaping, into the adjacent material. However, this process of diffusion – of electrons into the p-type material and holes into the n-type – does not go unopposed. The fixed positive charges in the left-hand block will create an electric field that tends to pull the escaping electrons back; this field is increased by the holes percolating into the n-type material. These holes also experience a "tug", from the fixed negative charges in their original half, and the electrons that have migrated over the barrier. We can actually list four separate phenomena operating at the join of our two slabs:

(1) Creation and annihilation of electron-hole pairs by thermal fluctuations.
(2) Conductivity (carrier drift prompted by electric forces).
(3) Diffusion (carriers trying to smooth out the charge density).
(4) Electrostatic processes (due to the fixed charges).

After a while, this complicated physical system will settle down into an equilibrium state in which there is a concentration of fixed charges either side of the junction (Figure 8.4).

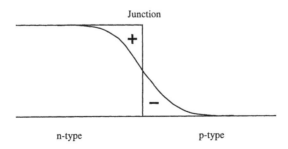

FIGURE 8.4
The equilibrium charge distribution for an np junction.

The central region is actually depleted of charge carriers and is referred to as the *depletion* region. The density of the fixed charges in this region is not quite mirror-symmetric, as P and B have different effective masses, but we will treat them as the same. We can add the signed densities of electrons, holes, and the fixed charges to obtain the net charge density in the device (Figure 8.5).

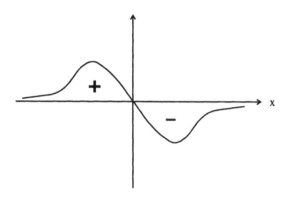

FIGURE 8.5
The net charge density.

The physical situation in equilibrium, it must be stressed, is not static. There are currents flowing due to diffusion. However, in equilibrium, the current flowing to the right cancels that to the left, resulting in no *net* current flow.

Let us see what happens if we apply a voltage across this system. We have two choices as to how we hook up our battery: we can connect the positive terminal to either the n-type or the p-type material. Let us consider connecting it to the p-type material first. If you think about it, you should be able to see that the effect will be to reduce the opposition to current flow caused by the fixed charges in the depletion region – the positive potential on the right will attract the electrons from the n-type material into the p-type. As the voltage increases, more and more electrons are able to diffuse across the boundary, and more and more holes, of course, can go in the opposite direction. Put bluntly, if we wire it right, our "device" conducts madly! (An important rider to this bald statement is that it is *essential* for the maintenance of a current that "external" free electrons be drawn into the n-type part of the material from the point of contact with the battery to continually replenish the flow. This is necessary because many of the "indigenous" carriers in the semiconductor will recombine with their opposite charges once over the boundary.)

What happens if we apply the voltage in the opposite direction? Now it gets interesting – we find that *the material does not conduct!* Why so? Well, any free electrons in the n-type material can happily go left, away from the junction region, and free holes in the p-type can go right, flowing out of the semiconductor and into the circuit. However, the application of the voltage has increased the height of the potential barrier across the depletion zone, in fact, to the point where *electrons in the p-type material cannot traverse it*. (Needless to say, neither can any electrons that might be sourced from the point of contact with the battery.) Similarly, holes cannot maintain a current to the right, so after an initial blip, the current just drops off. There are too few free carriers available in the right places to sustain it. We say that the voltage *reverse-biases* the junction: in the current flow condition, the junction is said to be forward-biased. We call this device a *junction diode* and it has the fundamental property that it conducts in one direction but not the other.

Is there absolutely no current when the junction is reverse-biased? Well, not quite – there will be some flow due to thermal electron-hole pair creation at the junction. We wait for it to happen, it happens, and the electron scoots off in one direction, the hole in the other. The magnitude of the current created will clearly be temperature dependent – it actually increases exponentially with temperature – and largely independent of the applied reverse voltage. If we wish, we can aid the thermal process in a reverse-biased diode by creating electron-hole pairs ourselves. The thermal current is typically so small that, if we do create any pairs, we can easily detect them over the thermal background. How we go about producing pairs depends on the magnitude of the semiconductor's band gap. In gallium arsenide (GaAs), for example, we can create pairs utilizing photons (in fact, this process is quite efficient even in Si). Naturally, the reverse is true: if we flip the electric field so that the diode becomes "forward-biased" and current can flow freely, electrons and holes move toward each other and annihilate, producing photons (in GaAs) or phonons (in Si). In this way, we can make semiconductor lasers and LEDs.

Recall that electrons and holes annihilate at a rate n_i^2. If this were zero, then in the reverse-biased case there would be no current since, by (8.1), n_e and n_h would have to vanish also! However, in the forward-biased case, current could flow by filling up the p-type material farther and farther from the junction with electrons. If the applied voltage were to flip again, these electrons would have to all be laboriously brought back, and the diode would no longer prevent current flowing the wrong way – the device would be acting like a large capacitor. Annihilation solves this problem by letting the electrons in the p-type material fill in holes rather than be stored at increasingly great distances from the junction. So, in fact, the diode would not really work without this annihilation process.

Let us return to the case when the field allows current to flow fully. It is possible to calculate the current I that flows through the diode as a function of the applied voltage V. The math isn't exceptionally difficult, but we will not go into it here (see suggested reading). The relationship turns out to be non-linear:

$$I(V) = I_0[\exp(q\ V / kT) - 1] \tag{8.2}$$

where V is the effective potential difference across the device – the voltage after allowing for a voltage in the "wrong" direction caused by the fixed charges at the boundary – and q is the magnitude of the charge on the carriers (in Coulombs) and I_0 is a constant. The graph of I against V can now be drawn (Figure 8.6).

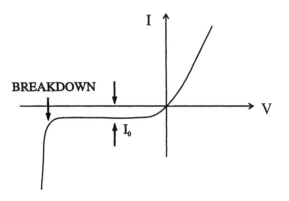

FIGURE 8.6
Variation of diode current with voltage.

In the real world, $I(V)$ cannot just keep on growing exponentially with V; other phenomena will come into play, and the potential difference across the junction will differ from that applied. Note also that the current trickle that exists in the reverse-biased case catastrophically increases (negatively) at a certain voltage, the so-called *breakdown* voltage. This can vary from as few as five volts to hundreds and can actually be exploited in some situations to limit the voltages in a circuit.

The next step in this chapter is to look at the operation of another semiconductor device, the famous *transistor*. Now, transistors come in all shapes and sizes, and those that are deployed extensively in VLSI chips are quite different from those that are used by hobbyists. As an example of this latter type, we mention very briefly the venerable *npn bipolar junction transistor*, one of several historical antecedents of more modern transistor devices. This transistor is formed by sandwiching an extremely thin slice of p-type material between two of n-type material (hence "npn"). The various slabs are denoted the base, the collector, and the emitter, as shown in Figure 8.7.

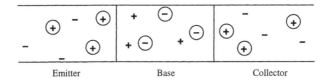

FIGURE 8.7
Structure of the npn bipolar junction transistor.

The "base" received its name from the fact that the transistor was built of this material: the terms "emitter" and "collector" both derive from prehistoric vacuum technology. Note the relative thinness of the base to the slabs on either side of it – this feature is essential to the transistor's correct operation. What this device is, if operated properly, is an *amplifier*: small changes in the current to the base are amplified at the emitter. It can also act as a switch and can be used in all of the transistor circuits discussed thus far in this book. However, bipolar transistors are not the most commonly used transistors in modern VLSI chips and I will therefore not discuss the (complicated) mode of operation of this transistor at the electron and hole level here. Good discussions can be found in many standard texts (see suggested reading). Instead, we will take a look at the type of transistor that *is* most commonly employed in VLSI systems – the sort of transistor that is usually built onto silicon chips. This is the MOSFET, an acronym for Metal Oxide Semiconductor Field Effect Transistor.

8.1.2 The MOSFET

We begin with a sketch of the structure of the MOSFET (and will worry about how to actually build such a device on a chip later).

The bulk of a silicon chip consists of a slab of lightly doped silicon, the substrate onto which the transistors and whatnot are laid down. If the doping is of the *p*-type, we are dealing with so-called *nMOS* technology; if the substrate is n-type, we have *pMOS*. We will focus on nMOS, as shown in Figure 8.8. We can identify three ports for our MOSFET: the *gate* is a

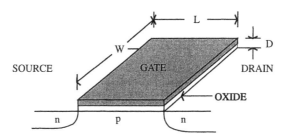

FIGURE 8.8
The MOSFET.

conducting layer of "polysilicon" (a substance rather like a metal), which is separated from the silicon by a thin layer of non-conducting oxide. To either side of the gate, also separated from it by the oxide layer, are two n-type *diffusion layers*, materials so-called because they have doping diffused into them (as illustrated): they are referred to as the *source* and *drain* – as opposed to the emitter and collector of the npn device. These diffusion layers also conduct.

The transistor works as easy as pie. From our discussion of the np junction diode, we can see roughly what's happening inside the device before we apply any external voltages. Hie substrate material is lightly p-type and is usually kept grounded. The substrate forms a diode-like structure with the n-type layers – in essence, the MOSFET is built from two back-to-back diodes. As before, an equilibrium state will arise in which there are depletion regions at the n-p junctions with very few electrons and holes diffusing across them. So, the source and drain are effectively cut off from one another and if a voltage is put across them no current will flow. However, if we now put a positive voltage on the gate, things are different. The effect of such a voltage is to attract electrons to the underside of the oxide (they won't conduct into the gate because the oxide layer is an insulator). These electrons chiefly come from the source and drain – the positive potential having lowered the barriers at the depletion layers that prevented their free movement. The electrons under the oxide form what is called an *inversion layer*. If we *now* put a voltage across the source and drain, we *will* get a current – the inversion layer is essentially a channel that allows electrons to flow freely between the two contacts. So we have a fantastic device – a switch! The voltage that controls it is the gate-source voltage V_{gs} – the bigger this is, the more charge carriers there are under the gate, and the more current that can flow. Note, however, that if $V_{gs} < 0$, the MOSFET will not conduct. In such a case, electrons are repelled from beneath the gate, and an inversion layer cannot be formed.

Actually, we have been a little simplistic here. Simply having $V_{gs} > 0$ does not automatically allow a current to flow. It is necessary for V_{gs} to exceed a certain minimum voltage, the *threshold* voltage V_{th}, before this happens (typically, V_{th} is of the order of $0.2V_{DD}$, where V_{DD} is the supply voltage, say five volts). Because of this, it turns out to be convenient to define a shifted gate-source voltage:

$$V'_{gs} = V_{gs} - V_{th} \qquad (8.3)$$

in terms of which the condition for current flow is $V'_{gs} > 0$. We can actually design our MOSFET to make this threshold voltage either positive or negative. A negative threshold voltage is obtained by doping the silicon slab so that there is a thin conducting layer of n-type semiconductor under the gate, connecting the source and the drain. Transistors with $V_{th} > 0$ such as we discussed above are called "enhancement mode" transistors: if $V_{th} < 0$, they are called "depletion mode" transistors. (Depletion mode transistors turn out to be useful for fabricating resistors in nMOS VLSI, as we shall see later.)

Let's make all this a bit more quantitative. Suppose we want to find the drain-source current I_{ds} for given gate-source and source-drain voltages. Those readers not interested in details can skip this as we will not need it subsequently – it's just a nice bit of physics! We can consider the gate/oxide/silicon sandwich to be a capacitor, modeled by two conducting plates of area $A = WL$ (where W is the width of, and L the distance between source and drain), separated by a material of depth D and permittivity ε. Let us denote the charge on this capacitor (i.e., the charge under the gate) at any time by q_g. We can calculate the capacitance C_g for this system, using the well-known general formula for a parallel plate capacitor $C = \varepsilon\, A/D$. We have:

$$C_g = \varepsilon\, WL / D. \qquad (8.4)$$

Using the standard relationship between the voltage across a capacitor and the charge stored in it, we can write:

$$q_g = C_g\, V'_{gs} \qquad (8.5)$$

Let us suppose first that the drain-source voltage V_{ds} is small. We know that the current I_{ds} is just the charge under the gate divided by the time it takes for the electrons to drift from the source to the drain. This is a standard result in electricity. How long is this drift time? Drawing on engineering practice, we can write the drift velocity v_{drift} in terms of the "mobility" μ of the charge carriers as $v_{drift} = \mu\, E$, where E is the electric field across the drain/source. E is easily seen to be given by V_{ds}/L. We can now straightforwardly find the drift time, which we denote by τ :

$$\tau = L / \mu E = L^2 / \mu V_{ds} \qquad (8.6)$$

Combining this expression with that for the charge q_g in (8.5), we find that the current (charge divided by time) is given by:

$$I_{ds} = (\mu \varepsilon W / LD) V'_{gs}\, V_{ds} \qquad (8.7)$$

Our calculation has been a little simplistic: this result strictly only holds for small values of the source-drain voltage V_{ds}. However, we can see that as long as V_{ds} is fairly small, the transistor has the interesting property that *the current through it is proportional to the applied voltage*. In other words, it effectively functions as a *resistor* (remember $V = IR!$), with the resistance proportional to $(1/V'_{gs})$.

As the drain-source voltage increases, matters become more complicated. The drift velocity of the charge carriers depends upon the electric field E, and this in turn is determined by both V_{ds} and V'_{gs}. However, it so happens that, if V_{ds} gets too big, the current across the transistor actually becomes *independent* of V_{ds}, a phenomenon known as *saturation* – at this point, the current becomes proportional to $(V'_{gs})^2$. We can understand this strange phenomenon better by the use of a fluid model analogy (described in more detail in the book by Mead and Conway). If you like water and gravity better than electricity, you should like this!

Let us envisage two reservoirs of water, separated by a partition. We will actually take this water to be without inertia – a little like honey – so it will flow slowly, and not splash about all over the place. To begin, consider the state of affairs shown in Figure 8.9 where the "water" is on either side of a movable central partition:

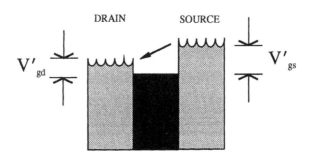

FIGURE 8.9
Fluid analogy for the forward-biased MOSFET.

The diagram has been annotated with MOSFET-related words to force the analogy on the reader. In this case, there will clearly be a flow of water from the "source" to the "drain": in the transistor, this equates to a current, with the partition playing the role of the potential barrier the charge carriers have to overcome. In this situation, the transistor is said to be "forward-biased". The height of the water column above the partition on the right represents V'_{gs}, of the left column above the partition, V'_{gd}, and so on. Note that the precise nature of the flow across the partition will depend on the level of the "water" in the drain reservoir. Consider now a second state shown in Figure 8.10.

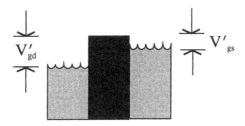

FIGURE 8.10
Fluid analogy for the back-biased MOSFET.

In this case, there will be no flow at all: for the MOSFET, this represents the "back-biased" case, where $V_{gs} < 0$. Third, consider the case of saturation (Figure 8.11).

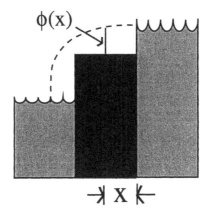

FIGURE 8.11
Fluid analogy for the case of saturation of the MOSFET.

In this instance, where the level of the drain reservoir is below the level of the partition, the water from the source will simply "waterfall" into the drain, at a rate *independent of the actual drain-level*. In the MOSFET, this behavior would begin to occur when the "drain reservoir" and the "partition" were at the same height, for example, when $V_{ds} = V'_{gs}$. At such a point, the current flow will become constant, irrespective of V_{ds}.

We can implement this analogy physically to gain some insight into the process of saturation. This is an alternative derivation of the magnitude of the current flowing through the MOSFET. Let us take Figure 8.11 to define a voltage $\varphi(x)$ beneath the gate, which, due to the energy level structure of Si, happens to be proportional to the number of free electrons under the gate. Now the electric field under the gate is proportional to the derivative of this voltage $(d\varphi/dx)$. Hence, since the current must depend on the density of electrons multiplied by the electric field (which controls the electron drift velocity), we must have:

$$I = K\phi(x)d\phi/dx = (K/2)d/dx(\phi^2) \tag{8.8}$$

where K is a constant. Now if you think about it, you should be able to see that the current I has to be a constant, independent of x. This means that the function φ^2 must be linear in x and thus $\varphi \propto \sqrt{x}$. We therefore obtain the general expression:

$$I = (K/2L)(\phi^2[0] - \phi^2[L]) \tag{8.9}$$

However, at saturation we have $\varphi(0) = V'_{gs}$ and $\varphi(L) = 0$ so that we recover the quadratic dependence of I on V'_{gs} that we flagged earlier. What happens if we are not at saturation? Will we rederive (8.7)? Actually, no. We get something better. In the unsaturated case, $\varphi(L) = V'_{gd}$, the gate-drain voltage, and we find that:

$$I \propto (1/2)(V'_{gs} + V'_{gd})V_{ds} \tag{8.10}$$

where $V_{ds} = V'_{gs} - V'_{gd}$. Why is this an improvement? It removes an anomaly in our earlier expression (8.7). In that expression, the current I was strangely asymmetric between source and drain. Intuitively, one might prefer to replace it with an average over the gate-source voltage and the gate-drain voltage – that is just what (8.10) amounts to. For the record, we give here a plot of the variation of the current with the various voltages we have considered in this section (Figure 8.12).

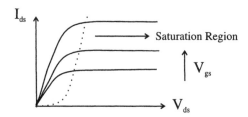

FIGURE 8.12
Current-voltage variations for MOSFET.

Another question we can ask about the MOSFET is: in the state for which no current flows, how good an insulator is the device? You will remember that the junction diode permitted a small thermally induced current to flow even when reverse-biased. Let us briefly discuss this effect in the transistor. Recall Figure 8.10, the "water" diagram for the back-biased case.

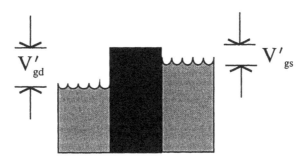

The probability of an electron jumping from the source to the drain must be proportional to the probability for it to have enough thermal energy for it to overcome the potential barrier V'_{gs}, namely, $\exp(-qV'_{gs}/kT)$. Thus, the forward current will also be proportional to this factor. There will also be a backward current – we can similarly see that this will be proportional to $\exp(-q[V'_{gs} + V_{ds}]/kT)$. There will therefore be an overall current given by

$$I = (constant).\exp\left(-qV'_{gs}/kT\right)\left[1-\exp\left(-qV_{ds}/kT\right)\right] \qquad (8.11)$$

when $V'_{gs} < 0$. So, if we switch from a current-flow to a back-biased case, the current will not switch off instantly (note that the temperature of the device is likely to be higher than room temperature). However, if we have $kT \sim 1/40$ electron-volt, then the current turns off quite quickly for circuitry run at about 5 volts. We can amend Figure 8.12 to include the reverse current (Figure 8.13).

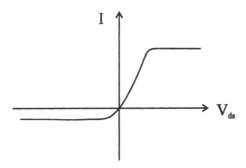

FIGURE 8.13
Transistor current as a function of source-drain voltage.

Let us take stock and summarize what we have learned about the properties of MOSFETs. First, rather than redraw the physical picture of Figure 8.8
every time we want to discuss the device and draw circuits involving it,
we'll need a diagrammatic symbol. There are many such symbols for
the differing varieties of transistors encountered in VLSI technology.
Figure 8.14 illustrates the fairly common convention for the MOSFET that
we shall adopt.

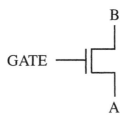

FIGURE 8.14
Conventional MOSFET symbol.

This can represent either an n- or p-type transistor. (We will show later
how to amend this figure to indicate which – this is necessary in CMOS
technology which uses both types.) The rule for both is that they act
like switches – when the gate voltage V_g is positive enough, they conduct. For the n-type MOSFET, we have the following rules: the most
negative of A and B is called the source, the other called the drain; and
if the gate voltage is more positive than a certain threshold voltage, V_{th},
above the source voltage the device conducts – the switch is "closed"
and current flows. For a p-type device, the most positive of A and B
is the source and the transistor conducts if the voltage on the gate is
more *negative* than a certain threshold *below* the source voltage. We also
defined two modes of operation for a MOSFET – the *enhancement* and
depletion modes. The former is the case when the threshold voltage, V_{th},
> 0 for n-type and < 0 for p-type. In depletion mode, it is the other way
around. Now, a nice feature of depletion mode MOSFETs is that, if V_g =
V_s (the source voltage), the device always conducts. Thus, if we directly
connect the source to the gate so that each is automatically at the same
voltage, we find our transistor acting no longer as a switch but as a
resistance (Figure 8.15).

Why should we want to do this? It's a matter of simple economics and
design. Implementing a standard resistance on a chip is both expensive
and takes up a lot of space, neither of which are obstacles to the use of

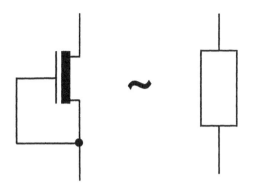

FIGURE 8.15
Depletion mode MOSFET operating as a resistor.

depletion mode MOSFETS.[*] (Note, incidentally, that this trick would not work with an enhancement mode device.)

Problem 8.1 I will now set you a problem – I will actually give you the answer shortly, but you might like to work through the math to get some practice on the theory of the internal guts of these things. The question has to do with capacitance. In the diode, not only was the current a non-linear function of the voltage but so was its capacitance. This is also the case for the MOSFET. Now, although most of the capacitance is in the oxide, the overall set up turns out to be actually highly non-linear – and quite interesting! The problem I am about to give you is designed to illustrate the nature of this non-linearity.

We model the electrode contact at the gate in the MOSFET as shown in Figure 8.16. Suppose we have a large mass of lightly doped p-type material – in principle, this material should be of infinite depth (measured

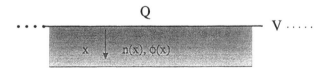

FIGURE 8.16
Schematic model of the gate-semiconductor electrode contact.

---

[*] With the advent of CMOS technology (see later), nMOS depletion mode transistors are now rarely used: they are replaced by p-channel enhancement mode transistors. [Editor]

vertically in the diagram). On top of this, we place a metal plate, also of very large extent (but now in the horizontal plane):

We put a positive voltage V on the plate and, as a result, negative charge carriers are attracted to its underside. Deep into the material, where the electrostatic forces due to the plate are negligible, the number of negative carriers will just equal the number of doping ions. However, because of the voltage on the plate, the density of carriers near to it will be greater, falling off to a constant as we go further away. If we label the depth into the material by x, we can define a position-dependent carrier density $n(x)$. We can also define a resulting electrical potential within the material, $\varphi(x)$. Finally, let us denote by n_0 the initial positive doping concentration. Now, the question I want you to answer is: how much charge is there on the electrode? Put differently, what is the capacitance of this physical set up?

To help you practice, I'll give you some hints. First, you have to take $\varphi(0)$ (i.e., at the electrode) to be the plate voltage V, and take the idealized value $\varphi(\infty)$ to be zero. I will hand you on a plate a relationship between $n(x)$ and $\varphi(x)$ resulting from thermodynamical considerations:

$$n(x) = n_0 \exp\left[q\phi(x)/kT\right] \tag{8.12}$$

where q is the charge on the negative carriers, and T is the temperature, as usual; n_0 is a constant. Another essential relationship is that between the rate of change of $\varphi(x)$ as we go deeper into the material (in other words, the electric field within the semiconductor) and the charge density on the plate, Q. Note that defining a charge *density* here is important – it would be meaningless to discuss the *total* charge for an *infinite* plate. We find (cf. (8.4) or by Gauss' theorem) the result that at $x = 0$:

$$\partial\phi / \partial x = Q / \varepsilon \tag{8.13}$$

where ε is the *permittivity* of the doped material and determines how rapidly the electric field drops off with distance from the plate. Using the standard Poisson equation $\partial^2\varphi/\partial x^2 = -\rho(x)$, where $\rho(x)$ is the charge density within the material (which you can find in terms of $n(x)$ and n_0) and integrating using the boundary conditions, you should find the result of the form of (8.14):

$$Q = V[2(e^V - V - 1)/V^2]^{1/2} \tag{8.14}$$

in some set of units ($kT/q = 1$ and $n_0 q\varepsilon = 1$). The rather odd appearance of a V^2 within the square root, which you might think ought to cancel with the V outside it, is necessary to get the correct sign for Q. Now you can see by comparison with the standard formula defining capacitance, $Q = CV$, that the capacitance of this system displays an extremely non-linear relationship with the plate voltage, V. To my knowledge, this property is not much exploited in VLSI – although there have been recent applications in "hot-clocking" (which we discuss later).

Thus far, we have considered an isolated MOSFET device on a silicon substrate. The next stage in our journey into the heart of VLSI is to take a look at how these transistors might actually be put together on chips to make logic circuits. We now come to real machines!

8.1.3 MOSFET Logic Gates and Circuit Elements

To build logic circuits we need to be able to build logic gates, and we have already seen, in Chapter 2, how to do this using generic "transistors". We use the same approach with MOSFETs. Consider what happens when we hook up a transistor to a supply voltage, V_{DD} across a resistance as shown in Figure 8.17.

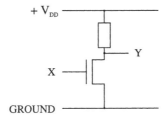

FIGURE 8.17
Inverter circuit.

We will take our transistor to be of the nMOS variety, operating in enhancement mode. (There are many types of VLSI design, and we cannot consider all of them – it makes sense to focus on one in particular.) If terminal X (the gate) is near zero, then the transistor is an insulator, and the output voltage at Y is near the supply voltage, V_{DD}: we interpret this state of affairs as meaning that the output Y is at logical 1. However, if X is near V_{DD}, then the transistor conducts. If we suppose it conducts much better than the resistance, then Y is near zero: this state we equate with logical 0. As a rule, we do not operate between these extremes, except perhaps temporarily. This single MOSFET device therefore operates as a NOT gate (an inverter), as we saw in Chapter 2, since it just flips the input signal.

We can follow Chapter 2's lead for the other canonical gates. For example, the NAND (NOT AND) gate is built as follows (Figure 8.18).

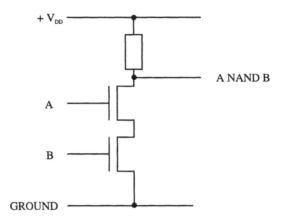

FIGURE 8.18
The NAND gate realized by MOSFETS.

In this system, both inputs *A* and *B* must be logical 1 for the output *Y* to be 0. To get the AND gate, we obviously just tag an inverter onto the output. To remind yourself of how to get a NOR gate, check out Chapter 2!

One can build other useful elements onto chips using MOSFETs apart from logic gates. Consider the matter of resistors – as I stated earlier, it turns out to be expensive and area-consuming to put standard forms of resistor onto silicon chips so it is normal practice to employ depletion-mode transistors in this role. Thus, in nMOS technology, the MOSFET structure of our inverter would actually be that shown in Figure 8.19.

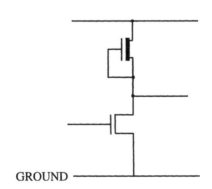

FIGURE 8.19
nMOS inverter with MOSFET resistance.

Now there is another essential property of MOSFETs that is not evident from strictly logical considerations. This is their behavior as *amplifiers*. Consider what happens if we place two inverters in sequence (Figure 8.20).

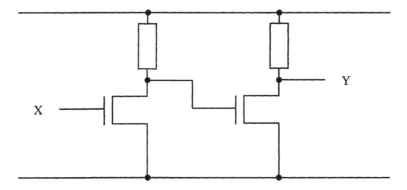

FIGURE 8.20
"Follower" circuit.

From a logical viewpoint, this is a pretty trivial operation – we have just produced the identity. We're not doing any computing. However, from the viewpoint of machinery, we have to be careful; transistors dissipate energy, and one might naively think that the output of a chain of devices such as this would ultimately dwindle away to nothing as the power dropped at each successive stage. This would indeed be disastrous! However, this clearly doesn't happen: the input current to the second transistor may drop slightly, but it will not be enough to alter the mode of operation of this transistor (i.e., conducting or not), and the output Y will still be pulled up to the supply voltage (or down to ground, whichever is appropriate). In other words, the output will always represent a definite logical decision, being relatively insensitive to minor power fluctuations along the chain. This circuit is an extremely effective so-called "follower", which jacks up the power or impedance behind the line (if you like, it is a double amplifier). In a sense, we can control the whole dog just by controlling its tail. Needless to say, this amplifying property is crucial to the successful operation of circuits containing thousands or millions of transistors, where we are constantly needing to restore the signals through them. The presence of amplification is essential for any computing technology.

With VLSI, as with other areas of computing, we are often concerned with matters of timing. In this regard, it is interesting to ask how fast an inverter can go. That is, if we switch the input at the far left of a chain of connected inverters, what happens at the output on the right? The switching certainly won't be instantaneous: the output of each transistor must feed the input of another and charge up its gate, and this will take time. Each gate voltage must be changed by some value V with the gate having some effective capacitance C_g, say. If we can find how long the process takes, and maybe think up ways of speeding it up, we might be able to get better machines. We can shed some light on this process by examining the circuit depicted in Figure 8.21, in which we have explicitly inserted a capacitor to represent the gate capacitance.

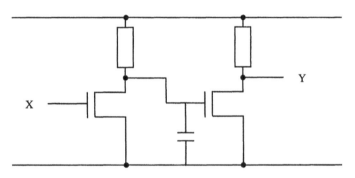

FIGURE 8.21
Effective electrical analog of a follower circuit.

Suppose the accumulated charge needed for a decision (i.e., for the gate voltage to be adequate for the transistor to switch) is Q. Then, $Q = C_g V$. How fast can we deliver this charge or take it away? First, note that the state $X = 1$ does not correspond to the first transistor's output being exactly at ground; the transistor will have a certain minimum resistance (which we call R_{min}) resulting in a slight voltage drop across the device. Now, it is a standard result in electronics that the discharge time is determined by the product $R_{min}C_g$, assuming an analogy with the standard RC circuit shown in Figure 8.22.

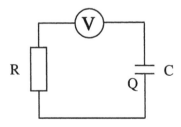

FIGURE 8.22
Equivalent RC circuit.

Again, from standard circuit theory, the charge Q on the capacitor at time t, $Q(t)$, obeys

$$Q(t) \propto \exp\left(-t / R_{min}C_g\right) \equiv \exp(-t / \tau), \tag{8.15}$$

with $\tau = R_{min}C_g$.

Clearly, if we were interested solely in getting the inverter to go faster, then we could achieve this by decreasing both R and C, something we could do by making the circuit smaller. However, there is a limit to this: recall that, even in an inactive state, electrons from the source and drain nonetheless seep a small distance into the silicon substrate of the MOSFET. As we shrink the device down, these carriers drift closer and closer to the opposite pole, until there comes a point where they actually *short-circuit* the region under the gate, and we will get a current flowing without having to manipulate the gate voltage. When this happens, it is back to the drawing board: a redesign is now needed, as the transistor will no longer work the old way. This is a nice example of how Nature places limitations on our technology!

So what do we do if we want to build smaller machines? Well, when the rules change, redesign, as I have said. Consider, for example, the case of aeronautical engineering with incompressible air and low-speed aircraft. A detailed analysis concluded that propeller-based machines would not work for speeds in excess of that of sound: there was a "sound barrier". To get a faster-than-sound plane, it was necessary to go back pretty much to square one. At this moment in time, we have yet to find a fundamental limit on sizes for Si computers – there is no analog of the sound barrier. This problem is an instance of how thinking differently from everyone else might pay dividends - you might blunder into something new! Currently, state-of-the-art devices have $RC \approx 10$ picoseconds. By the time you have managed to reduce this significantly, you'll probably find that others will have undercut you using some other technology! This actually happened with super-conducting computing devices: as researchers were working in this area, its advantages were continually disappearing as advances were made in conventional VLSI technology. This sort of thing is quite a common occurrence.

Thus far in this chapter, we have reviewed the structure of various semiconductor devices used in computing but have so far had little to say on the practical limitations in this area. We address some examples of this now, beginning with a discussion of the important topics of heat generation and power loss in computers.

8.2 Energy Use and Heat Loss in Computers

In Chapter 5, we pointed out that a typical transistor dissipates some $10^8 kT$ in heat per switch. This is a phenomenal amount – if we could get it down by a factor of ten or 100, we could simplify our machines considerably just by getting rid of all the fans! One particularly annoying problem with the nMOS technology we have discussed up to now is that even in the steady state of a MOSFET's operation – when X=1 (Y=0), say, and the transistor is merely *holding* this value, not changing it – *current flows continuously*. So even if our transistors aren't doing anything, they're throwing away power! Obviously, any technology that offers the hope of more economical behavior is worth exploring; and the *complementary metal oxide semiconductor* (CMOS) technology that we will look at in this section is just such a technology.

8.2.1 The CMOS Inverter

In the CMOS approach, we employ a mixture of n-type and p-type MOSFETs in our circuitry. The way in which we combine these to make a standard CMOS inverter is shown in Figure 8.23. As with the nMOS inverter, logical 1 is held to be near +V, for some voltage V, but logical 0 is not at ground but can be chosen to be at −V.

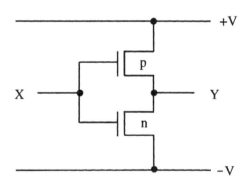

FIGURE 8.23
The CMOS inverter.

To indicate the doping type of each MOSFET (n or p) we have followed the convention of writing the appropriate letter adjacent to its symbol. Note that the nMOS depletion mode transistor has effectively been replaced by a conventional p-channel transistor. Is this circuit worth building? Yes, for the following reason. Suppose the input X is positive. Then the n-type

MOSFET has its gate voltage above that of the source and it conducts: the p-type device, on the other hand, is reverse biased and therefore *doesn't* conduct. The output Y is pulled down to -V. Now switch X to zero. As you can see from Figure 8.23, the upper transistor now conducts and the lower doesn't; the voltage Y rises to the supply. So far, nothing new – this is just the standard operation of an inverter. However, this circuit has a novel feature: specifically, after the transition occurs, *no current flows through the circuit!* The route to −V is cut off by the insulating n-type MOSFET. (I'll leave it to the reader to see what happens when the input is switched back again.)

This is a remarkable property. In a CMOS inverter, no energy is required to *hold* a state, just to *change* it.[*] The CMOS inverter can also serve as a useful simple "laboratory" for investigating some of the energetics of logic gates. The matter of how much energy is required for a logical process was considered in the abstract in Chapter 5, but it is obviously important to get a handle on the practicalities of the matter. We would, unsurprisingly, like our devices to use the very minimum of energy to function – and to this end we will have to consider the amount of energy required to make a decision, the time taken in switching, the reliability of our components, their size, and so forth. Let's start by considering in more detail the electrical behavior of a CMOS inverter as part of a chain. This will enable us to also examine the amplification properties of CMOS devices. To proceed, we will employ a simplified (and none too accurate) model due to Mead and Conway.

In this model, we treat the two transistors simply as controlled resistors. We thus have the following equivalence (Figure 8.24).

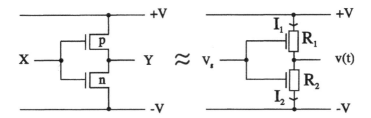

FIGURE 8.24
Simplified model of a CMOS inverter.

---

[*] Strictly speaking, there will be a small current flowing through the reverse-biased transistor, but we largely neglect this in our considerations. [RPF]

This CMOS device is to be visualized as one in a linear sequence. The input at X is fed in from the previous gate: the output at Y is to be considered the input to the next gate, which has an effective capacitance to ground of C, say (which we take to be a constant, although this isn't strictly true). Ultimately, we want to examine the behavior of the output voltage as we vary the gate voltage, $V_g(t)$, at X – i.e., as we perform a switch. Let us first consider the simple case where we keep the voltage to the I input gate constant. This will prompt a flow of current. What will the final, equilibrium voltage at the output be? Denote the

currents through the transistors by I_1 and I_2, and define the difference between them (that is, the current that transfers charge to any subsequent component connected to Y) to be $I = I_1 - I_2$ The voltage at Y is a function of time, say v(t). Let us also take the charge accumulating at Y to be Q(t). From standard circuit theory, we know that:

$$dQ / dt = I_1 - I_2 = Cdv(t) / dt \tag{8.16}$$

and from (8.11), we can see that for small drain-source voltages the currents I_n are given by:

$$I_n \approx V_{ds\,n} / R_n \tag{8.17}$$

where the interpretation of $V_{ds\,n}$ is obvious, and the effective resistances R_n are given by

$$R_1 = R_0 \exp(qV_g / kT), R_2 = R_0 \exp(-qV_g / kT). \tag{8.18}$$

Note that we have sneakily removed all signs of the threshold voltage in V_g – we are considering our devices to be somewhat ideal.

If we now combine the basic equations for I_1 and I_2 given below:

$$I_1 = (V - v) / R_1 \,, I_2 = (v + V) / R_2 \tag{8.19}$$

with (8.16) and (8.18), and setting $V_T = kt/q$, we can straightforwardly derive a differential equation relating C, v(t), and R:

$$Cdv / dt = -(2V / R_0)\sinh(V_g / V_T) - (2v / R_0)\cosh(V_g / V_T). \tag{8.20}$$

So, if we keep the voltage on the gate fixed, what is the equilibrium value at the output, that is, the value it has when everything has settled down? Well, when everything has stopped sloshing about, dv/dt = 0, and we see directly that the equilibrium value, v_e say, is given by:

$$v_e = -V \tanh\left(V_g / V_T\right) \tag{8.21}$$

where v_e is a constant. Since V_g/V_T is typically a large positive number or a large negative number, the equilibrium voltage asymptotically approaches $+V$ or $-V$.

We can use this result to analyze the amplification properties of a chain of CMOS inverters. Suppose we vary the gate input slightly, say let $V_g \rightarrow V_g + \delta V_g$. In response to this, the output will vary by some amount, which we denote $\delta v_e = A.V_g$. In response to this, the output of the gate fed by v_e, v' say, will itself vary, by $\delta v' = Av_e = A^2 V_g$; and so on down the chain. Clearly, if this CMOS device is to work, it must be the case that the magnitude of this factor $|A|$ is *greater than one*: if it were not, then any change of input at the left hand of the chain would not propagate all the way through and eventually peter out. The amplification factor A is the slope of the graph of v_e against V_g at the origin, $V_g = 0$ (Figure 8.25).

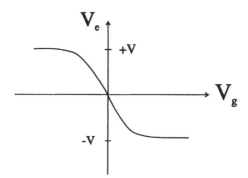

FIGURE 8.25
Gate versus output voltages for the CMOS inverter.

The slope at the origin is $-V/V_T$ (as you can show). Hence, we only need our DC supply voltage to exceed the order of 1/40th of a volt for the chain to work. In practice, of course, the supply voltage is much higher (say five or six volts) so we see that the amplification is quite significant. The output voltage is an extremely sensitive function of the input since small input changes are magnified many times at the output.

Problem 8.2 Here are some problems, not easy, for you to try. So far, we have considered the equilibrium behavior of a CMOS circuit. What I'd like you to do now is analyze its behavior in time, by solving (8.20) to find how long it takes the output to switch if we switch the input. The general solution, for which V_g is an arbitrary function of time, is obviously too difficult, so assume in your calculation that the input voltage switches *infinitely rapidly*. Next, consider the dissipation of energy in the inverter. I stated earlier that, while it is a useful qualitative idealization to think of no current flowing through the circuit in equilibrium, this is not actually the case (indeed, our previous calculation presumes otherwise!). The reverse-biased transistor just has a very high resistance. This results in a small perpetual power loss, which you can find using the standard electrical formula for the power dissipated by a voltage drop V across a resistance R: V^2/R, where R is the "non-conducting" resistance (alternatively, you could use I^2R, where I is the leakage current). There will also be power loss in the switching process – this occurs when we dump the charge on the gate through the (now conducting) resistance. You should find the energy lost during switching to be $2Cv_e^2$. Also, what is the time constant τ of the effective gate capacitance?

Although we are interested in CMOS technology chiefly for what it can tell us about the energetics of VLSI, for completeness I will briefly digress to illustrate how CMOS inverters can be used to construct general logic gates. Consider the implementation of a NAND gate – remember, if we can build one of these, we can build everything. A NAND gate then results from the arrangement shown in Figure 8.26.

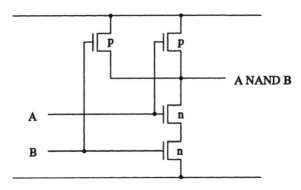

FIGURE 8.26
NAND gate realized in CMOS.

Let us see how this works. Recall that, for a NAND gate, the output is zero if both inputs are one, and one for all other inputs. That is clearly what will occur here: the output voltage in Figure 8.26 can only be pulled down to $-V$, i.e., logical zero, *if both of the lower transistors conduct*. This can only occur when both inputs are positive. If either input is negative, the respective transistor will fail to conduct, and the output voltage will stay at $+V$.

Let us return to the matter of energy dissipation in CMOS devices. In practice, the energy dissipated per switch is of the order of $10^8 kT$. This is very big so here is an opportunity for people to make a splash in the engineering world: *there is no reason why it should be so high.* Obviously, the voltage must be a certain size depending on the technology implemented in our devices, but this is not a fundamental limitation, and it should be possible to decrease the energy dissipated. (Remember our analysis in Chapter 5 where we saw that $kT\log2$ was theoretically attainable.) Let us discuss what can be done in this area.

Consider what actually happens in the switching process. Before we make the switch, there is a voltage on the input capacitance and a certain energy stored there. After we switch, the voltage is reversed, but the energy in the capacitance is the *same* energy. So we have done the stupid thing of getting from one energy condition to the same energy condition by dumping all the juice out of the circuit into the sewer, and then recharging from the power supply! This is rather analogous to driving along the highway at great speed, slamming on the brakes – screeech! – until we come to a halt; and then pushing the car back up to speed again in the opposite direction! We start off at 60 miles an hour, and we end up there, but we dissipate an awful lot of energy in the process. Now, in principle, it should be possible to put the energy of the car into (say) a flywheel and store the energy. Then, having stopped, we can get started again by drawing power from the flywheel rather than from a fresh source. We shouldn't have to throw the energy away. Is there some parallel in VLSI to this flywheel?

One suggestion is to store the energy in an inductance, the electrical analog of inertia. We build the circuit so that the energy is not thrown away but stored "in a box" so that we can get it out again subsequently. Is this possible? Let's see. To explain the concept of inductance, I'll turn to another useful analogy using water. Those of you who are electrically minded are used to analogies between water and electricity: those more comfortable with mechanics than electricity will also find water is easy! Imagine we have the arrangement shown in Figure 8.27, consisting of a large water-holding vessel with a couple of pipes leading into it.

Each pipe is connected to an essentially bottomless reservoir (not shown), into or from which water can flow – this flow is regulated by a valve on each pipe. The analogy here is that the pipes plus valves represent the transistors, and the water in the reservoirs is charge from the power supply just waiting to be dumped through them. The upper reservoir corresponds to the voltage $+V$, the lower to voltage $-V$, and the height of the water in the tank can be interpreted as the voltage through which the charge will be dumped. To keep the analogy meaningful, the valves are rigged so that if one is open (conducting), the other is closed (insulating). To model the switching process in an inverter, we open and close the valves in this system and see what happens.

The initial condition is that shown in Figure 8.27, with the upper valve closed and the lower valve open. The water sits at some equilibrium level. Suppose we now switch the system by closing the lower valve and opening the upper (corresponding to a negative gate voltage). The water from the upper reservoir rushes in – sploosh! – filling up the tank until a new equilibrium depth is reached. In the process, there is noise, friction, turbulence, and whatnot, and energy is dissipated. There is a power loss. Eventually, everything settles down to a fresh equilibrium point. We now want to go back to our initial situation, so we switch again, opening the lower valve and closing the upper. Down comes the water level, dissipating energy in a variety of ways, until the water in the tank reaches its original height. We are back where we started, but we have used up a heck of a lot of energy in getting there!

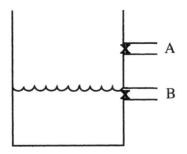

FIGURE 8.27
Water analogy for the CMOS switch.

We would like to alter this set-up so that we don't lose so much energy every time we switch. One way we could do this is as follows. We put another tank next to the first and join the two with a tube containing a valve (Figure 8.28).

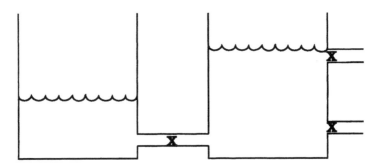

FIGURE 8.28
Energy-saving analogy for the CMOS switch.

Suppose we have the upper valve open so the water level of our original tank is as shown in Figure 8.28. If we now close the upper valve and open up the valve into the adjoining tank, the water goes splashing through the connecting tube into the new tank. When the water level reaches its maximum height in this tank, we close the valve. If we were to just leave the adjoining valve open, the water would slosh back and forth between the two tanks and eventually settle down into a state where the height in both tanks was the same. In this case, the pressure would be equalized but this finite time to stability results from the fact that water has inertia. When the valve is first opened, the water level reaches a height in the new tank that is higher than what would be the equilibrium value if we let the system continue sloshing about. Likewise, the initial level in our original tank will be lower than its equilibrium value. By closing the valve just after this high point is reached, we have actually managed to catch most of the water, and hence its potential energy, in the new tank. Not all of it, of course – there will be losses due to friction, and so on, and we might have to top the new tank up a bit. But now if we want the energy of the water back, we just have to open up the adjoining valve to the adjoining tank when the right-hand tank is at a low ebb.

To implement this in silicon we need the electrical analog of this and that means we need the analog of inertia. As I've said, for electricity this is inductance. One way to implement the above idea can be seen by considering the following circuit (Figure 8.29).

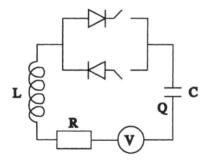

FIGURE 8.29
An inductive circuit.

This circuit contains a capacitor, an inductance, a resistance, and two "check valves", based on diodes. When one of the switches is closed, the diode ensures that the current can only flow one way, mimicking the one-way flow of water through the two pipes in the water model. You should be familiar with the basic equation defining the behavior of the circuit:

$$Ld^2Q/dt^2 + RdQ/dt + (Q/C) = V \qquad (8.22)$$

where V is the voltage across the circuit. I will leave it to you to see if you can implement this sort of idea using CMOS as the basis. Unfortunately, it turns out that it is extremely difficult to make appreciable inductances with silicon technology. You need long wires and coils and there's no room! So, it turns out that this is not a practical way of getting the energy losses down. However, that need not mean we have to abandon the basic idea – a very clever thing we can try is to have just one inductance, off the chip, instead of many small ones, as in one per switch.

8.2.2 Hot-Clocking

Here is a completely different, and very clever, way to get the energy dissipation down. It is a technique known as *hot-clocking*. In this approach, we try to save the energy by *varying the power supply voltages*. How and why might something like this work? Let's return to our water analogy. Earlier, we saw that if we opened the upper valve while the level of water in the tank was low, then we would lose energy as water flooded in from above and cascaded down. Where we are going wrong is in *opening the switch while there was a difference in water level*. If we do that, we will unavoidably lose energy. In principle, however, there are other ways of filling tanks that aren't nearly so wasteful. For example, suppose we have a tank to which is attached a single switched pipe, at the end of which is a water reservoir. If we fill the tank by the gradual process shown in Figure 8.30, opening the switch and moving the pipe up the tank as it pours so that it is always at the height of the water, then we will dissipate no energy:

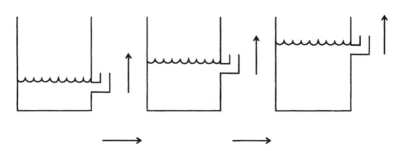

FIGURE 8.30
Non-dissipative filling of a tank.

Of course, we would have to perform the operation infinitesimally slowly to *completely* avoid a dissipative waterfall (this type of argument was used frequently in Chapter 5). However, it is clear that if we could move things so slowly, then we could really get the energy loss down as long as we never opened the switch when there was a difference in level between the pipe and the head of water in the tank. There is an analogous principle in electricity: *never open or close a switch when there's a voltage across it.* But that's exactly what we've been doing!

Here's the basic principle of hot-clocking. Consider the amended inverter circuit in Figure 8.31, where two pn diodes have been inserted.

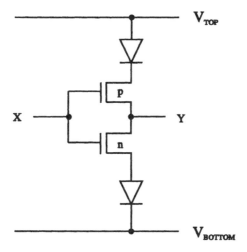

FIGURE 8.31
Sample "hot-clocking" circuit.

In Figure 8.31, the upper and lower voltages V_{TOP} (= V, say) and V_{BOTTOM} (= $-V_{TOP}$) are *not* to be considered constant: they can, and will, vary, so watch out! We will define the two main states in which they can be in as the *quiescent* state, which corresponds to the upper voltage being negative and the lower positive, and the *hot* state, the inverse of this, with the upper voltage positive and the lower negative. (These designations are arbitrary – we could just as well have them the other way around.) The principle of operation of this device is this. Suppose we start in the quiescent state, so the upper voltage in Figure 8.31 is negative and has X positive (= +V). Then, the p-MOSFET is open, the nMOSFET is closed, and no current flows (there is no voltage across the n device). In fact, even if X is negative, no current will flow due to the rectification property of the diode. So we can switch the input willy-nilly while in the quiescent state – the circuit is quite insensitive to the input voltage. This clearly leaves us free to choose our initial state for Y: we will take this to be positive.

Now, we let the voltages go hot – we gradually turn them around. Now a positive voltage gradually grows across the bottom diode, which conducts. This draws the output Y down to that of the lower voltage (which is now negative). The energy dissipation as this occurs is small as the resistance of the diode is low. When this lower voltage bottoms out and things have settled down, we switch back to the quiescent state again: the output Y would like to revert to its previous value but cannot, as the diode prevents any current from flowing. We can change X, that is, make a switch, as we please once in this stable state. It is necessary to run the first part of the cycle, when Y changes, rather slowly; the second stage, the return to quiescence, can be performed rapidly.

Now the output of Y must feed another gate. Clearly, we cannot use it while it is changing so the voltage cycle of the next gate must take place somewhat out of phase, with a different power supply (rather like a two-phase clock). It is possible, as is common with flip-flops, to have the second signal simply the inverse of the first, and hence use just the one supply – but this is dangerous, and slightly confusing, as going back to quiescence allows Y to vary a little. It is safer to design conservatively, with two separate power supplies. We can exhibit this diagrammatically by plotting the voltage changes of the two supplies (Figure 8.32).

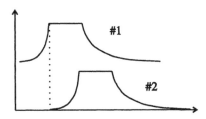

FIGURE 8.32
The supply voltages.

Note that the leading edge of each pulse is more leisurely than the trailing edge, reflecting the differing times of switching in the two stages. Let us also point out that these power supplies are universal to the entire chip or chips: otherwise, we could see that the amount of energy required to vary the supply voltage would effectively offset any savings we might make. We store outflowing energy in the power supply machinery.

Let's go back to the diode arrangement and calculate the energy lost during the switch. Let's suppose that the "rise time" we are allowing for the supply voltage to shift is t. The charge that we have to move during the change is $Q = CV$ and hence the current that will flow is (on average) just $Q/t = CV/t$. If we further suppose that the resistance we encounter in the diodes when we close them is R, a small quantity similar to that of the transistors, then the rate of energy loss, i.e., the power loss, is just $P = I^2R = Q^2R/t^2$. Hence, the total energy loss in switching is:

$$\Delta E = Pt = C^2V^2R/t = \left(CV^2\right)\left(CR\right)/t = \left(CV^2\right)\tau/t, \qquad (8.23)$$

where τ is the time constant of the original, naive CMOS inverter circuit. Also, recall that CV^2 was the energy loss during switching in that circuit. Therefore, we see that the energy loss multiplied by the time in which this loss takes place is the same for both the old and the new circuits. This is suggestive of a general relationship of the form:

$$(Energy\ loss)\left(Time\ of\ loss\right) = Constant \qquad (8.24)$$

for each switching step or simple logical operation. This expression certainly appears to be in sympathy with the findings of Chapter 5: the slower we go, the less energy we lose. In actual circuits, the clocks are much slower than the transistors (e.g., by a factor of 50 to one), and so clocking enables us to save a great deal of energy in our computations. Unfortunately, such is the current obsession with speed that full advantage is not being taken of the opportunities that power savings might offer. Yes, the machines would be slower, and bigger because of the extra components, but this might be offset by the fact that they would be cheaper to run, and there would be considerably less need for all the pumps and the fans and so forth needed to keep the things cool!

Now, although I used diodes in my example of Figure 8.31, I ought to point out that a more realistic set-up, if we don't want to use too many different types of component, is that shown in Figure 8.33, in which the diodes are replaced by transistors.

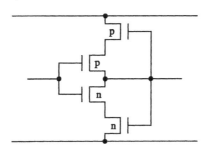

FIGURE 8.33
"Diode-less" hot-clocking circuit.

We have looked at just one of the so-called "hot-clocking" methods for reducing energy dissipation. These techniques (developed largely at Caltech[*]) allow clock lines to deliver power but were not originally intended for trading time for energy. Let me finish this section by pointing out that hot-clocking is a fairly recent development, and there are still many unanswered questions about it – so you have a chance to actually do something here, to make a contribution! The circuit I drew was my own, different from others that have been designed and built, and I'm not sure if it has any advantages over them. But you can check out all manner of ideas. For example: what if the supply voltage was AC, i.e., sinusoidal? Could we perhaps use two power supplies, both AC and out of phase? Why not let the voltage across the logic elements be AC? Perhaps we could define two states, one in phase with the power supply (logical one) and one out of phase (logical zero). There are many opportunities, and perhaps if you delved further and kept at it, you might uncover something interesting.

8.2.3 Some General Considerations and an Interesting Relationship

One of the central discoveries of the previous section, which might be general, is that the energy needed to do the switching, multiplied by the time used for this switching, is a constant – at least for resistive systems. We will call this constant the "dissipated action" (a new phrase I just made up). Now the typical time constant τ of an inverter is of the order of 0.3 ns, which is pretty small. Does it have to be so tiny? Well, yes, if we want to go as fast as possible. But we can approach the matter from a different angle. Because of delays on the lines, and because each element might have to feed others, and so forth, the actual clock cycles used are a hundred times greater in length – you can't have everything changing too quickly, or you'll get a jam. Now it is not obvious that we cannot slow the inverter down a bit – if we do so, it is not necessarily true that we will lose time overall in our computation in proportion to this reduction. Since this is unclear, it is interesting to find out exactly what is the value of our constant, which we shall write as $(Et)_{sw}$.

One way to do this is to work out the value of the constant for a specific switch for which it is directly calculable. We will therefore focus on the fastest possible switch and evaluate it for this – this is as good as any other choice. Let's first recap our basic equations. Our switch, a single transistor, will have a certain capacitance C_g, and we put a voltage V_g on it, and

---

[*] A 1985 paper on 'Hot-Clock nMOS' by Chuck Seitz and colleagues at Caltech ends with the following acknowledgment: "We have enjoyed and benefitted from many interesting discussions about 'hot-clocking' with our Caltech colleagues Alain J. Martin and Richard P. Feynman." [Editor]

hence a charge $Q = C_g V_g$. This gives us a switching energy $E_{sw} = C_g V_g^2$. Now $\tau = C_g R$, so $(Et)_{sw} = C_g^2 V_g^2 R = Q^2 R$, that is, the square of the charge needed to make the switch work multiplied by the minimum resistance we get when the switch is turned on. (You can also understand this in terms of power losses, working with currents.) Now we're naturally interested in asking what this dissipated action constant is for our ordinary transistors. We want to know this to see if, by redesigning such devices, we can get it down a bit, and perhaps use less energy or less time.

In order to proceed with the calculation, which is rather easy, we will need some physical parameters. First, the electron charge $e = 1.6 \times 10^{-19}$ C. Also, at room temperature, we have $kT/e = 1/40$ Volt. Using the kinetic energy relationship $(1/2)mv^2 = (3/2)kT$ (where m is the effective mass of the electron), we can define a "thermal velocity" v_{th} – this turns out to be roughly 1.2×10 cm/s; We also need some of the properties of weakly doped silicon material: the electron carriers have surface channel mobility $6 = 800$ cm^{2}V^{-1}s^{-1}, and a mean free path $l_{col} = 5 \times 10^{-6}$ cm. As with our earlier analysis of the MOSFET, we take the silicon under the gate to be L cm in length, W in width: in 1978, a typical value for L was 6 microns, falling to 3 by 1985[7] (Figure 8.34).

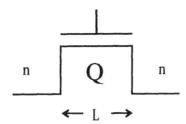

FIGURE 8.34
The simple MOSFET.

Suppose we have electrons sloshing about under the gate, and we impose a force F on them, for a time τ_{col}. This latter quantity we take to be the average time between electron collisions, which is a natural choice given the physics of the situation. There is an intuitively satisfying relationship between the mean free path and the collision time: $l_{col} = v_{th}\tau_{col}$. Now, from standard mechanics, at the end of this time an electron will have gained a momentum $mv_D = F\tau_{col}$ where the velocity v_D is the "drift velocity" in the direction of the force and is quite independent of (and much smaller than) v_{th}. Since mobility is defined by the relation, $v_D = \mu F$, we have $\mu = \tau_{col}/m$. Now, take the current flowing under the gate to be I. We have $I = Q/$(time of passage across the gate) $= Q/(L/v_D) = (Q/L).$ $(\mu e).(V_{ds}/L)$. However, the source-drain voltage $V_{ds} = IR$, so we have, for the resistance, $R = L^2/(Q\mu e) = mL^2/(Qe\tau_{col})$. (Incidentally, the effective mass of an electron moving through Si is within 10% of its free mass, so we can take m to be the latter.) Now, using our expression for the dissipated action in terms of Q and R, and using the relationships we have derived, we find:

$$(Et)_{sw} = N(L/l_{col})^2 (3kT)\tau_{col} \tag{8.25}$$

where we have introduced the number N of (free) electrons under the gate, $N = Q/e$. Now focus on the last two factors on the right-hand side of (8.25) – $3kT$ is an energy, of the order of the kinetic energy of an electron, and τ_{col} is a time, the time between collisions. Maybe it will help us to understand what is going on here if we define the product of these terms to *itself* be a dissipated action – just that dissipated during a single collision. This isn't forced on us: we'll just see what happens. Let us call such an action $(Et)_{col}$. Then we have:

$$(Et)_{sw} = N(L/l_{col})^2 (Et)_{col} \tag{8.26}$$

So we find that the (Et) that we need for the whole switch is larger than the Et for a single collision by two factors. One is the number of electrons under the gate, and the other is the ratio of the width of the gate to the mean free path. Taking L to be 6 microns (hence L/l_{col} to be about 100), and the number of electrons N under the gate to be about 10^6, we find:

$$(Et)_{sw} = 10^{10}(Et)_{col} \approx 10^{10}kT\tau_{col}. \tag{8.27}$$

This ties in with what we have quoted before. Now this is an *awful* amount, and we would certainly hope that we can improve things somehow! Why is this number so large? We know from the considerations of Chapter 5 that it in no way reflects a fundamental energetic limit. What can we do to get it down a bit?

Of course, all of our calculations thus far have been rooted in the conventional silicon VLSI approach – so perhaps what we ought to do is step outside that technology and look at another. Let us take a more general, and somewhat abstract, look at this question. Suppose that you design for someone a beautiful switch, the fundamental part of a computational device, which has a certain switching energy E_{part} and corresponding switching time t_{part}. Now you give this guy a pile of these parts, and he proceeds to build a circuit with them. But he does this in a most absurdly inefficient manner. He does this as follows (this might all seem a bit abstract at first but bear with me). First, he connects up, say, p switches in parallel, and hooks them all up to the same input (Figure 8.35).

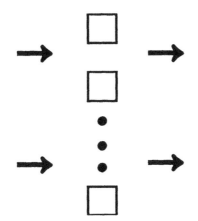

FIGURE 8.35
A possible parallel connection of fundamental parts.

These switches all operate simultaneously, the signal propagating from left to right. Clearly, the energy dissipated in switching all of these parts is $E_{sys} = pE_{part}$, and the time for it to occur is just $t_{sys} = t_{part}$. In other words, $(Et)_{sys} = p(Et)_{part}$. This is ridiculous, given that they all give out the same answer.

Next, the guy does something even dumber and connects up some parts in series as well, in chains s parts long (Figure 8.36).

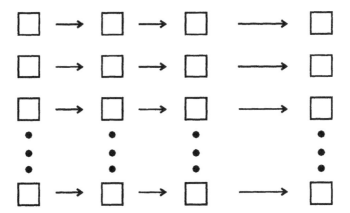

FIGURE 8.36
A serial connection of parts.

This really is dumb! Each switch in the chain just inverts the previous one, so all he has overall is a simple switch, effectively no better than the parallel arrangement he started out with! Yet compared to that, now $(Et)_{sys}$ $= ps^2(Et)_{part}$, as you should be able to see. So, of what relevance is all this? Well, an electron colliding is rather like a *1-electron switch*, with which we can associate a quantity $(Et)_{part} = 3kT\tau_{col}$. We can consider such a collision to be the fundamental operation. Now all the electrons jiggling beneath the gate are doing the same thing, bumping into one another, drifting and so forth, and so we can consider them to be operating effectively in *parallel*, with the number of parallel parts $p = N$, the number of electrons beneath the gate. Of course, one collision is not sufficient to account for the whole of an electron's activity between poles – the actual number of hits, on average, is $(L/l_{col}) \sim s$, using the serial analogy here. So we can actually interpret our result for $(Et)_{sw}$ in terms of the crazy handiwork of our engineer: $10^{10} = ps^2$! Surely room for improvement?

Okay – so how *can* we improve on this? First, is it completely silly to put things in parallel? Not at all: it's good for accuracy. It might be the case that we are working with parts that are extremely sensitive and which can easily be flipped the wrong way by thermal fluctuations and whatnot. Putting such parts in parallel and deciding the output on the basis of averaging, or by a majority vote, improves system reliability. If we have a part whose probability of malfunctioning is 1/4, then with just 400 of these in parallel, we can guarantee that the chance of the system spitting out a wrong answer is about 1 in 10^{18} – wow! And what about putting parts in *series*? Well, I've thought a lot about this, but have yet to come up with *any* resulting advantage. It wouldn't help with reliability – all it does is increase the lag. In fact, I can see no reason for having anything other than $s = 1$.

Problem 8.3 In our electron model, s = 1 would correspond to getting the fundamental ratio (L/l_{col}) down to unity. An interesting question arises if we actually take this notion seriously. In fact, I would like you to consider the most extreme case, that where the mean free path of the electrons below the gate is *infinite*: in other words, they suffer no collisions. Analyze the characteristics and behavior of such a device. Sure, on first impression, such a device could never function as a switch – it would always conduct. But we have forgotten about *inertia*: in order to conduct, the electrons have to speed up and change their speed, and can only start at zero; so there is a certain density of charge beneath the gate anyhow. In fact, this whole analysis, with infinite mean free path, was originally made for vacuum tubes, and these certainly worked. Therefore, a switch of this kind can be devised and analyzed – it's just that we can't do it with silicon (in which the electrons can be thought of as moving through some sort of "honey").

Generally, however we do it, we should make every effort to increase the mean free path and decrease L. There is a factor of 100 to be found in $(Et)_{sw}$ (not 10^4, because if we change the mean free path, we change τ_{col} as well). Current hardware design stinks! The energy loss is huge and there is no physical reason why we shouldn't be able to get that down at the same time as speeding things up. So go for it – you're only up against your imagination, not Nature.

An obvious suggestion is to simply reduce the size of our machines. We can make good gains this way. Let us scale L by a factor of $\alpha < 1$: $L \to \alpha L$. We then find $(L/l_{col})^2 \to \alpha^2(L/l_{col})^2$ as l_{col} does not scale. The number of electrons under the gate scales with area: $N \to \alpha^2 N$. Hence, we arrive at the result:

$$(Et)_{sw} \to \alpha^4 (Et)_{sw} \tag{8.28}$$

This is excellent scaling behavior, and though we cannot trust it down to too small values of α, it shows that simply shrinking our components will be advantageous.

The (Et) ideas I've put forward here are my own way of looking at these things and might be wrong. The idea that (Et) might be a constant is very reminiscent of the Uncertainty Principle in quantum mechanics, and I would love to have a fundamental explanation for it, if it turns out to be so. There is certainly room for you to look into such questions to see if you can notice something. Anything you can do to criticize or discuss these ideas could be valuable. If nothing else, because the simple relationship:

$$Power = E/t = (Et)/t^2, \tag{8.29}$$

shows that reducing the dissipative action (Et) should reduce the power loss from faster machines.

8.3 VLSI Circuit Construction

We now come, at last, to discuss the actual physical technology underlying VLSI. How are transistors actually made? How do we, being so big, get all this stuff onto such tiny chips? The answer is: very, very cleverly – although the basic idea is conceptually quite simple. The whole VLSI approach is a triumph of engineering and industrial manufacture, and it's a pity that ordinary people in the street don't appreciate how marvelous and beautiful it all is! The accuracy and skill needed to make chips are quite fantastic. People talk about being able to write on the head of a pin as if it is still in the future, but they have no idea of what is possible today! We can now easily get a whole book, such as an encyclopedia or the Bible, onto a pinhead – rather than angels! In this section, we will examine, at a fairly simplistic level of analysis, the basic processes used to make VLSI components. We shall once again focus solely on nMOS technology.

8.3.1 Planar Process Fabrication Technology

The process all begins with a very pure crystal of silicon. This material was known and studied for many years before an application in electronics was found, and at first, it tended to be both rare and, when unearthed, riddled with impurities – nowadays, in the laboratory, we are able to make it extremely pure. We start with a block of the stuff, about four inches square* and deep, and we slice this into thin wafers. Building integrated circuits on this substrate involves a successive layering of a wafer, laying down the oxide, polysilicon, and metals that we need according to our design. Remember from our earlier discussion that the source and drain of a MOSFET were n-type regions seeded into, rather than grafted onto, lightly doped p-type Si material – it is important to keep in mind that the silicon wafer we are using is actually this p-type stuff. To see the sort of thing that goes on, we'll explain in some detail the first step, which is to create and manipulate the non-conducting oxide layer on the silicon that will ultimately play a role in constructing the insulation layer under the gate of a transistor. We start by passing oxygen over the surface of the wafer, at high temperature, which results in the growth of a layer of silicon oxide (SiO_2). This oxide layer is shown in Figure 8.37. We now want to get rid of this oxide in a selective fashion. We do this very cunningly. On top of the oxide, we spread a layer of "resist", an organic material that we bake to make sure it stays put. A property of this resist is that it breaks down under ultraviolet light, and we use this property to etch an actual outline of our circuitry on the wafer. We take a template – a "mask" – and lay this over the material. The mask comprises a transparent material

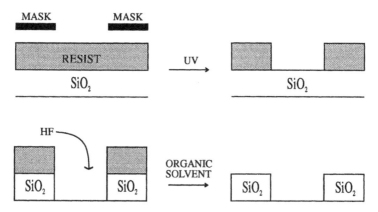

FIGURE 8.37
The first stages of chip fabrication.

---

* Cylinders of silicon 12 inches in diameter are now the industry norm. [Editor]

overlaid with an ultraviolet opaque substance, occupying regions beneath which, on the chips, we will want SiO_2 to remain. (Usually, the mask will repeat this pattern over its area many times, enabling us to produce many chips on one wafer, which may be cut into separate chips later.) We next bombard the wafer with UV light (or X-rays). The affected resist that is not shielded by the opaque regions of the mask breaks down and can be sluiced off. This exposes channels of SiO_2 that we can now remove by application of a strong acid, such as hydrofluoric acid. The beauty of the resist is that it is not removed by the acid so it protects the layer of SiO_2 beneath it – unlike the stuff we've just sluiced off – that we want to keep in place. After this stage, we have an upper grid of resist, under which lies SiO_2, and beneath this a bared grid of the original silicon. We now apply an organic solvent to the wafer, which removes the resist and leaves the underlying oxide intact. The result is, if you like, a layer of oxide with "silicon holes" in it (Figure 8.37).

That is the first of several steps. Step two involves laying down the basic material for any *depletion mode* transistors that may be required in the circuit (for use as resistors, for example). Such transistors differ in their construction from enhancement mode devices by having a shallow layer of n-type Si strung beneath the gate between the source and drain (Figure 8.38).

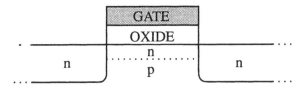

FIGURE 8.38
The depletion mode transistor.

Such a transistor is perpetually *closed* and current can always flow unless we place a negative charge on the gate to stem this current flow and open the switch (hence, $V_{th} < 0$, as stated earlier). To put such transistors on the chip, it is necessary to lay down their foundations before we go any further: this entails first delineating their gate regions and then creating a very thin region of n-type doped Si over these areas. To do this, we cover the wafer with resist again and place on it a mask whose transparent regions represent the depletion areas. Once again, we blast the wafer with UV or X-radiation, and this time we are left with a wafer comprising a covering of resist, dotted among which are spots of exposed silicon substrate. These open areas we dope with phosphorus, arsenic, or antimony to create the required depletion region. The resist prevents these ions from penetrating into the rest of the silicon. This done, we wash off the remaining resist.

The next layer to be taken care of is the *polysilicon* (polycrystalline silicon) layer. Recall that highly doped polysilicon conducts well, although not as well as a metal, and will be used to construct, among other things, the gates of transistors. As these gates are separated from the substrate by a thin layer of insulating oxide (see Figure 8.8), it should come as no surprise to you that before we do anything with our polysilicon, we have to coat the wafer with another thin layer of oxide as we did initially. As before, we do this by heating the wafer in oxygen (note that this will leave the depth of oxide across the wafer uneven). The wafer is then coated in polysilicon and another mask overlaid, this time designed to enable us to remove unwanted polysilicon. Having done this, we have to build the drains and sources (and, generally, the diffusion layer) of our transistors – and we do this by doping all of the remaining silicon appropriately (i.e., with phosphorus). We achieve this by removing any oxide that is not lying under the polysilicon and mass-doping the exposed Si regions. The depletion layers beneath the polysilicon are protected and will not be additionally contaminated.

We can now see how an enhancement mode transistor will arise from this process. To draw a diagram, we will adopt the conventions for the various layers shown in Figure 8.39 (the conventions in most common usage are actually color-coded).

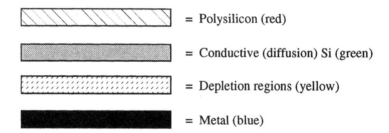

FIGURE 8.39
Conventions for chip paths.

We have added one more layer here – that of metal.* This layer comprises the "flat wires" we use to carry current a sizeable distance, in preference to polysilicon or the diffusion layer. (The power supply is usually drawn from metal paths.) It will also be necessary to add contact points to enable the current to flow freely between layers, as required. With this convention, we can draw an enhancement mode transistor as in Figure 8.40.

---

* Modern process technologies can now support more than 10 metal layers. [Editor]

FIGURE 8.40
Schematic diagram for enhancement mode transistor.

The transistor is just the crossing point of a polysilicon path and a diffusion path! Of course, the two paths do not cross in the sense of making physical contact – there is a layer of insulating oxide between them.

A full inverter requires a resistance in series. As we discussed earlier, we use a depletion mode transistor for this task. The inverter circuit is shown in Figure 8.41.

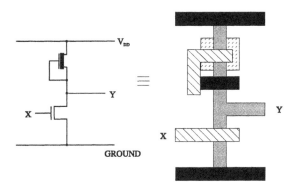

FIGURE 8.41
The full inverter.

You will note that I have included here the power supply and ground lines, both of which are metal paths. It is necessary for the fabrication process to leave patches of the diffusion paths exposed at the point where the metal crosses to ensure an electrical contact. These features you cannot see from a vertical picture. (The actual circuit is not laid out wholly flat as in Figure 8.41; it's all built on top of itself, in a clever, tight little box. See Mead and Conway for more details.) A similar procedure is necessary if we want to, say, use the source or drain of a transistor as the input to another gate – we then have to connect a diffusion path to a polysilicon path. Obviously, some kind of direct contact is needed; otherwise, we would find a capacitor or transistor where the lines cross. We can use a so-called "butting contact" where we overlay a direct diffusion/poly contact with metal, as shown in Figure 8.42.

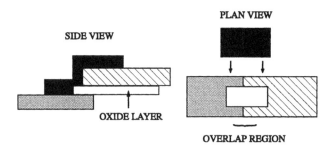

FIGURE 8.42
Polysilicon-diffusion layer contact.

To give an illustration of a more involved logic unit, we will look at the NAND gate. To make this, all we need to do is take our previous circuit and cross the diffusion path with another polysilicon path to make another transistor (Figure 8.43).

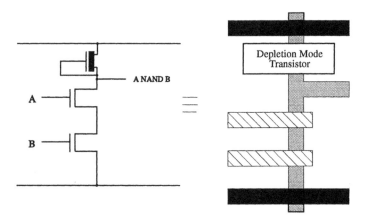

FIGURE 8.43
The NAND gate.

Note that in this circuit the polysilicon paths extend a little way beyond the diffusion path at each of the two transistors. Why? Well, there are many design rules governing precisely how we should arrange the various paths on a chip with regard to each other, how big the paths must be, and so on, and I'll briefly list some here. (For a fuller exposition of these 'lambda-based' design rules, see Mead and Conway). Let us begin by defining a certain unit of length, λ, and express all lengths on the chip in terms of this variable. In 1978, λ was about 3 microns; by 1985, it had fallen to 1 micron, and it falls further as time progresses. The minimum width for the diffusion and polysilicon paths is 2λ. The metal wire, how- ever, must be at least 3λ, across, to counter the possibility of what is known as "electromigration", a phenomenon whereby atoms of the metal tend to drift in the direction of the current. This can be a seriously destructive effect if the wire is especially thin (Figure 8.44).

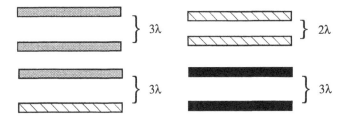

FIGURE 8.44
Silicon chip path widths.

Again, these are minima: the paths can be wider if we desire. Another set of rules pertains to how closely we can string wires together. Conducting paths cannot be placed too near each other because of the danger of voltage breakdown, which would allow current to crisscross the circuit (Figure 8.45).

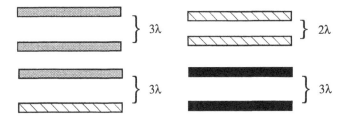

FIGURE 8.45
Silicon chip path separations.

Metal paths (blue) can go on top of poly (red) and diffusion ones (green) without making contact. Where red crosses green, as we've said, there is a transistor. It is important with such devices that the poly line forming the gate extends over the edge of the diffusion region to prevent a conducting path forming around it resulting from shorting the drain to the source. We usually require an overlap of at least 2λ to allow for manufacturing errors (Figure 8.46).

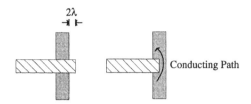

FIGURE 8.46
Rules for a transistor.

We must also consider the connections between levels. If we are hooking a metal line to another path, we must be sure the contact is good (the contact is typically made square). To ensure this, we do not just place the metal in contact with the path, area for area, but must have at least a distance λ of the path substance surrounding the contact to prevent leakage through the metal and into the surroundings. This is true whether we are connecting to poly, diffusion, or metal lines (Figure 8.47).

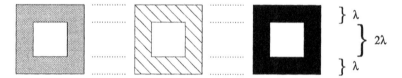

FIGURE 8.47
Rules for contacts.

8.3.2 Circuit Design and Pass Transistors

To actually make a specific circuit, we would design all of the necessary masks (typically enormously complex) and send them to a manufacturer. This manufacturer would then implement them in the construction process we have described to provide us with our product. There is a standard heuristic technique for drawing out circuits, one which tells us the topology of the layout, but not its geometry – that is, it tells us what which paths are made of, and what is connected to where; but it does not inform us as to scale, i.e., the relevant lengths of paths and so on. For example, the drawing (the so-called "stick figure") for the NAND gate is shown in Figure 8.48 (in which we have also indicated the new linear conventions we adopt for each type of path):

This tells us all the important interconnections in the circuit but if we were to actually trace the final physical product, the actual scaling of the respective parts might be radically different. This latter need not concern

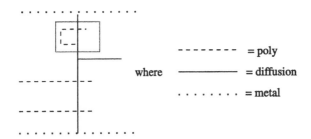

FIGURE 8.48
"Stick figure" for the NAND gate.

us here and we will adopt the stick figure approach in what follows, when we want to take a look at some specific circuits. To make things simpler still, we can sometimes deploy a kind of "half and half" shorthand, in which we represent sub-circuits on the chip by black boxes (a common enough procedure). So, for example, if we had a simple chain of inverters, it would be easier, rather than drawing the entire transistor stick figures over and over, to use the scheme of Figure 8.49

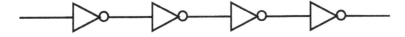

FIGURE 8.49
Simplified circuit diagram for chain of inverters.

where the triangles are just the conventional symbols for inverters, and the line convention is as explained in Figure 8.48.

A common type of circuit is the shift register. We represent this in Figure 8.50 as a doubly clocked inverter chain, crossed by polysilicon paths.

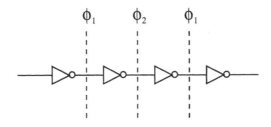

FIGURE 8.50
A shift register.

The two (complementary) clock pulses are sent down polysilicon lines, and where these cross a diffusion line, they form what is known as a *pass transistor*, so-called because it only allows a current to flow from source to drain (i.e., from left to right in the above picture) if the gate is forward-biased. This occurs whenever the clock pulse to the polysilicon line is on. At the next pulse, the next inverter in the chain switches and will hold its new value until the next clock pulse. The reader should be able to make contact with our discussion of clocked registers in Chapter 2 to figure out how Figure 8.50 works. It is a simpler, more accessible arrangement than a bunch of flip-flops and logic gates. Note, incidentally, that we can close such an arrangement (i.e., make it go "in a circle") if we want to use it as a memory store.

8.3.3 Programmable Logic Arrays

With programmable logic arrays (PLAs), we come on to examine the issue of *if-then control* in machines – that is, the matter of how, given a certain set of input data, the machine should determine what it does next. For example, "if such-and-such is zero, then stop" or "if both bits are 1, then carry 1". Abstractly, there is information coming out of some part of the machine that will tell us what we're to do next. This information hits some "sensors" (my own word, not the technical one), which tell us our present state. Once we know this, we can act on it, for example, by telling an adder to add or subtract. This instruction, or more generally, set of instructions, will take the form of data coming out on a set of lines (Figure 8.51).

SENSING   CONTROL   INSTRUCTIONS

FIGURE 8.51
A generic control device.

The first stage in designing a device to do this is, obviously, to know what set of instructions is associated with a given sensory set. This is pretty straightforward. For example, we might represent the instructions as in Table 8.1.

What this means is as follows. Each row in the left-hand column represents some configuration of bits on the sensor lines (of which there are five in this example). The corresponding rows on the right represent the bits sent out along the instruction lines (six, in this case), given the sensor set on the left. In this column, a 1 might mean "do something if the input from this line is 1" – such as "add" or "switch on light" – while a 0 might mean do nothing or do something else – "leave state X as it is"

TABLE 8.1

Example Instruction Set for a Control Device

Sense lines					Instructions					
1	2	3	4	5	a	b	c	d	e	f
1	1	0	1	0	1	0	1	1	0	1
1	0	0	1	1	1	0	1	1	0	1
0	1	1	0	1	0	1	1	0	1	0
			•						•	
			•						•	
			•						•	

or "switch off light". A very direct, and very inefficient, way of making a control system would be to simply store this table in memory, with the sensing lines as memory addresses, and the control lines as the contents of these addresses. Thus, we would separately store the actions to be performed for all possible combinations of sense lines. Since the contents of this memory are to be fixed, we might as well store everything on a read-only memory (ROM) device. The only potential hitch in this otherwise straightforward procedure arises from timing: it is conceivable that some instructions could leave the ROM device before the rest, changing the state of the machine and confusing the sensing. The effect of this might be fed back into the ROM before it has completely dealt with its previous sense set. This would be pretty bad if it happened but is usually avoided (you should be way ahead of me here) by deploying clocked registers at each end of the memory to ensure that the retrieval and use of an instruction occur at different times (Figure 8.52).

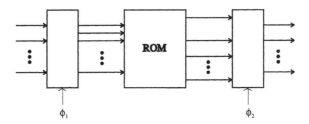

FIGURE 8.52
Clocked ROM control system.

When φ_1 is on, the sense lines feed through to the memory, which looks up the corresponding control signals. These latter signals cannot get out because φ_2 is off. Only when we can be sure that everything has settled down – that all the sensing information is in and that the instruction set has been chosen – do we switch φ_2 on. φ_1 has meanwhile gone off to freeze the memory input. With the external clock on, the instruction set can now get out and reach the rest of the machine without affecting the memory input. And so it goes on.

Thus, we see that control can be very, very simple. However, we are dissatisfied with this kind of approach because we would also like to be efficient! As a rule, stuffing our memory with 2^n entries is somewhat extravagant. Often, for example, two or more given input states will result in the same output state, or we might always filter a few sense lines through a multiple-OR gate before letting them into the ROM. This would leave us with a high degree of redundant information in our table, and naturally enough, we find ourselves tempted to eliminate the ROM completely and go back to basics, developing a circuit involving masses of logic gates. This was how things were done in the early days, carefully building immensely complicated logic circuits, deploying theorems to find the minimum number of gates needed, and so forth, without a ROM in sight. However, these days the circuits are so complex that it is frequently necessary – given the limitations of human brain power! – to use a ROM approach. But there are intermediate cases for which a ROM is not necessary because the number of possible outputs is small enough to enable a much more compact implementation using just a logic circuit – the setup is not *too* complicated for us to design. To illustrate one such instance, we shall examine a PLA, something we first encountered in Chapter 2. This is an ordered arrangement of logic gates into which we feed the sense input, and which then spits out the required instruction set. Ideally, such an array would exhibit no redundancy. In a "black box" scheme, a generic PLA would have the form shown in Figure 8.53.

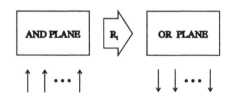

FIGURE 8.53
The generic PLA.

As can be seen, the PLA comprises two main sections: the "AND-plane" – formed exclusively from AND gates – and the "OR-plane" – formed exclusively from ORs. The planes are connected by a bridge of wires, which we label R. The inputs are fed into the AND-plane, processed, and fed into the OR-plane by the R-wires. A further level of processing then takes place and a signal emerges as output from the OR-plane. This output is the set of "what next" instructions corresponding to the particular input.

Let us consider a case where we have three input lines, A, B, and C, and four output lines, $Z_1 \ldots Z_4$. Each input, before being fed into the AND-plane, is split into two pieces – itself and its complement, for example, A and NOT A. We now have a device that can manipulate each signal with NOT, AND, and OR – in other words, it can represent any logical function whatsoever. Let us pick a specific PLA to show the actual transistor structure of such an array. We have three inputs telling us the state of certain parts of the machine and four possible outputs – pulses that will shoot off and tell the machine what to do next. Now suppose that the output Zs are to be given in terms of the inputs according to the following Boolean functions ($\vee$ = OR; $\wedge$ = AND; $'$ = NOT):

$$Z_1 = A$$

$$Z_2 = A \vee \left(A' \wedge B' \wedge C \right)$$

$$Z_3 = B' \wedge C' \qquad (8.30)$$

$$Z_4 = \left(A' \wedge B' \wedge C \right) \vee \left(A' \wedge B \wedge C' \right)$$

It is not immediately obvious that the particular Boolean functions of A, B, and C that we need to calculate the Zs can be written as the product of a series of ANDs followed by ORs. However, it is in fact the case, as we demonstrated for the general logical function in Chapter 2. In this instance, an acceptable output R_i from the AND-plane must only involve the ANDs and NOTs of A, B, and C. Thus, we can define the R_i as:

$$R_1 = A, R_2 = B' \wedge C', R_3 = A' \wedge B' \wedge C, R_4 = A' \wedge B \wedge C', \qquad (8.31)$$

and it is now straightforward to see that the Z-outputs can be written purely in terms of OR operations (or identities) on these Rs:

$$Z_1 = R_1$$

$$Z_2 = R_1 \vee R_3$$

$$Z_3 = R_2 \qquad (8.32)$$

$$Z_4 = R_3 \vee R_4$$

It is a general result that any Boolean function can be factorized in this way. The PLA for this function is shown in Figure 8.54.

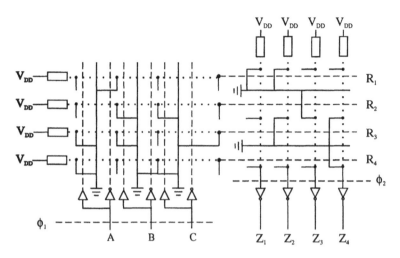

FIGURE 8.54
Circuit diagram for a PLA.

I will leave it to you to work out at the electronic level how this circuit gives us the advertised transformation!

As a rule, some 90% of the structure of a PLA is independent of its actual function. In consequence, PLAs are usually constructed by overlaying a standard design with select additions. For example, the above circuit results from taking the generic AND-plane and changing it into the circuit we want by the judicious addition of several diffusion paths in the right places (Figure 8.55).

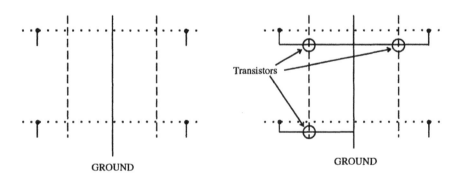

FIGURE 8.55
A generic AND-plane and the amended form.

This is a very practical approach. Of course, if you wanted more lines you would have to look up in a manufacturer's catalog which core arrays were available. Incidentally, note, from Figure 8.54 that the generic OR-plane is essentially the AND-plane rotated through a right angle.

Problem 8.4 Let me now give you an interesting problem to solve. This actually arose during the design of a real device. The problem is this: we would like to switch, that is, exchange, a pair of lines A and B by means of a control line, C. We are given C and its complement C' – they come shooting in from somewhere and we don't care exactly where – and if the control C is hot, then A and B change places: if C is cold, they don't. This is a variant of our old friend the Controlled Exchange. The circuit diagram we will use is that of Figure 8.56.

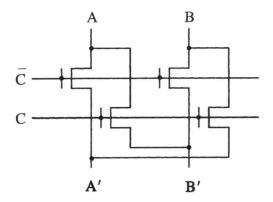

FIGURE 8.56
An exchange circuit.

To reiterate the rules: $C = 0 \Rightarrow A' = A, B' = B; C = 1 \Rightarrow A' = B, B' = A$. You should be able to see how it all works. Here is what I want you to do:

(a) Draw a stick figure with the correct conventions for diffusion, poly, and metal (hint: the inputs A and B are fed in on metal lines).

(b) Draw a legitimate layout on graph paper, obeying the λ design rules.

(c) This circuit can easily be amended to allow for more $A, B...$ inputs simply by iterating its structure (and extending the C, NOT C lines). Suppose now that we have eight input pairs coming in from the top. There are only 14 λs available horizontally for each pair, and 16 or 20 extra λs on the borders for about 132λ total width. But we are allowed 150λ deep. Now we want the As and Bs to go out of the circuit in metal too. Can you design it? You may assume more Cs from the left if you want.

8.4 Further Limitations on Machine Design

It doesn't take much thought to realize that one of the most important components of any computer is *wire*. We're so used to treating wires – more generally, transmission paths, including polysilicon lines – in an idealized way that we forget they are real physical objects, with real physical properties that can affect the way our machine needs to be designed. In this final section, I'd like to look at two ways in which wires play an important role in machine design. The first relates to how wire lengths can screw up our clocking, the so-called "clock skew" problem; the second to an even simpler issue, the fact that wires take up space, and that when we build a computer, we'd better make sure we leave enough room to get the stuff in!

8.4.1 Clock Skew

Let's return to our discussion of clocking the general PLA. Remember, we employed two clock pulses, φ_1 and φ_2, taking the general form (Figure 8.57).

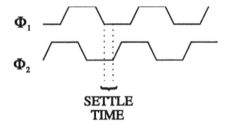

FIGURE 8.57
The PLA clock pulses.

The idea is that we feed data into the PLA while φ_1 is on, and then let things settle down for a while – let the logic gates go to work and ready their outputs, and so on. This is the reason for introducing a delay time, and not simply having the two clocks complementary. Then, we switch on φ_2, and during this time we allow the data to spew out. This sounds all very straightforward and simple.

However, in a real machine, there can be problems. For a start, charging up the gates of circuit elements takes a nonzero time, and this will introduce delays and time lags. Also, of course, the clock signals are current pulses sent along *wires* – metal, polysilicon, whatever – and *these pulses will take a finite time to travel*. A clock pulse sent along a short wire will reach the end before a pulse sent along a long wire. We can actually model a simple wire in an interesting way as an infinite sequence of components

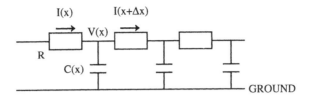

FIGURE 8.58
An infinite-limit model of a simple wire.

as shown in Figure 8.58 (which in the finite case could be taken as modeling a chain of pass transistors).

We have a line of resistors interspersed with capacitors. If we assume we have infinitely many small capacitors and resistors, bunched up infinitely closely, then we effectively have a wire, with a resistance per unit length of R, and a capacitance per unit length of C. Now what we want to do is to load up one end of the line (which needn't be metal – it could be polysilicon) and wait for the signal to propagate along to the other end. Let the distance along the wire from the origin be x. At each junction, we can define a potential $V(x)$, and a current flowing into it, $I(x)$. Taking the limit as $\Delta x \to 0$, elementary math and electricity give us the set of equations:

$$\partial I / \partial x = -C \partial V / \partial t \qquad (8.33)$$

$$\partial V / \partial x = -IR \qquad (8.34)$$

$$\partial^2 V / \partial x^2 = RC \partial V / \partial t = \tau \partial V / \partial t \qquad (8.35)$$

defining $\tau = RC$. Equation (8.35) is an example of the *diffusion equation*. Charge flows in at one end and diffuses through the system. The general form of the solution in terms of Green's functions is well-known. With our boundary conditions the solution is:

$$V(x,t) = \exp\left(-x^2 \tau / 4t\right) \qquad (8.36)$$

It is easy to see from this that if the overall length of the wire is X, then the time to load the wire scales as X^2. For 1 mm of polysilicon, this time comes to 100 ns. For 2 mm, it is 400 ns. This is a pretty lousy line, especially if you're more used to transmission lines for which the loading time is proportional to the distance. Metal, however, has such a low resistance that the load time is relatively much shorter – so if you want to send a signal any great distance, you should put it on metal.

The issue of clocking is of such importance to computing (indeed, much more important than you'd think given how little I've talked about it) that we are naturally encouraged to explore other avenues, other ways of controlling our information flow. The problem with the standard way, so-called synchronous clocking – the only type we've considered so far – is that in designing our machine we have, at each part of the system, to allow for the "worst case scenario". For example, suppose we have to take an output from a complex adder that could take anywhere from, say *t* units of time up to 5*t* to show. Now, even if the output zips through after *t* units, we still have to put our machine on hold for at least 5*t* just on the off chance that we get a slow decision. This can lead to severe time inefficiencies. Now, another way to design machines – although one which is not yet used commercially – is an "asynchronous" method: we *let the adder control the timing*. Let it tell us when it's ready! It carries out its computation, and then sends a signal saying it's ready to send the data. In this way, the timing is controlled by the computing elements themselves, and not a set of external clocks.

Interestingly enough, a little thought will show you that even synchronous systems have asynchronous problems of their own to solve. For example, consider what happens if such a machine has to accept data from a keyboard or another machine hooked up to it. Keyboards don't know anything about the "right time" to send in the data! We have to have a *buffer*, a little box that lets data into the machine only if the machine clocks are in the right state. It has to make a decision: whether to accept the data right now or to wait until the next cycle, as the data came in too late. The fact that a decision has to be made introduces the theoretical possibility of a hang-up caused by the data coming in at *just* such a time that the buffer is not quick enough to make a decision – it can't make its mind up. It's a fascinating problem, and one well worth thinking about.

8.4.2 Wire Packing: Rent's Rule

Up until now we've been discussing transistors, VLSI, and this and that – and we think that's the hard part of machine design. But whenever you get to the end of a big design, and you set out to build it, you'll discover that all the algorithms and so forth that you've worked out are not enough – something always ends up getting in the way. That something is *wires*. We look at that now.

I would like to emphasize that wires represent a real problem in system design. We've discussed one difficulty they cause: timing problems resulting from the finite time it takes to load them. But another problem is that the space needed for the wiring, connecting this chip to that and the other[*] is greater than that needed for the functioning components, like transistors! Now there is no guarantee that wires will forever reign supreme: with optical fibers, for example, we can send multiple messages down single wires by using light of differing frequencies. People occasionally break down and begin to dream, having brilliant ideas such as that of building a machine, by analogy with our broadcasting system, in which each component radiates light of a particular color (say via a LED), which is broadcast throughout the machine to be picked up and acted on by frequency-sensitive components. However, at this moment in time, the predominant method of current transmission is via wires, and I'd like to spend some time discussing them. Specifically, I want to address the question of how *much* wire we might need for a generic design.

Now there's very little I can say here about wire-packing – they're just wires, after all – but it turns out that there is an empirical rule, *Rent's Rule*, which purports to shed some light on this question. It's a curious rule, and I can't really vouch for how accurate it is in general, but it appears to be the case based on the *experience* of IBM. Here's how it goes. Let us suppose we have a unit, like a circuit board, and suppose further that we can segregate elements on the unit into "cells" – not too big, not too small. These cells could be individual chips, for example. Now suppose that:

(1) Each cell has t pins, or terminals,

(2) N cells make up a unit, and

(3) The number of terminals, or output pins, on our unit is T.

Needless to say, these numbers have to be interpreted with a certain latitude. Let us suppose we try to connect everything up so that the components talk a lot, that is, we try to minimize the wire length by packing. Then Rent's Rule states that:

$$T = tN^r \tag{8.37}$$

where $0.65 \leq r \leq 0.70$. (Since this inequality is only approximate, we will take $r = 2/3$.) In other words, it claims to relate the number of wires leading to and from the unit ($= T$) to the density of cell packing on the unit ($= N$). A naive first question to ask might be: why not just $T = tN$? Well, for the obvious reason that many of the wires will be *internal* to the unit (Figure 8.59).

---

[*] I am not now concerned so much with the "wires" on the chips, but those connecting chips together – real bunches of wires that interfere with how closely chips can be stacked and so on. [RPF]

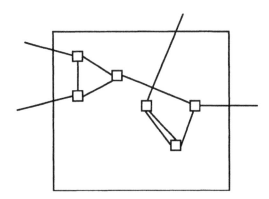

FIGURE 8.59
Schematic depiction of fundamental cells on a board.

We can see how an expression such as that in (8.37) might arise by moving "up" in our hierarchy of units and cells. We have considered units on which cells were joined together. We now consider *units* joined together. So let us imagine that we have a bigger unit, a "super-unit", the cells of which are the units bearing the original cells. Suppose this super-unit contains M units. Now, because we have set no fundamental level of analysis, there must be some consistency of scaling between these two situations. Let the number of terminals on the super-unit be T_s. Clearly, each of the M units will have T terminals. Then, Rent's Rule would say:

$$T_S = TM^r \tag{8.38}$$

However, returning to our initial level of analysis, we can treat the super-unit as comprising NM of the original cells, each of which has t pins. Using Rent's Rule again, we get:

$$T_s = t(NM)^r \tag{8.39}$$

Clearly, using (8.37), we see that (8.38) and (8.39) agree so that Rent's Rule has the correct scaling properties. This is very important.

Note that this treatment tells us nothing about the value of r (although it should be obvious from the form of the rule and the discussion following (8.37) that r would have to be less than 1). Where does this exponent come from? Well, you should remember that the value that was chosen was derived from experience, and this experience must have been influenced by problems of geometry in designing and connecting up logic circuits. That is, while it might be enticing to think that there is some neat logical reason for the value of r, that it might drop out of a pretty mathematical treatment, it's possible that it's an artifact of conventional design approaches. But for the moment, with this caveat in mind, let's assume it *is* true in the general case and see what it might teach us about wire packing.

Let's go back to the two-dimensional case. Suppose we have a square board, of side length L cm, say. Let this be the unit. We pack it with cells, each of length l cm; we can write the number of cells on the board as $N = (L/l)^2$ (Figure 8.60).

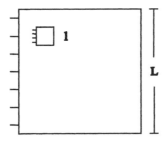

FIGURE 8.60
A general two-dimensional unit.

Now suppose that there is a restriction on how many terminals we can fit on each of the cells – that we can only place them so closely together. Let the maximum number of pins per cm on a cell perimeter be s_c. Suppose that there is also a minimum pin separation for the board terminals, with the maximum number per cm of perimeter being s_B. Rent's rule then becomes:

$$T = \left(4s_B L\right) = tN^r = \left(4s_C l\right)(L/l)^{2r} \tag{8.40}$$

and we have:

$$s_B = s_c (L/l)^{2r-1}. \tag{8.41}$$

It is clear from this that if $r > 1/2$ then, as we increase L, we need more and more pins per inch on the perimeter to take care of all the junk inside it. Therefore, we'll eventually get a jam. So as we build the machine bigger, the wiring problem becomes more serious. At the heart of this is the fact that the length of the perimeter varies as the square root of the area but the number of terminals (according to Rent) goes as the (2/3)rd power, a much faster scaling factor. A big incompressible mess of wires is unavoidable, and we have to increase the spacing between cells, leading to more boards, and increasing the spacing between boards, and so on, to make room. Now interestingly enough, if we were to rework this argument in three dimensions, rather than two, we get a different result: in 3D, we replace the perimeter with the surface area (length²) and the area with the volume (length³). Clearly, the former scales as the latter to the (2/3)rd power, the same as the number of terminals! So, in 3D, we could just make it – we could always use the same density of pins over the surface, and we wouldn't get into a wire hassle. The problem with this sort of 3D design, of course, is that for anyone to look at it – to see what's going on – they have to be able to get inside it, to get a hand or some tools in. At least with two dimensions, we can look at our circuits from above!

Still assuming the validity of Rent's Rule, we can ask another interesting question. What is the distribution of wire lengths in a computer? Suppose we have a big, two-dimensional computer, a board covered in cells and wires. Some of the wires are short, maybe going between adjacent cells, but others may have to stretch right across the board. A natural question to ask is: if we pick a wire at random, what is the chance that it is of a certain length? With Rent's Rule, we can actually have a guess at this, after a fashion. Return to the two-dimensional case shown in Figure 8.60, and now take L to be the side-length of some arbitrary unit on the board. We can consider any wires connecting cells within this unit to other cells within it to be less than L in length. This is not strictly true, of course, as we might have diagonals. However, if we just deal with orders of magnitude, we shall assume we can neglect this subtlety. There will also be wires going out of this unit and hooking up to other units on the board. We take these to be longer than L. From Rent's Rule, we can calculate the number of wires of length greater than L – this will be T, the number of terminals on the unit, given in this case by:

$$T = t\left(L/l\right)^{2r}. \tag{8.42}$$

We can now calculate the probability that a random wire will have a length greater than L. It is just the right-hand side of (8.42) divided by the total number of wires on the unit. This is easily seen to be $\propto t(L/l)^{2}$. So, if the probability of a wire having length greater than L is $P(L)$, we clearly have:

$$P(L) \propto \left(\frac{L}{l}\right)^{2r-2} \quad \text{or } P(L) \propto 1/L^{2/3}, \tag{8.43}$$

using $r = 2/3$ in Rent's expression.

We can take these statistics further. Introduce the probability density $p(L)$, which is defined in the standard way – the probability of finding the wire length to lie between L and $L + \delta L$ is $p(L)\delta L$. Then we have:

$$P(L) = \int_L^\infty \rho(L')dL' \tag{8.44}$$

with

$$\rho(L) = dP / dL \propto 1 / L^{5/3}. \tag{8.45}$$

Let us compute a quantity of particular interest, the mean wire length. By conventional statistical reasoning, this is:

$$\int_l^{L_{max}} L\rho(L)dL \Big] / \Big[\int_l^{L_{max}} \rho(L)dL \tag{8.46}$$

Note that we have tinkered with the limits of integration in (8.46); if we let the length L range from zero to infinity, then the numerator gives us trouble at its upper limit (infinity), as the integrand is of positive dimension in L, and the denominator gives us trouble at its lower limit (zero), as its integrand is of negative dimension in L. We hence set an upper limit for L, L_{max}, and also set a natural lower limit, the cell-size l. The reader can perform the integrals in (8.14) to obtain the mean wire length. The answer is:

$$2l \left(L_{max} / l \right)^{1/3}. \tag{8.47}$$

Note that this quantity is divergent: the bigger our machine (its size being given roughly by L_{max}), the bigger the mean wire length. No surprise there. However, note how it is the *cell size, l,* that is calling the shots in (8.47); the mean length scales half as quickly with machine size as it does with cell size (which is equivalent to cell spacing in our model). If we space our cells a little further apart, the size of the machine must balloon out of proportion.

It used to be said in the early 1980s that a good designer, with a bit of ingenuity and hard work, could pack a circuit in such a way as to beat Rent's Rule. But when it came to the finished product, something always came up – extra circuits were needed, a register had to be put here, an inductance there – and, when the machine was finally built, it would be found to obey the Rule. When it comes to the finished product, Rent's Rule holds sway, even though it can be beaten for specific circuits. Nowadays, we have "machine packing programs", semi-intelligent software that attempts to take the contents of a machine and arrange things to minimize the space it takes up.

9

The Future of Computing beyond Moore's Law

CONTENTS

9.1 Introduction

9.1.1 Overview

Society has come to depend on the rapid, predictable, and affordable scaling of computing performance for consumer electronics, the rise of "big data" and hyperscale data centers (e.g., Google, Facebook), scientific discovery, and national security. There are also many other parts of the economy and economic development that are intimately linked with these dramatic improvements in information technology and computing such as avionics systems for aircraft, the automotive industry (e.g., self-driving cars), and smart-grid technologies. The approaching end of lithographic scaling threatens to hinder the continued health of the four-trillion-dollar

DOI: 10.1201/9781003358817-9

electronics industry, impacting many related fields that depend on computing and electronics. Furthermore, the loss of exponentially improving energy efficiency for electronics technologies in the face of substantial market growth will create an energy crisis that would make it very difficult to meet energy efficiency targets necessary to address climate change. At the time of writing this chapter, data centers alone account for 8% of worldwide energy consumption [1] and are projected to reach 20–30% of worldwide electricity [2] if substantial energy efficiency improvements cannot be realized as depicted in Figure 9.1.

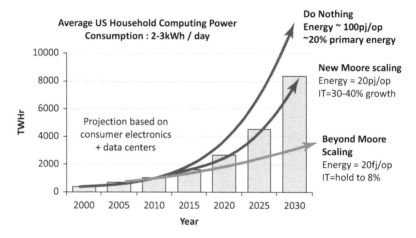

FIGURE 9.1
Datacenters already account for 8% of worldwide power consumption and are projected to grow to 25–30% of worldwide power if we cannot substantially improve the energy efficiency of future microelectronic systems.

Moore's law [3] is a techno-economic model that has enabled the information-technology (IT) industry to nearly double the performance and functionality of digital electronics roughly every two years within a fixed cost, power, and area. This expectation has led to a relatively stable ecosystem (e.g., electronic design automation tools, compilers, simulators, emulators) built around general-purpose processor technologies such as the x86, ARM, and Power instruction set architectures. However, within a decade, the technological underpinnings for the process Gordon Moore described will come to an end as lithography gets down to the atomic scale. At that point, it will be feasible to create lithographically-manufactured devices with dimensions nearing atomic scale, where only a dozen or fewer silicon atoms are present across critical device features. This will therefore be a practical limit for implementing logic gates for digital computing [4]. Indeed, in 2015, the International Technology Roadmap for Semiconductors (ITRS), which has tracked the historical improvements

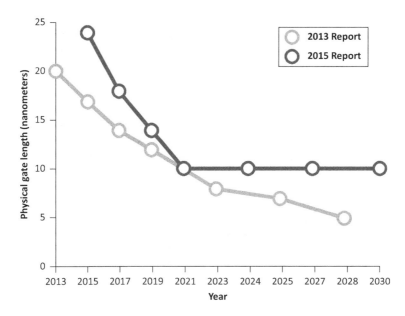

FIGURE 9.2
The most recent ITRS report predicts transistor scaling will end in 2021 (a decade sooner than was predicted in 2013). (Figure from ITRS.)

in lithography for the past 30 years, projected no improvements beyond 2021, as shown in Figure 9.2, and has subsequently disbanded having no further purpose. The classical technological driver that has underpinned Moore's law for the past 50 years is failing [5] and is anticipated to flatten by 2025 as shown in Figure 9.3. Evolving technology in the absence of Moore's law will require investment *now* in computer architecture and the basic sciences (including materials science) to study candidate replacement materials and alternative device physics to foster continued technology scaling.

9.1.2 Multiple Paths Forward

To address this daunting problem both in the intermediate and long term, a multi-pronged approach is required: evolutionary for the intermediate (ten-year) term and revolutionary for the long (ten-to-twenty-year) term strategy. Timing needs for the intermediate term will require an evolutionary approach based on achieving manufacturing technology advances allowing the continuation of Moore's law with current CMOS technology. This will rely on new computing architectures and advanced packaging technologies such as monolithic 3D Integration (building chips in the third dimension) and photonic co-packaging to mitigate data movement

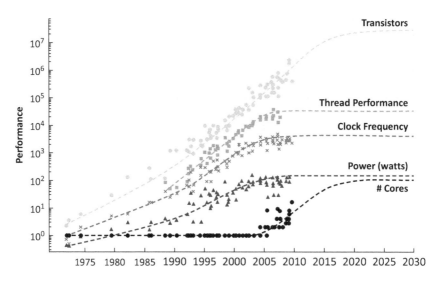

FIGURE 9.3

Sources of computing performance have been challenged by the end of Dennard scaling in 2004, all additional approaches to further performance improvements end in approximately 2025 due to the end of the roadmap for improvements to semiconductor lithography. (Figure from Kunle Olukotun, Lance Hammond, Herb Sutter, and Mark Horowitz, and extended by John Shalf.)

costs [6, 7]. The long-term solution requires fundamental advances in our knowledge of materials and pathways to control and manipulate information elements at the limits of energy flow, ultimately capable of achieving less than 1 attojoule/operation (atto = 10^{-18}), which would be six orders of magnitude less energy use than in today's devices. As we approach the longer term, we will require groundbreaking advances in device technology going beyond CMOS, requiring fundamentally new knowledge of control pathways as well as new system architectures and programming models to allow the energy benefits of scaling to be realized. Using the history of the silicon fin field-effect transistor (FinFET) as an example, it takes about ten years for an advance in basic device physics to reach mainstream use. Any new technology will therefore require a long lead-time and sustained R&D of one to two decades. Options abound, the race outcome is undecided, and the prize is invaluable. The winner will not only influence chip technology but will also define a new direction for the entire computing industry as well as many other industries that have come to depend heavily on computing technology.

There are numerous paths forward to continue performance scaling in the absence of lithographic scaling as shown in Figure 9.4. These three axes represent different technology scaling paths that could be used to extract additional performance beyond the end of lithographic scaling.

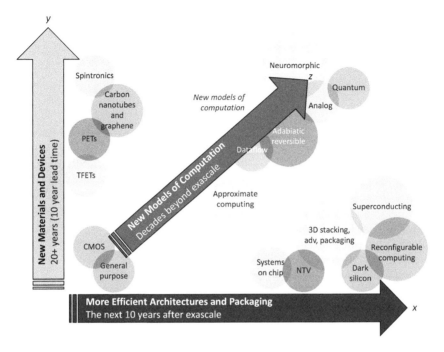

FIGURE 9.4
There are three potential paths forward to realize continued performance improvements for digital electronics technology.

The near-term focus will be on the development of ever-more specialized architectures and advanced packaging technologies that arrange existing building blocks (the horizontal axis of Figure 9.4). In the mid-term, emphasis will likely be on developing CMOS-based devices that extend into the vertical dimension and are based on improving materials and transistors that will enhance performance by creating more efficient underlying logic devices. The third axis represents opportunities to develop new models of computation such as neuro-inspired or quantum computing, which solve problems that are not well addressed by digital computing.

9.2 The Complementary Role of New Models of Computation

Quantum, neuromorphic, and artificial intelligence (AI) technologies have garnered increasing attention recently due to the rapid pace of recent improvements. Despite the rapid influx of funding into such technologies, it is important to understand that they are not replacement technologies for digital electronics as we currently understand them. They certainly expand computing into areas where digital computing is deficient. Digital computing is well known for providing reproducible and explainable calculations that are accurate within the precision limit of digital representation. Neural network models of AI and machine learning (ML) algorithms have substantially improved our ability to recognize patterns in *big data* and automate data mining processes over traditional pattern recognition algorithms, but they are less reliable for handling operations that require a precise response and reproducibility (and even explainability for that matter). Quantum computing algorithms will expand our ability to solve some combinatorially complex problems in polynomial time, but they will not be particularly useful for applications such as word processing or graphics rendering, for example. It is quite exciting and gratifying to see computing expand into new spaces, but it is equally important to recognize the complementary role that digital computing plays in our society, which will not and cannot be replaced by these emerging modes of computation as illustrated notionally in Figure 9.5.

Much of advanced architecture development and many new startup companies in the digital computing space are targeting the AI/ML market because of its explosive market growth rate. Growth markets are far more appealing business opportunities for companies and venture capital as they offer a path to rapid profit growth, whereas a large market that is static invites competition that slowly erodes profits over time. As a result, there is far more attention paid to technologies that are seeing a rapid rate of expansion, even in cases where the market is still comparatively small.

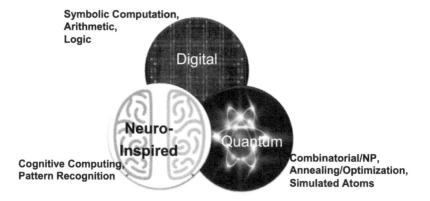

FIGURE 9.5
Quantum and AI/neuro-inspired technologies offer approaches to expand computing into new areas where traditional digital computing is weak. They are complementary technologies and they are not a replacement technology for digital electronics.

So, although interest in quantum and AI/ML computing is currently superheated due to market opportunities, it is still urgent to advance digital computing even as we pursue these new computing directions.

9.3 Architectural Specialization

9.3.1 The Challenges for Specialized Hardware

In the absence of any miraculous new transistor or other device to enable continued technology scaling, the most promising tool left to a computer architect for extracting continued performance improvements is to use transistors more efficiently by specializing the architecture to target scientific problems. In the near term, the most practical path to continued performance growth will be architectural specialization in the form of many different kinds of accelerators. We believe this to be true because historically it has taken approximately ten years from the introduction of a new transistor concept demonstrated in the lab to become incorporated into a commercial fabrication process. The report for the US Office of Science and Technology Policy (OSTP) by myself and Robert Leland surveyed the landscape of potential CMOS-replacement technologies and found many potential candidates [7], but no obvious replacements were demonstrated in the laboratory at this point. We are therefore already a decade too late to resolve this crisis by finding a scalable post-CMOS path forward. The only hardware option for the coming decade will be architectural specialization and advanced packaging for lack of credible alternatives. In the past, when competing against an exponentially improving general-purpose computing ecosystem, it was very difficult to compete using hardware specialization due to long lead-times and high development costs. However, Thompson and Spanuth's [8] article evaluating the economics of Moore's law points out that the current tapering of Moore's law improvements makes architecture specialization a credible and economically viable alternative to continued reliance on fully general-purpose processors. However, it should be recognized that such a path will have a profound effect on algorithm development and the programming environment.

Overall, there is strong consensus that the tapering of Moore's law will lead to a broader range of accelerators and specialization technologies than we have seen in the past three decades. Examples of this trend already exist – in smartphone technologies, which contain dozens of specialized accelerators co-located on the same chip [9]; in hardware deployed in massive data centers, such as Google's Tensor Processing Unit (TPU) that accelerates the Tensorflow programming framework for machine learning tasks; in-field programmable gate arrays (FPGAs) in the Microsoft cloud used for Bing search and other applications; and in a vast array of other deep learning accelerators. Industry is already moving forward with production implementation of diverse acceleration technologies in the AI and ML markets (e.g., Google TPU [10], Nervana's AI architecture [11]), and in other forms of compute-in-network acceleration for mega-data centers (e.g., Microsoft's FPGA Configurable Cloud and Project Catapult for FPGA-accelerated search [12]). Even before the explosive growth in the AI/ML market, system-on-chip (SoC) vendors for embedded, Internet of Things (IoT), and smartphone applications were already pursuing specialization to good effect. Sophia Shao et al. [9] from Harvard University tracked the growth rate of specialized accelerators in iPhone chips and found a steady growth rate for discrete hardware accelerator units. This grew from around 22 accelerators for Apple's 6th generation iPhone SoC to well over 40 discrete accelerators in their 11th generation chip. Companies engage in this practice of developing such diverse heterogeneous accelerators because the strategy works!

There have also been demonstrated successes in creating *science-targeted* accelerators such as D.E. Shaw's Anton computer, which accelerates molecular dynamics (MD) simulations by nearly 180 times over contemporary HPC systems [13], and the GRAvity PipE (GRAPE) series of specialized accelerators for cosmology and molecular dynamics [14]. Heterogeneity and diversity of architecture are nearly inevitable given current architecture trends. Indeed, even major industry groups such as the IEEE Heterogeneous Integration Roadmap (HIR) [15] are betting their economic future on this approach by building pre-competitive roadmaps for advanced packaging technologies that will deliver an extremely heterogeneous future. This

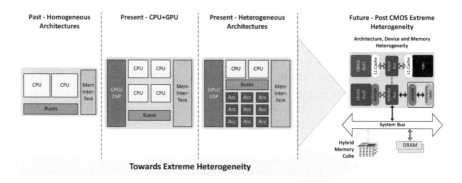

FIGURE 9.6
Architectural specialization and extreme heterogeneity are anticipated to be the near-term response to the end of classical technology scaling. (Figure courtesy of Dilip Vasudevan from LBNL.)

trend toward co-packaging of diverse *extremely heterogeneous* accelerators is already well underway, as shown in Figure 9.6.

Specialization is therefore the most promising near-term approach for continuing to provide the year-on-year performance increases required by all users of scientific computing systems, but specialized architectures also need a well-defined target application. This creates a particular requirement for the sciences to focus on the unique aspects of scientific computing for both analysis and simulation. Recent communications with computing industry leaders suggest that post-exascale HPC platforms will become increasingly heterogeneous environments. Heterogeneous processor accelerators – whether they are commercial designs (e.g., evolutions of current GPU or CPU technologies), emerging reconfigurable hardware, or bespoke architectures that are customized for specific science applications – will therefore optimize hardware and software for particular tasks or algorithms. This will achieve performance and/or energy efficiency gains that would not be realizable using general-purpose approaches. These long-term trends in the underlying hardware technology (driven by the physics) will create daunting challenges for maintaining the productivity and continued performance scaling of HPC codes on future systems.

9.3.2 Programming System and Software Challenges

New software implementations, and in many cases new mathematical models and algorithmic approaches, are necessary to advance the science that can be done with new architectures. This trend will not only continue but will also intensify; the transition from multicore systems to hybrid systems has already caused many code teams to re-factor and redesign their implementations. However, to take the next step to systems that exploit not just one type of accelerator but a full range of heterogeneous architectures will require more fundamental and disruptive changes in algorithm and software approaches [16]. This applies to the broad range of algorithms used in simulation, data analysis, and learning. New programming models or low-level software constructs that hide the details of the architecture from the implementation can make future programming less time-consuming but they will not eliminate or in many cases even mitigate the need to redesign algorithms. Key elements of a path forward include:

- Understanding the impact of proposed architectures on current mathematical kernels and algorithms and using this knowledge to steer the HPC hardware deployment choices through feedback in an iterative codesign process.

- Redesigning current algorithms in response to proposed architectures; hardware choices should be based not only on current algorithms but on the potential performance of new algorithms and even new science use cases.
- Developing advanced programming environments that ease the implementation of these new algorithms and numerical libraries and are able to generate code for these diverse, heterogeneous accelerators.

Applied mathematics is critical to our ability to codesign application and science-relevant accelerators. There are two categories of applications that will need to be redesigned to run effectively in a heterogeneous accelerated environment. In the first type, a single computational motif or kernel is paramount, such as stencil computations with fixed spatial patterns. In this case, there is likely to be a single best choice of hardware design. Most of the success stories regarding specialized architectures fall into this category. The advances in numerical methods can be encapsulated in numerical libraries, such as SPIRAL for FFTs [17], the Basic Linear Algebra Subprograms (BLAS) [18] composable primitives for dense linear algebra, SuperLU [19] for sparse iterative solvers, and GraphBLAS [20] as a composable set of primitives for graph analytics applications, and application frameworks, such as AMReX [21], to make these advances broadly available to the community. These numerical libraries define a handful of simple primitives that can be composed in many ways to implement a myriad of complex algorithms. Currently, this enables vendors to hand-tune a small number of primitive kernels to get maximum performance on their platform for a broad range of applications but this same approach could be used to package dedicated hardware accelerators built around the interfaces to these primitives in a more accessible manner. The second, more complex type is that in which solving the science problem requires fundamentally heterogeneous operations. The heterogeneous operations can be staggered, as one might envision in a data pipeline, so that the data moves through the pipeline as different operations are performed on it. In this scenario, the data may also be moving physically in steps from source to destination, making the use of different architectures for different stages of the computation transparent and separable. For example, in a simulation of a time-evolving state or any iterative solution procedure, each step may contain multiple heterogeneous sub-steps, with each step repeated multiple times, perhaps with different relative (i.e., dynamically changing) costs of the components. No single specialized architecture will be ideal for all stages, suggesting an architectural layout that allows a single code to exploit multiple specialized components. Existing hybrid CPU/GPU systems already allow this, and applications are being re-factored to use this capability. The generalization of this current trend of offloading different algorithmic components to different specialized architectures will not only continue but become more important.

To make this future of diverse accelerators usable and accessible, the first case example will require the codesign of new compiler technology and domain-specific languages (DSLs) designed around the requirements of the target computational motifs (the 13 motifs that extended Phil Colella's original 7 Dwarfs of algorithmic methods [22]). The higher levels of abstraction and declarative semantics offered by DSLs enable more degrees of freedom to optimally map the algorithms onto diverse hardware than traditional imperative languages that over-prescribe the solution. Because this will drastically increase the complexity of the mapping problem, new mathematics for optimization will need to be developed, along with better performance introspection (both hardware and software mechanisms for online performance introspection) through extensions to the roofline model [23]. The use of machine learning and AI technologies will be essential to enable the analysis and automation of dynamic optimizations.

9.3.3 Data Movement Challenges

Extracting more compute performance alone may not be sufficient to realize performance gains in future systems. A potential complication for future digital technologies is that the cost of data movement, and not necessarily computing, already dominates electrical losses and could undermine any potential improvements in the overall compute energy efficiency if not addressed. Since the end of Dennard scaling in 2004, a new technology scaling regime has emerged. As elegantly described in David A. B. Miller's article [24, 25], according to the laws of electrical resistance and capacitance, the intrinsic energy efficiency of a fixed-length wire does not improve appreciably as it shrinks in size with Moore's law improvements in lithography. In contrast, the power consumption of transistors continues to decrease as their gate size, and hence capacitance, decreases. Since the energy efficiency of transistors is improving as sizes shrink, and the energy efficiency of wires is not improving, we have come to a point where the energy needed to move data exceeds the energy used to perform the operations on those data, as shown in Figure 9.7. This leads to extreme bottlenecks and heterogeneity in the cost of accessing data because the costs to move data are strongly distance dependent. Furthermore, although computational performance has continued to increase, the number of pins per chip has not improved at a similar rate [26]. This leads to bandwidth contention that leads to additional performance non-uniformity. The natural consequence of this technological limitation is an increased heterogeneity in data movement and nonuniform memory access (NUMA) effects so long as copper/electrical

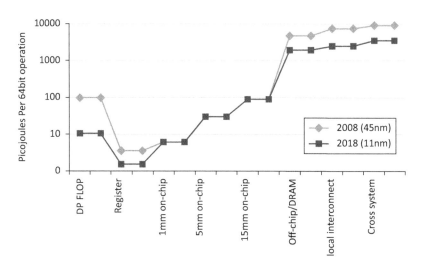

FIGURE 9.7
As lithography has improved, the energy efficiency of wires has not improved as fast as the efficiency of transistors. Consequently, moving two operands just 2 mm across a silicon chip consumes more energy than the floating-point operation performed upon them. (Data for this graph was taken from the 2009 ITRS report and updated using 2015 ITRS data in 2016.)

communication is being utilized. Data locality and bandwidth constraints have long been concerns for application development on supercomputers but recent architecture trends have exacerbated these challenges to the point that they can no longer be accommodated with existing methods such as loop blocking or compiler techniques. Future performance and energy efficiency improvements will require more fundamental changes to hardware architectures, more advanced packaging approaches, and new algorithm designs.

The most significant consequence of these assertions is the impact on scientific applications that run on current HPC systems, many of which codify years of scientific domain knowledge and refinements for contemporary computer systems. To adapt to computing architectures beyond 2025, developers must be able to reason about new hardware and determine what programming models and algorithms will provide the best blend of performance and energy efficiency into the future. Even our theory of complexity for numerical methods is based on counting the number of floating-point operations, which fails to account for the order of complexity of compulsory data movement required by the algorithm. Ultimately, our theories about algorithmic complexity are out of step with the underlying physics and cost model of modern computation. Future systems will express more levels of hierarchy than we are accustomed to in our existing programming models. Not only will there be more levels of hierarchy, but it is also likely that it will become important to optimize the topology of communication. Programmers are already facing NUMA performance challenges within the node but future systems will see increasing NUMA effects between cores within an individual chip package in the future [27, 16]. Although it will become important to optimize for the topology of communication, current programming models do not express information needed for such optimizations, and current scheduling systems and runtimes are not well equipped to exploit such information were it available. Overall, our current programming methodologies are ill-equipped to accommodate changes to the underlying abstract machine model, which would break our current programming systems. An article about the Programming Abstractions for Data Locality (PADAL) workshop by Unat et al. [28] outlines the current state of the art in data locality management in modern programming systems and identifies numerous opportunities to greatly improve automation in these areas.

New algorithms favoring less data movement or higher arithmetic intensity, such as communication-avoiding and high-order operators, are already being developed. In addition, data-centric programming abstractions must be built into new partitioned global address space (PGAS) programming systems in order to communicate algorithmic information about data locality to the underlying software system. These capabilities are even more crucial for heterogeneous architectures where different accelerators have different memory/communication speeds. More complex algorithms increase the challenges of performance modeling, and tools such as the Roofline model need to be improved to take heterogeneity into account. Although applied mathematicians must lead the effort to re-factor core simulation and analysis algorithms, they will need to be working as part of collaborative teams containing algorithm, application, software, computer architecture, and performance analysis expertise. Looking ahead, we expect to demonstrate algorithmic redesign of simulation algorithms that target multiple specialized architectures and refine the software prototypes to the point that they can transition to production release and adoption by leading-edge facilities.

9.3.4 Photonics and Resource Disaggregation

Architectural specialization is creating new data center requirements such as support for emerging accelerator technologies for machine-learning workloads and rack disaggregation strategies that will push the limits of current interconnected technologies. While the latest high-throughput processor chips with many CPU/GPU cores are intrinsically capable of carrying extremely demanding computing tasks, they do not have the necessary off-chip bandwidth for full and efficient utilization of their resources. To address this challenge, we must overcome the bandwidth density limitations of current electrical packages. An alternative to this future is to explore co-integration of photonic technologies that do not suffer from this data movement distance constraint. Photonic interconnect technologies have been proposed to address this critical data-movement challenge because of their well-known bandwidth density and energy efficiency advantages. However, system-wide energy efficiency and performance gains cannot be attained by simple one-to-one photonic replacement of existing links and switches. Co-packaging of photonics as in-package devices for photonic multi-chip modules (MCMs) has been offered as a compelling approach [29]. The in-package bandwidth densities due to the extremely high pin-density enabled by copper pillar or solder microbump technologies are very well matched to photonic technologies. Whereas photonic technologies are currently often sold on the basis of higher bandwidth and energy efficiency (e.g., lower picojoules per bit), these emerging workloads and technology trends will shift the emphasis to other metrics such as bandwidth density (as opposed to bandwidth alone), reduced latency, and performance consistency. Such metrics cannot be accomplished with device improvements alone but will require a systems view of photonics in computing platforms.

9.4 CMOS Replacement: Inventing the "New Transistor"

9.4.1 The Challenge

The development of new devices (e.g., a better transistor or digital logic technology) could greatly lower the energy consumed by logic operations. However, the development of such a "new transistor" will require fundamental breakthroughs in materials, and their suitability for use in future computing devices must be evaluated in the context of circuits and full system architectures. This will enable us to determine how to make the best use of such new devices and whether efficiency improvements at device scale can translate to delivered improvements to applications at chip and system scale. An integral dimension of this challenge is combining these two primary paths with other promising avenues such as 3D integration and novel memory technologies, as well as overcoming packaging and integration challenges arising from incorporating new materials or technology improvements. Taking information and metrology from such studies will guide the development of any new post-CMOS transistor and logic technologies. A prior article written by Robert Leland and myself for the US Office of Science and Technology Policy (OSTP) in 2013, and then re-released as an IEEE Computer article in 2015 [7], surveys the many different technology options that are currently available and scores those opportunities. Current state-of-the-art electronic devices are limited by what is termed "Boltzmann's Tyranny." Statistical thermodynamics (primarily the probability distribution function that an electron will occupy a particular energy state) sets an ultimate limit on the switching efficiency for electronic devices that depend upon a potential barrier, which is the basis of operation for all modern transistors. A conventional transistor is limited by Boltzmann's tyranny to $k_B Tln(10)/qk_B Tln(10)/q$, or 60 mV, to create a ten-times change in current – commonly expressed as "60 mV per decade" on a logarithmic scale – as a fundamental theoretical limit for switching performance for a theoretically *perfect* switch that uses a potential barrier as its mode of operation. The limits are stark and they are physically fundamental. Nikonov and Young's 2016 article [30] reviews the challenges of "Boltzmann's Tyranny" for electronic devices, and also describes quantitatively just how far these technology options are from being clear candidates for completely replacing our present CMOS technology.

9.4.2 Deep Codesign to Accelerate the Pace of Discovery

Typically, new electronic devices, such as new transistors or memory elements, are evaluated in isolation at a physical level, which fails to capture the architectural-level impact of the device. Existing hardware design tools do not adequately account for the benefits and limitations of future devices. *It is essential that architects and system designers, in particular, have metrics they can use to reason out the impact of each new device on architectures and designs and their complex interactions with existing technologies. They do not have that now.* To correct this shortfall we need urgently and immediately to explore the specialized architectural design space efficiently and systematically in combination with emerging device technologies to avoid stalling performance scaling while waiting for radical new technologies to mature. The ability to guide the development of future devices requires evaluation of their performance based on the ultimate outcomes they achieve for target applications. The value of new and novel materials or new device technologies is not currently understood in a system context. Performance and behaviors in a system context are not currently understood at a device or materials level. True codesign to advance future systems containing novel devices and materials requires feedback that spans all layers, from atomic-scale materials to large-scale complex systems to meet the needs of emerging scientific applications as diagrammed in Figure 9.8.

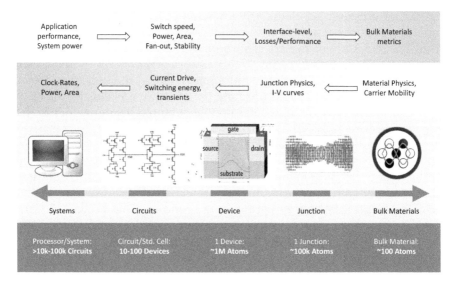

FIGURE 9.8
A deep codesign framework to accelerate the discovery of CMOS replacement technologies must include research and development at all scales from the design of new materials to processor and system design.

Only with codesign to cover this broad space, with consideration of manufacturing challenges can we expect to make progress in all areas cohesively to bring about real change to the IT energy outlook. Further, the output of this work will provide a path to sustaining exponential growth in computing capabilities to enable new scientific discoveries and maintain economic vitality in all segments of the computing market, from IoT and consumer electronics to data centers and supercomputing. Figure 9.8 shows an early prototype of an integrated approach spanning research in fundamental materials science to architectures, circuits, and full system architectures, with the intent to dramatically accelerate the discovery process for future transistors. Our vision is to develop a codesign framework that integrates the physical layers and the logical layers of complex systems so that it can propagate quantitative information to guide decisions for the development of better solutions. The codesign framework would enable us to develop unified materials/device/circuit/system electronic design automation and simulation tools to reduce the development timeline for mission-critical science. The long-term solution requires fundamental advances in our knowledge of materials and pathways to control and manipulate information elements at the limits of energy flow. We will require groundbreaking advances in device technology going beyond CMOS that arise from fundamentally new knowledge of control pathways, as well as advances in system architectures and programming models to allow the energy benefits of scaling to be realized. A complete workflow will need to be constructed, linking device models and materials to circuits. These circuits can be evaluated through the efficient generation of specialized hardware architectural models so that the different advances in basic technologies can be compared for their benefits to ultimate system performance. The architectural simulations that result from this work will yield a better understanding of the performance impact of these emerging approaches on target applications. It would also enable early exploration of new software systems that would make these new architectures useful and programmable.

In the longer term, the research community can expand the modeling framework to include non-traditional computing models and accelerators, such as neuro-inspired architectures and quantum accelerators, as components in our simulation infrastructure. We will also need to develop the technologies to automate aspects of the algorithm/architecture/software-environment system codesign process so that developers can evaluate their ideas early in future hardware development cycles. Ultimately, we need to close the feedback loop from the software all the way down to the device level in order to make software an integrated part of this infrastructure.

9.5 Revisiting Reversibility

In his lectures, Feynman spoke eloquently about the possibility of using reversible computing to go below the Landauer limit [31] for the minimum energy expended on a logical operation. As Moore's law, based on CMOS technology and shrinking transistors, slows down, it is a good time to revisit the alternatives. The Landauer limit of $k_B T ln(2) k_B T ln(2)$ energy expended by a logic operation is a brilliant merger of the second law of thermodynamics with Shannon's limit for information representation. All digital computation as we know it today is subject to this limit, and we are still several orders of magnitude away from that limit even with the most cutting-edge CMOS technologies. However, moving electrons through wires to communicate information is fundamentally irreversible and is not treated by this analysis. Even with current CMOS-based digital electronics, there is consensus that data movement now consumes more power than actual computing operations [32]. The present analysis of reversible computation does not address the issue that resistive losses during communication between the compute elements are fundamentally irreversible.

This is not to say that reversible computing is a dead-end. On the contrary, the analysis highlights the fact that foundational physics and technology needed to successfully implement this approach will have to be fundamentally different from what we have been using up until now. Semiconductors are out of the picture. New physics and supporting materials that can enable the transfer of information without loss will be essential. For example, reversible computing using superconducting technology [33] is one possible approach. Other opportunities with low loss include optical reversible computing [34], nano-structured thermodynamic devices that exhibit isentropic (reversible adiabatic) behavior [35], and even possible exotic quasiparticles such as skyrmions [36] in topologically engineered materials such as topological insulators. This calls out again for a deep codesign approach covering physics, materials, and new theories of computation. Feynman's exploration of reversibility says that there *is* no lower limit for the energy loss of logic operations but it may require decades of exploration to find the right physics and materials to overcome the engineering obstacles to a successful implementation. However, Feynman's analysis causes me to regard with great skepticism any claims that we have reached the end of the line for improving computing performance and efficiency, and to resist any calls to stop looking for better approaches that will continue to sustain us in the future. Although Moore's law as we know it today is coming to a close, Moore's law is an ambition to continue scaling of computing performance into the future. Feynman's analysis of reversible computing has moved the fundamental energy loss limit far beyond what we can contemplate achieving at present. It is certainly the end of conventional computing as we know it, but Feynman shows us that this is not the end of computing. There remains plenty of room at the bottom!

9.6 Conclusion

Semiconductor technology has a pervasive role to play in future energy, economic, and technology security. To effectively meet societal needs and expectations in such a broad context, any new devices and computing paradigms must be economically manufacturable at scale and provide an exponential improvement path [7]. The realization of these requirements could necessitate as substantial a technological shift as the transition from vacuum tubes to semiconductors. Making such a transition will require not years but decades to achieve. Whether the semiconductor roadmap has ten or twenty years of remaining vitality, researchers must begin now to lay a strategic foundation for change.

Bibliography

1. International Energy Agency. *Gadgets and Gigawatts: Policies for Energy Efficient Electronics*. International Energy Agency (IEA), 2009.

2. Nicola Jones. How to stop data centres from gobbling up the world's electricity. *Nature*, 561(7722):163–167, 2018.

3. Gordon E. Moore. Cramming more components onto integrated circuits. *Electronics*, 38(8), April 1965.

4. Chris Mack. The multiple lives of Moore's law. *IEEE Spectrum*, 52(4):31–31, 2015.

5. I.L. Markov. Limits on fundamental limits to computation. *Nature*, 512(7513):147–154, 2014.

6. Robert Colwell. The chip design game at the end of Moore's law. In *2013 IEEE Hot Chips 25 Symposium (HCS)*, pages 1–16, 2013.

7. John M. Shalf and Robert Leland. Computing beyond Moore's law. *Computer*, 48(12):14–23, 2015.

8. Neil C. Thompson and Svenja Spanuth. The decline of computers as a general purpose technology. *Communications of the ACM*, 64(3):6472, February 2021.

9. Yakun Sophia Shao, Sam Likun Xi, Vijayalakshmi Srinivasan, Gu-Yeon Wei, and David M. Brooks. Co-designing accelerators and soc interfaces using gem5-aladdin. In *49th Annual IEEE/ACM International Symposium on Microarchitecture, MICRO 2016*, Taipei, Taiwan, October 15–19, 2016, pages 48:1–48:12. IEEE Computer Society, 2016.

10. N. Jouppi, C. Young, Nishant Patil, David A. Patterson, Gaurav Agrawal, R. Bajwa, Sarah Bates, Suresh Bhatia, N. Boden, Al Borchers, Rick Boyle, Pierre luc Cantin, Clifford Chao, Chris Clark, Jeremy Coriell, Mike Daley, M. Dau, J. Dean, Ben Gelb, T. Ghaemmaghami, R. Gottipati, William Gulland, R. Hagmann, C.R. Ho, Doug Hogberg, J. Hu, R. Hundt, D. Hurt, J. Ibarz, A. Jaffey, Alek Jaworski, Alexander Kaplan, Harshit Khaitan, Daniel Killebrew, Andy Koch, N. Kumar, Steve Lacy, J. Laudon, James Law, Diemthu Le, Chris Leary, Z. Liu, Kyle A. Lucke, Alan Lundin, G. MacKean, A. Maggiore, Maire Mahony, K. Miller, R. Nagarajan, Ravi Narayanaswami, Ray Ni, K. Nix, Thomas Norrie, Mark Omernick, Narayana Penukonda, A. Phelps, J. Ross, Matt Ross, Amir Salek, E. Samadiani, C. Severn, G. Sizikov, Matthew Snelham, J. Souter, D. Steinberg, Andy Swing, Mercedes Tan, G. Thorson, Bo Tian, H. Toma, Erick Tuttle, Vijay Vasudevan, Richard Walter, W. Wang, Eric Wilcox, and D. Yoon. In-datacenter performance analysis of a tensor processing unit. In *2017 ACM/IEEE 44th Annual International Symposium on Computer Architecture (ISCA)*, pages 1–12, 2017.

11. Jeremy Hsu. Nervana systems: Turning neural networks into a service resources startups. *IEEE Spectrum*, 53(6):19, 2016.

12. Adrian M. Caulfield, Eric S. Chung, Andrew Putnam, Hari Angepat, Jeremy Fowers, Michael Haselman, Stephen Heil, Matt Humphrey, Puneet Kaur, Joo-Young Kim, Daniel Lo, Todd Massengill, Kalin Ovtcharov, Michael Papamichael, Lisa Woods, Sitaram Lanka, Derek Chiou, and Doug Burger.

A cloud-scale acceleration architecture. In *2016 49th Annual IEEE/ACM International Symposium on Microarchitecture (MICRO)*, pages 1–13, 2016.

13. David E. Shaw, J.P. Grossman, Joseph A. Bank, Brannon Batson, J. Adam Butts, Jack C. Chao, Martin M. Deneroff, Ron O. Dror, Amos Even, Christopher H. Fenton, Anthony Forte, Joseph Gagliardo, Gennette Gill, Brian Greskamp, C. Richard Ho, Douglas J. Ierardi, Lev Iserovich, Jeffrey S. Kuskin, Richard H. Larson, Timothy Layman, Li-Siang Lee, Adam K. Lerer, Chester Li, Daniel Killebrew, Kenneth M. Mackenzie, Shark Yeuk-Hai Mok, Mark A. Moraes, Rolf Mueller, Lawrence J. Nociolo, Jon L. Peticolas, Terry Quan, Daniel Ramot, John K. Salmon, Daniele P. Scarpazza, U. Ben Schafer, Naseer Siddique, Christopher W. Snyder, Jochen Spengler, Ping Tak Peter Tang, Michael Theobald, Horia Toma, Brian Towles, Benjamin Vitale, Stanley C. Wang, and Cliff Young. Anton 2: Raising the bar for performance and programmability in a special-purpose molecular dynamics supercomputer. In *Proceedings of the International Conference for High Performance Computing, Networking, Storage and Analysis*, SC '14, page 4153. IEEE Press, 2014.

14. Itta Ohmura, G. Morimoto, Y. Ohno, Akifumi Hasegawa, and M. Taiji. Mdgrape-4: A special-purpose computer system for molecular dynamics simulations. *Philosophical Transactions. Series A: Mathematical, Physical, and Engineering Sciences*, 372(2021), 2014.

15. William Chen and Bill Bottoms. Heterogeneous integration roadmap: Driving force and enabling technology for systems of the future. In *2019 Symposium on VLSI Technology*, pages T50–T51, 2019.

16. H. Johansen, L.C. McInnes, D. Bernholdt, J. Carver, M. Heroux, R. Hornung, P. Jones, B. Lucas, A. Siegel, and T. Ndousse-Fetter. Software productivity for extreme-scale science report on DOE workshop, 2014.

17. Christopher K. Turnes and Justin K. Romberg. Spiral FFT: An efficient method for 3-d FFTS on spiral MRI contours. In *ICIP*, pages 617–620. IEEE, 2010.

18. L. Susan Blackford, Antoine Petitet, Roldan Pozo, Karin Remington, R. Clint Whaley, James Demmel, Jack Dongarra, Iain Duff, Sven Hammarling, Greg Henry, et al. An updated set of basic linear algebra subprograms (BLAS). *ACM Transactions on Mathematical Software*, 28(2):135–151, 2002.

19. Xiaoye Sherry Li, James Demmel, John Gilbert, Laura Grigori, and Meiyue Shao. *SuperLU*. Springer US, Boston, MA, pages 1955–1962, 2011.

20. Jeremy Kepner, Henning Meyerhenke, Scott McMillan, Carl Yang, John D. Owens, Marcin Zalewski, Timothy Mattson, Jose Moreira, Peter Aaltonen, David Bader, et al. Mathematical foundations of the graphblas. In *2016 IEEE High Performance Extreme Computing Conference (HPEC)*, September 2016.

21. W. Zhang et al. AMReX: A framework for block-structured adaptive mesh renement. *Journal of Open Source Software*, 4(37):1370, 2019.

22. Krste Asanovic, Ras Bodik, Bryan Christopher Catanzaro, Joseph James Gebis, Parry Husbands, Kurt Keutzer, David A. Patterson, William Lester Plishker, John Shalf, Samuel Webb Williams, and Katherine A. Yelick. The landscape of parallel computing research: A view from Berkeley. Technical Report UCB/EECS-2006-183, EECS Department, University of California, Berkeley, December 2006.

23. Samuel Williams, Andrew Waterman, and David Patterson. Roofline: An insightful visual performance model for multicore architectures. *Communications of the ACM*, 52(4):6576, April 2009.

24. David A.B. Miller. Optical interconnects to electronic chips. *Applied Optics*, 49(25):F59–F70, September 2010.

25. D.A.B. Miller. Rationale and challenges for optical interconnects to electronic chips. *Proceedings of the IEEE*, 88(6):728–749, 2000.

26. M. Horowitz, Chih-Kong Ken Yang, and S. Sidiropoulos. High-speed electrical signaling: Overview and limitations. *IEEE Micro*, 18(1):12–24, 1998.

27. Peter M. Kogge and John Shalf. Exascale computing trends: Adjusting to the "new normal" for computer architecture. *Computing in Science and Engineering*, 15(6):16–26, 2013.

28. Didem Unat, Anshu Dubey, Torsten Hoefler, John Shalf, Mark Abraham, Mauro Bianco, Bradford L. Chamberlain, Romain Cledat, H. Carter Edwards, Hal Finkel, Karl Fuerlinger, Frank Hannig, Emmanuel Jeannot, Amir Kamil, Jeff Keasler, Paul H.J. Kelly, Vitus Leung, Hatem Ltaief, Naoya Maruyama, Chris J. Newburn, and Miquel Perics. Trends in data locality abstractions for HPC systems. *IEEE Transactions on Parallel and Distributed Systems*, 28(10):3007–3020, 2017.

29. Nathan C. Abrams, Qixiang Cheng, Madeleine Glick, Moises Jezzini, Padraic Morrissey, Peter O'Brien, and Keren Bergman. Silicon photonic 2.5d multi-chip module transceiver for high-performance data centers. *Journal of Lightwave Technology*, 38(13):3346–3357, 2020.

30. D.E. Nikonov and I.A. Young. Overview of beyond-CMOS devices and a uniform methodology for their benchmarking. *Proceedings of the IEEE*, 101:2498–2533, 2016.

31. R. Landauer. Irreversibility and heat generation in the computing process. *IBM Journal of Research and Development*, 5(3):183191, July 1961.

32. Shekhar Borkar. Role of interconnects in the future of computing. *Journal of Lightwave Technology*, 31(24):3927–3933, December 2013.

33. Naoki Takeuchi, Yuki Yamanashi, and Nobuyuki Yoshikawa. Reversible computing using adiabatic superconductor logic. In Shigeru Yamashita and Shin-ichi Minato, editors, *Reversible Computation*, pages 15–25. Cham. Springer International Publishing, 2014.

34. Huanan Li, Lucas J. Fernández-Alcázar, Fred Ellis, Boris Shapiro, and Tsampikos Kottos. Adiabatic thermal radiation pumps for thermal photonics. *Physical Review Letters*, 123(16):165901, October 2019.

35. T.E. Humphrey and H. Linke. Reversible thermoelectric nanomaterials. *Physical Review Letters*, 94(9):096601, March 2005.

36. Yan Zhou. Magnetic skyrmions: Intriguing physics and new spintronic device concepts. *National Science Review*, 6(2):210–212, 10, 2018.

10

Feynman on Artificial Intelligence and Machine Learning

Eric Mjolsness

CONTENTS

10.1 Introduction

Richard Feynman's role in sparking the development of quantum computing is well known, as well as his affinity for calculating machines. Less well known is that he was seriously interested in artificial intelligence (AI), and in neural networks in their 1980s incarnation as an approach to artificial intelligence, a subject led at Caltech by John Hopfield, who had recently arrived as a full professor.

The Physics of Computation course began as a collaborative offering between Feynman, Hopfield, and Carver Mead, the pioneer of very large-scale integration (VLSI) computer chip design, who was also interested in neural computation as a new hardware paradigm. Feynman thought quite a bit about how to achieve progress in AI that had eluded the mainstream symbolic approach, through neural networks instead, particularly in their capabilities for pattern recognition and machine learning (ML).

DOI: 10.1201/9781003358817-10

As a physics graduate student at Caltech from 1980–1985, I was very fortunate to have many interactions with Richard Feynman, some of which bear on his interest in neural networks and artificial intelligence. To me, these years at Caltech seemed like a time of great intellectual ferment in multiple overlapping directions. What follows is reconstructed from decades-old fallible memory. Others will almost surely remember things a bit differently.

Today, neural networks are so successful that it is becoming hard to understand how non-mainstream such thinking was at the time. Neural networks had lost the evolutionary race with von Neumann computer architectures in the late 1950s and early 1960s, despite pioneering hardware work by Frank Rosenblatt, Bernard Widrow, and others. Enthusiasm and funding for neural network research had also suffered from the reception (but not the actual text) of the book on "Perceptrons" by Marvin Minsky and Seymour Papert, which included negative results on what we could now call single-layer feed-forward neural networks without hidden units. But in the early 1980s, John Hopfield, Geoffrey Hinton, Terry Sejnowski, Kunihiko Fukushima, and others were trying to revive neural networks, with new quantitative dynamical algorithms and architectures. Feynman was interested (along with other physicists) because of the physics-like characteristics of this approach. The essence of a neural network could be described in quantitative terms, mostly by equations, at a higher level of abstraction than computer code. Feynman certainly knew Hopfield's work well; how much he knew about other neural network research is harder to say because he preferred to re-invent everything independently, from first principles.

At MIT, Marvin Minsky was a pioneer of artificial intelligence of the classic non-neural "symbolic" variety, and Feynman and Minsky were friends. As Feynman became interested in AI, somehow a deal was struck by which Minsky recommended that Papert's former doctoral student, Gerald Sussman, also prominent in mainstream symbolic AI as well as programming languages, should visit Caltech on an extended sabbatical where he could interact with Feynman on AI … and also execute his plan to build a cost-effective special purpose "digital orrery", a planetary orbit computer. So, Gerry Sussman collaborated on the Physics of Computation course too, delivering lectures on hardware, programming using his own Scheme dialect of Lisp, and AI.

The role in which I knew Feynman best was as a teaching assistant for the Physics of Computation class in several iterations from 1983–1985. Also, as a physics/neural network PhD student of John Hopfield, I was deputized to push the buttons and flip the tapes on a tape recorder for many of Feynman's lectures. Feynman was democratically minded, and very often after class, he would continue the discussion with his TAs and a changing cast of visitors, including Sussman, at the Institutes' unpretentious cafeteria a few steps away. (US universities have evolved; the cafeteria has become more pretentious.) Sometimes we would instead go to the Caltech Atheneum, the faculty club that also welcomed graduate students. Another TA for later iterations of this course was the remarkable physics graduate student Mike Douglas, then master of the digital orrery software and knowledgeable in AI, and later of string theory renown. Needless to say, the lunch discussions were fascinating and sometimes spectacular. The TAs sometimes also met Feynman in his office. I heard his voice ringing in my ears for years afterwards.

It is well known that Feynman had a leading role in organizing the computing effort at Los Alamos from 1943–1945, during the Manhattan Project in World War II. The "computers" of the day were human operators working on mechanical calculators, but the information flow among them had to be designed somehow. During lunchtime socializing of the 1980s, we heard previews of many of Feynman's Los Alamos and other stories that subsequently showed up in his later, memoir-like books. It is also well reported that in the mid-1980s, during the time of the Physics of Computation classes, he consulted for the Thinking Machines computer company. He worked on the bit-serial message-routing protocol for their highly parallel Connection Machine CM-1 computer and performed an analysis of its need for message overflow buffers. Also during this time, he undertook the quantum computing work described elsewhere in this book. These contemporaneous projects were described in the class. So his interest in computing was broad and long-lived.

Feynman had various dream projects that he didn't get time to finish. He died just a few years after giving these lectures, in 1988, at the age of 69. He wanted to think and visualize his way through the configuration space of quantum chromodynamics, the theory of the strong nuclear force, in order to deeply understand it the way he had visualized his way through several famous problems in statistical physics. And he wanted to train, shape, or perhaps evolve a neural network to achieve artificial intelligence, complete with vision and language. The neural network/learning machine (details were a bit sketchy) was apparently going to be person-like and female. "She" would master one skill after another, like a human baby, trained by doting scientists who would supply the right learning tasks in the right order.

Gerry Sussman, who was by far the most knowledgeable AI expert on the AI-skeptical Caltech campus during his visit, did not love the Feynman plan for neural network-style learning machines. He thought it was naive, harking back to the late 1950–early 1960 ideas that had not panned out. On the other hand, mainstream AI was just entering the "AI winter" phase of decreased confidence and funding, so Sussman could not win an argument on the future of AI outright, based on the accomplishments of the field at that time. Feynman in turn thought Sussman's "problem is he knows too much" about current AI research (he probably knew all of it, as the field was much smaller then), so Sussman couldn't see that the real progress was going to come from a completely different direction. These mutual diagnoses emerged only under fairly persistent questioning by the younger scientists present. In retrospect of more than 30 years, after the great strides in neural networks for computer vision and natural language of 2010–2020, Feynman looks to have had the decisive advantage in this symbolic vs. neural AI debate ... but as I will suggest, things could change again!

10.2 Physics-like Neural Networks of the 1980s

We can use Feynman's approach to statistical mechanics to understand Hopfield's neural networks and closely related networks. In particular, Feynman's variational method (e.g., [1], Section 3.4) allows us to understand key aspects of such neural networks. We will make a mathematical analogy between a network of idealized "neurons" that interact only with their neighbors in a network of connections, on the one hand, and a rigid lattice of simplified "atoms" that interact only with their neighbors in a solid material, on the other. The common theme is the use of real-number connection or interaction strength values T_{ij} that specify whether, and how strongly, and with what ± sign two network nodes (either neurons or atoms, indexed by integers i and j) influence one another. When $T_{ij} = 0$ then there is no network or lattice connection and no direct influence between nodes i and j, and for large systems we can imagine this is the most common case. It is the nonzero connection strengths that specify the "network". A major point of difference between neural networks and solid material models in physics will be that the connection strengths in a neural network are allowed to change under some dynamics of learning.

For the physics-based material model, we assume the Boltzmann probability distribution of equilibrium statistical mechanics where the probability of a state

$$p_{Boltz}\left(state\right) \propto e^{-\beta Energy(state)}$$

This proportionality makes sense because energy is extensive (additive over noninteracting subsystems), and probability is multiplicative over independent subsystems, such as networks that comprise unconnected subnetworks.

10.2.1 Neural Activation Dynamics

Next, we establish the conventional statistical mechanics context, exhibit Feynman's convexity inequality, derive a mean field theory (MFT) approximation, and apply it to neural network activation dynamics.

For a Boltzmann distribution, the free energy $F(\beta,T,h) = \langle H \rangle - TS$ of an equilibrium statistical mechanics model on classical atomic "spin" variables $s_i \in \{\pm 1\}$ satisfies

$$e^{-\beta F} = Z(\beta,T,h) = \sum_{\{s_i \in \{\pm 1\}\}} e^{-\beta H[s]}, \tag{10.1}$$

where for an Ising model, for example,

$$H_{Ising}[s] \equiv -\frac{1}{2}\sum_{i \ne j} T_{ij}s_i s_j - \sum_i h_i s_i. \tag{10.2}$$

In this notation, $\beta = 1/T$ is the inverse temperature, angle brackets denote thermal averages with respect to the Boltzmann probability $p_{Boltz}[s] = e^{-\beta H[s]}/Z$, and $S = -\langle \log p[s] \rangle$ is entropy. We use units in which Boltzmann's constant k_B is unity; otherwise, we would have $\beta = 1/k_B T$. The connection strength matrix $\boldsymbol{T}$ is symmetric and may be sparse and/or structured. Don't confuse the indexed connection matrix $\boldsymbol{T} = [T_{ij}]$ with the scalar temperature parameter T. To verify (10.1), it suffices to substitute p_{Boltz} into S. We could also interpret the $s_i \in \{\pm 1\}$ variables as idealized binary-valued neuron number i being "on" vs. "off", or "firing" vs. "not firing". Below we will seek energy-minimizing dynamics for "analog" real-valued neurons, whose real values are derived as probabilities of firing vs. not firing. This derivation proceeds by way of mean field theory.

Since the free energy F and the "partition function" Z may be hard to calculate, we compare them to their values in some related system (H_0, F_0) chosen to be easier to calculate:

$$Z = \frac{\sum_{\{s_i\}} e^{-\beta(H[s]-H_0[s])}e^{-\beta H_0[s]}}{\sum_{\{s_i\}} e^{-\beta H_0[s]}} e^{-\beta F_0} = \left\langle e^{-\beta(H-H_0)} \right\rangle_0 e^{-\beta F_0}$$

From the *convexity* of the exponential function, Feynman points out that

$$\left\langle e^{-\beta(H-H_0)} \right\rangle_0 \ge e^{-\beta\langle H-H_0 \rangle_0}$$

and therefore

$$e^{-\beta F} = Z = \left\langle e^{-\beta(H-H_0)} \right\rangle_0 e^{-\beta F_0} \ge e^{-\beta\langle H-H_0 \rangle_0}e^{-\beta F_0}$$

whence

$$F \leq F_0 + \langle H - H_0 \rangle_0 \text{ or } F \leq \langle H \rangle_0 - TS_0. \tag{10.3}$$

Of course, Feynman actually does this for path integrals and quantum systems. This key inequality, relating quantities (F and Z) that may be hard to calculate to quantities that are much easier to calculate, is sometimes attributed to Gibbs, Bogoliubov, and Feynman.

Now from (10.3), we can derive the (MFT) approximation by choosing an independent product distribution of all the classical spins, i.e.,

$$H_0 = \Sigma_i \, \mu_i s_i = -\Sigma_i u_i s_i$$

Then

$$TS_0 = -\Sigma_i u_i \langle s_i \rangle_0 - F_0 \text{ where } -\beta F_0 = \Sigma_i \log(2\cosh \beta u_i).$$

The variational bound can be calculated as

$$F \leq \langle H \rangle_0 - TS_0 = -\frac{1}{2}\sum_{i \neq j} T_{ij} \langle s_i \rangle_0 \langle s_j \rangle_0 - \sum_i (h_i - u_i) \langle s_i \rangle_0 - \frac{1}{\beta} \sum_i \log(2\cosh \beta u_i)$$

Calculating $\langle s_i \rangle_0 = -\partial F_0 / \partial u_i$, or equivalently minimizing this upper bound with respect to u_i, yields $v_i \equiv \langle s_i \rangle_0 = \tanh(\beta u_i)$ where $v_i \in (-1,1)$. Then we can eliminate u_i from the variational bound in favor of v_i. The bound becomes:

$$F \leq E_{MFT}[v] \equiv -\frac{1}{2}\sum_{i \neq j} T_{ij} v_i v_j - \sum_i h_i v_i + \sum_i \varphi(v_i) \tag{10.4}$$

where

$$\varphi(v) = \frac{1}{\beta}\left[v \tanh^{-1}(v) - \log\left(2\cosh\left(\tanh^{-1}(v)\right)\right) \right]$$

Since

$$\tanh^{-1}(v) = \frac{1}{2}\log\left(\frac{1+v}{1-v}\right)$$

for

$$v \in (-1,1),$$

we find

$$\varphi(v) = \frac{1}{\beta}\left[\left(\frac{1+v}{2}\right)\log\left(\frac{1+v}{2}\right) + \left(\frac{1-v}{2}\right)\log\left(\frac{1-v}{2}\right)\right] = -TS_{MFT}[v],$$

which is another $\sum p \log p$ entropy expression since $(1 \pm v)/2 \in [0,1]$ and they sum to 1.

The quality of the approximation of (10.4) is naturally measured by how close the right-hand side $E_{MFT}[v]$ comes to the left-hand side quantity F, with the best possible approximation of this class being obtained by minimizing $E_{MFT}[v]$ as a function of all the v_i. We could do this algorithmically, or we could consider some *dynamics* for v_i that seeks a minimum, and then implement that on a computer or more directly in analog electronic hardware.

So, in addition to fast equilibration of the vector $s = [s]$ of spins s_i, we now imagine a slower time-scale dissipative dynamics that act irreversibly to minimize the approximating free energy as modeled by $E_{MFT}[v]$. The derivative of the potential $\varphi(v)$ is $\tanh^{-1}(v)/\beta$, so $E_{MFT}[v]$ is minimized at

$$v_i = \tanh\left(\beta u_i \equiv -\beta \partial E_{Ising}[v]/\partial v_i\right)$$

so that

$$v_i = \tanh\left(\beta \sum_j T_{ij}v_j + \beta h_i\right) \tag{10.5}$$

at a local minimum. Once an energy-minimizing system has reached an isolated local minimum of the energy, it can go no further and must stop – it has reached a "fixed point" of the dynamics. Equation (10.5) is the fixed-point equation for Hopfield's "analog" (real-valued, $v_i \in (-1,1)$) neural network. Similarly, for MFT and $\{0,1\}$-valued "neurons" or "units", we obtain the entropy expression

$$\varphi(v) = \frac{1}{\beta}\left(v \log v + (1-v)\log(1-v)\right)$$

whence

$$v = g(u) = e^{\beta u}/\left(1 + e^{\beta u}\right))$$

$$= 1/\left(1 + e^{-\beta u}\right).$$

More generally, $\varphi(v) = \int^v g^{-1}(v)\,dv$ where g is the artificial neuron's (mono-tonic) activation function. Then the fixed-point equation becomes

$$v_i = g_i\left(\sum_j T_{ij}v_j + h_i\right). \tag{10.6}$$

One can also allow the potential function φ and the activation function $g = \varphi'^{-1}$ to depend on the variable index i as indicated in equation (10.6), while retaining the useful energy-minimizing optimization property of the network with

$$E[v] = -\frac{1}{2}\sum_{i\neq j}T_{ij}v_iv_j - \sum_i h_iv_i + \sum_i \varphi_i(v_i)$$

but either of these generalizations gives up the derivation from equilibrium MFT.

Hopfield also introduced a descent dynamics capable of finding these fixed points by energy minimization [2]:

$$v_i = g(u_i) \quad \text{and} \quad \tau_i\frac{du_i}{dt} + u_i = \sum_j T_{ij}v_j + h_i. \tag{10.7}$$

Input to the system can be provided through h or through the initial condition $v(t = 0)$. This architecture was applied to small combinatorial optimization problems, such as the Traveling Salesman Problem.

This variational MFT derivation generalizes naturally [3] from Ising "spins" $s_i \in \{\pm 1\}$ to Potts model discrete state variables, which take one of a fixed finite number of values $q_i \in \{1,\dots A\}$ that we can encode using $s_{ia} \in \{0,1\}$ and $q_i = \sum_a as_{ia}$, where $\sum_a s_{ia} = 1$. Then the potential energy function becomes

$$\varphi\left([v_{i*}]\right) = \frac{1}{\beta}\sum_a v_{ia}\log v_{ia} = -T\tilde{S}[v]$$

(the entropy of the ith soft choice distribution) and the activation function is:

$$v_{ia} = e^{\beta u_{ia}} / \sum_b e^{\beta u_{ib}} \quad \text{where} \quad u_{ia} \equiv \partial E_{Potts}[v]/\partial v_{ia}. \tag{10.8}$$

This is the well-known "soft-max" activation function, which softly (depending on parameter β) chooses among a finite set of alternatives to maximize u_{ia} over index a, automatically obeying the constraints $v_{ia} \geq 0$ and $\sum_a v_{ia} = 1$. Its usage and importance have steadily increased over time since it provides a differentiable choice over more than two alternatives. A further natural generalization is to the "soft-assign" fast dynamics which alternates row and column normalizations like (10.8) and converges to a solution satisfying the constraints of a doubly stochastic or "assignment" matrix: $v_{ia} \geq 0, \sum_a v_{ia} = 1$, and $\sum_i v_{ia} = 1$. The soft-assign architecture was applied to a number of semi-symbolic problems (e.g., [4]) bordering on the cognitive.

10.2.2 Neural Learning Dynamics

In a somewhat looser manner, we can use statistical mechanics ideas to obtain *learning dynamics* in neural networks. We allow the symmetric, real-valued connection matrix T to become dynamic by analogy with the activation spin variables, adding a potential function $\Phi(T_{ij})$ chosen here so the resulting activation function Φ^{-1} for T_{ij} will be linear:

$$H[s,T] = -A\sum_i h_i s_i - B\sum_{i<j} T_{ij}s_i s_j + B\sum_{i<j}\Phi(T_{ij}) \qquad (10.9)$$

$$\Phi(T) = \frac{1}{2c}T^2$$

where the parameters $A \geq 0$ and $B \geq 0$ have different relative sizes in learning mode vs. recall mode. In learning mode, $A \gg B$ is large so *s* tracks and binarizes the externally imposed input vector *h*, and T adapts to that input. At low temperature, energy is minimized and the matrix entry T_{ij} is driven down the gradient $\partial H / \partial T_{ij}$ toward $T_{ij} = cs_i s_j$. For example, one may use dissipative dynamics

$$\tau_T \frac{dT_{ij}}{dt} + T_{ij} = cs_i s_j. \qquad (10.10)$$

If members of a small set of input vectors $\{h\}$ are imposed repetitively, and the learning rate $1/\tau_T$ is low enough, then the binarized input vectors *s* will get averaged over the input population:

$$T_{ij} \rightarrow c\langle s_i s_j \rangle_{population}.$$

On the other hand, in recall mode, $B \gg A$ and the learning rate is zero, so $\dot{T} = 0$ and at low temperature the activation state *s* falls into whichever local minimum of $-\sum_{i<j} T_{ij}s_i s_j$ best aligns with the current input *h* or $s(t = 0)$.

   The resulting computational functionality is that of a "content address-able memory", which retrieves a learned memory by its similarity to an input vector, as was demonstrated computationally for a discrete-time version of this learning rule in [5]. It has similarities to the Boltzmann machine learning algorithm [6], and even to Rosenblatt's (1962) perceptron learning rule. Equations like (10.10) that model connections or "synapses" T_{ij} as strengthening or weakening depending on coincidence or correlation between the "neurons" they connect, also realize "Hebb's rule" for synaptic plasticity mediating learning as proposed in neurobiology.

Another important learning rule can be derived as follows, starting with (10.6) and/or (10.7). If we impose a sparsity pattern on the weights T by partitioning the neurons (units) into *layers* indexed by $l \in 0, \ldots L$, with nonzero connection weights only allowed between adjacent layers $(l-1, l)$ (or more generally from earlier to later layers), and if we further scale such weights by a factor of ε^l where $\varepsilon \to 0$, and allow a compensating layer-dependent $\beta \propto \varepsilon^l$ parameter in the activation functions $g_l(v_{il})$, *then* layer l effectively only receives input from layer $l-1$ and not from layer $l+1$. The result is a feed-forward multilayer neural network, often called a "multilayer perceptron" (MLP), with solution dynamics obtained simply by updating the fixed-point equation

$$v_{il} = g(u_{il}) \text{ where } u_{il} = \sum_{j=1}^{J_l} T_{ij}^{(l)} v_{jl-1} + h_{jl-1} = \sum_{j=1}^{J_l+1} T_{ij}^{(l)} v_{jl-1}, \qquad (10.11)$$

once for each layer in succession, ordered by increasing layer number. Here, the convention $v_{J_l+1l} \equiv 1$ and $T_{iJ_l+1}^{(l)} \equiv h_{jl-1}$ obviates the need for a separate learning rule for the biases h_{jl}.

Now learning can proceed by following a gradient with respect to the weights T_{ij}, which can be efficiently calculated. One imposes an objective function on the output activations v_{jL} in the last layer, such as $E(T) = \sum_p E_p(T)$ where $E_p(T) = \frac{1}{2} \sum_i (v_{iL} - y_{pi})^2$ and where p indexes input pattern vectors $x_i = v_{i0}^{(p)}$ and we have suppressed extra (p) indices on the activations. Now we just need to use the chain rule to find all the derivatives recursively, by assuming every u and v activation variable depends on all the weights of its own and earlier layers. By this means we will learn an approximate mapping (function) from input vector x to output vector y.

For any layer $l > 0$ we can differentiate u_{il} in (10.11) to find

$$\frac{\partial E_p}{\partial T_{ij}^{(l)}} = \frac{\partial E_p}{\partial u_{il}} \frac{\partial u_{il}}{\partial T_{ij}^{(l)}} \equiv \delta_i^{(l)} v_{jl-1}$$

(again suppressing p indices). We need to find the δ s. For output layer L in particular, we can differentiate E_p and v_{il} in (10.11) to find

$$\delta_i^{(L)} \equiv \frac{\partial E_p}{\partial u_{iL}} = \frac{\partial E_p}{\partial v_{iL}}\frac{\partial v_{iL}}{\partial u_{iL}} = \left(v_{iL} - y_{pi}\right)g'\left(u_{iL}\right). \tag{10.12}$$

For any other layer $l-1 \geq 0$ except the last one, we can differentiate (10.11) to find

$$\delta_i^{(l-1)} \equiv \frac{\partial E_p}{\partial u_{il-1}} = \sum_k \frac{\partial E_p}{\partial u_{kl}}\frac{\partial u_{kl}}{\partial u_{il-1}} = g'\left(u_{il-1}\right)\sum_k \delta_k^{(l)}T_{ki}^{(l)}$$

i.e.,

$$\delta_i^{(l-1)} = g'\left(u_{il-1}\right)\sum_k T_{ik}^{(l)transpose}\delta_k^{(l)} \tag{10.13}$$

in which the sum over k is a matrix multiplication in the (i,k) index space. This method is sometimes called the "generalized delta rule". It emerges from this calculation that the crucial error signal Δ_{il} travels *backwards* through the layers of the network by the transpose of each layer's weight matrix. Hence, the algorithm's name, "backwards propagation of error" or just "backpropagation".

Now we have the ingredients for continuous descent dynamics

$$\tau_T dT_{ij}^{(l)} / dt = -\eta_{cont}\partial E / \partial T_{ij}^{(l)} = -\eta_{cont}\sum_p \partial E_p / \partial T_{ij}^{(l)} = -\eta_{cont}\sum_p \delta_i^{(pl)}v_{jl-1}^{(p)}$$

or discretized in time to first order:

$$\Delta T_{ij}^{(l)} = \sum_p \Delta_p T_{ij}^{(l)} = -\eta\sum_p \delta_i^{(pl)}v_{jl-1}^{(p)} \tag{10.14}$$

where $\eta = \eta_{cont}\Delta t / \tau_T > 0$. The more cost-effective "stochastic gradient descent" version of this rule is just to update according to one input pattern p at a time as in (10.10):

$$\Delta T_{ij}^{(l)} = \Delta_p T_{ij}^{(l)} = -\eta\Delta_{il}^{(p)}v_{jl-1}^{(p)}.$$

Equation (10.13) is pregnant with consequences. It implies that a many-layered ("deep") network will have both products of many factors of g', each of which can be quite large or small, and many factors of $T^{(l)transpose}$ potentially resulting in a large matrix condition number. The resulting numerical problems fomented an accumulation of special methods or "hacks" to deal with them. One of the most common is weight decay: adding a regularizer $\frac{\lambda}{2}\sum_{lij}(T_{ij}^{(l)})^2$ to the objective E, whose negative derivative then shows up in the gradient descent dynamics for learning.

Besides the layer-stratified weight matrix $T(\varepsilon \to 0)$ mapping used above, there is another, more rigid way to map Hopfield analog neural networks into multilayer feed-forward neural networks. This is to "unroll" or "unfold" them by adding a discretized time index. For example, a forward Euler solution method for (10.7) could be expressed as

$$v_{ik} = g(u_{ik}) \text{ and } \tau_i \frac{u_{ik} - u_{ik-1}}{\Delta t} + u_{ik-1} = \sum_j T_{ij} v_{jk-1} + h_i + O(\Delta t^2)$$

and approximated by discrete-time dynamics

$$v_{ik} = g(u_{ik}) \text{ and } u_{ik} = \left(1 - \frac{\Delta t}{\tau_i}\right) u_{ik-1} + \frac{\Delta t}{\tau_i} \left(\sum_j T_{ij} v_{jk-1} + h_i\right) \qquad (10.15)$$

Even if $\tau_i \neq \Delta t$, (10.15) is very close to the form of (10.11), especially after absorbing the relative time factors into T and h. So this architecture supports a modified "delta rule" for learning by backpropagation of error. Alternatively, there are several slightly generalized forms of (10.10) that can provide a common generalization of (10.11) and (10.15), complete with a delta rule for backpropagation learning. For example, one can add extra linear units $(g_i(u_i) = u_i)$ to convey the $\boldsymbol{u}$ information of layer $l-1$ forward to layer l. The more substantial difference between (10.15) and (10.11) is actually one of specialization. In (10.15), all layers have the same connection matrix $T_{ij}^{(l)} = T_{ij}$, whereas in (10.11) they generally differ.

This reduction of parameters is one example of "weight sharing", whereby the full connection matrix (respecting an architecture's pattern of sparsity) is a function of some smaller number of parameters θ, and the chain rule is used to create a version of backpropagation that computes gradients with respect to θ from backpropagated gradients with respect to free weights T.

In this way, we obtain a family of gradient descent learning algorithms for Hopfield-style continuous time analog neural networks with feedback, a type of "recurrent neural network". Better ODE integration schemes will give rise to somewhat more complex learning algorithms. The method can be generalized to other differential equations with other kinds of parameters than neural networks. In the continuous limit $\Delta t \to 0$, we recover the "adjoint method" for locally optimizing parameters in a differential equation model, again with the training signals flowing backwards in time.

10.3 The AI/ML Spring

What has changed since the 1980s? And which changes would have resonated with Feynman?

Neural networks are usually specified with a combination of algebraic equations with free (learnable) parameters and illustrated by block diagrams showing how variables are connected to one another in these equations. The total amount of information to specify the architecture is usually very small indeed, compared with that required by a computer program aimed at a similarly nontrivial problem in perception, pattern recognition, or many other successful applications. Almost all of the information in a trained network comes instead from the data set used to train the network and is stored in the large number of learned parameters: the numerical "weights" by which one artificial neuron or "unit" influences another.

The field of neural networks has gone through several transformations since the 1980s. Some of the innovations that can be expressed in simple equations include:

- Fukushima's neurobiologically inspired "neocognitron" networks.
- Hopfield's optimizing or energy-minimizing networks (e.g., (10.7) above).
- Boltzmann (probabilistic) machines.
- Backpropagation of error through feed-forward multilayer perceptrons (variants and specializations of (10.11) and (10.13) above).
- Deep convolutional networks (essentially, a neocognitron network with backpropagation of error learning).
- "Soft-max" mixtures of experts.
- "Soft-assign" correspondence-optimizing networks.
- Regularization methods such as "weight decay" and random unit "dropout" promote successful generalization.
- Restricted Boltzmann Machines.
- Graphical probabilistic models.
- Latent semantic spaces.
- Variational autoencoders.
- Generative models based on diffusion.
- Long short-term memory.
- Residual networks.
- Tensor networks.
- "Transformer" architectures (based on soft-max "attention" equations incorporating (10.8) above).
- Graph neural networks.
- Multiscale/multigrid neural networks, and
- Large Language Models,

among many others.

Other kinds of learning architectures that are mathematical and specified by equations, without necessarily being neural networks, include:

- Self-organizing maps.
- Kernel methods.
- Manifold learning.
- Information geometry.

Often, practitioners encounter these architectures in the form of computer code or somewhat ill-defined diagrams. But what is essential to specify are the relatively compact equations, and ample training data. This combination sounds like theoretical and experimental physics. In an odd temporal coincidence, and as shown in this book, both Feynman and Hopfield wrote down Hamiltonians as simple equations describing non-traditional computing machines; Feynman for quantum computing (1986 and presented earlier in the Physics of Computation class), and Hopfield for neural networks (1982). The Boltzmann Machine of Ackley, Hinton, and Sejnowski (1985) also has a statistical physics Hamiltonian description.

Other aspects of neural network performance enhancement are more procedural than mathematical. One procedural approach to regularization is "early stopping" of a training optimization algorithm. Another example is Feynman's advocacy of a carefully designed sequence of learning tasks, related to what is known in animal training as "shaping". This occurs in current neural network practice as "transfer learning" and is very effective across many domains – although the sequence may be less important than the collection of learning tasks on which the same network is jointly trained ("multitask learning") – and also to "lifelong learning", which is more or less what it sounds like. And paradigms such as reinforcement learning and unsupervised learning each start with a different premise about how data from the world is to enter into the training process.

At this point, I must apologize to my many colleagues whose own great architectures and algorithms did not happen to get onto the foregoing lists. I am barging through decades of high-quality research and laying waste to nuance.

With these machine learning architectures and procedures, vast improvements have taken place in the decade 2010–2020. As is now widely known, these improvements have re-sculpted the classic AI areas of computer vision, natural language, robotics, game-playing, and recently even theorem-proving, which was the *ne plus ultra* of symbolic AI (Newell and Simon's Logic Theorist, 1956; Robinson's resolution-based theorem proving, 1965). Commercial and industrial applications are legion. There are now many domains in which it is not possible to be competitive without using machine learning. Even the practice of the hard sciences, including physics, is starting to feel an influx of AI/ML methods adapted to include scientific knowledge, under banners such as "physics-informed machine learning".

Some of the best-known and most influential recent successes in neural networks include AlexNet [7], which was a milestone in computer vision, AlphaGo [8] and AlphaZero (2017) for playing the intellectually demanding board games of Go and Chess, and the Generative Pre-Trained Transformer GPT-2 [9] that generates readable natural language text in response to natural language prompts. More recent large language models (LLMs) are even more impressive, but would benefit from the kind of validity-checking that more symbolic approaches to AI can provide. These examples go far toward answering classic challenges that were identified right from the beginning of AI. These include, for example, early computer vision at MIT, "Los Alamos chess" without bishops on a 6×6 board, and the "SHRDLU" and "ELIZA" text interaction programs – all running on computers of minuscule capacity by current standards. The notable recent successes involve neural network training of tens of millions to hundreds of billions of parameters, and they comprise a complete revolution in the field by comparison to AI capabilities in the mid-1980s. More importantly, using the kinds of techniques listed earlier, there is now a vast array of successful applications. An increasing fraction of high-technology industry, including the hardware and software that we all interact with every day, is dependent on neural networks and related machine-learning trainable models. Consequently, a very large amount of computing goes into their training every day.

What would Feynman have thought of all these advances? Remembering his commitment to having a personal, independent point of view on everything, and the impossibility of emulating a great mind, there can be little certainty on this point. Nevertheless, it seems to me that:

1. He would be very comfortable with the data-driven nature of the field, which was developing but far from universal in pre-1980s symbolic AI, and with the fact that serious validating computer experiments with some kind of "real data" are universally expected as part of the development of new techniques. This characteristic may be partly due to the large influx of physicists and other quantitatively minded people into the field.

2. However, Feynman reviewed experiments critically. In his classes and public lectures, he often expounded that in science "you are the easiest one [for you] to fool". I think he might be guarded about AI/ML methodological and sociological problems such as stopping upon success, selective reporting of best-among-competitors performance, community overtraining on benchmarks,

and related sources of what he called "too much happy talk" in science.

3. He would be enthusiastic about transfer learning, and about the particular neural network architectures that have led to success in vision, natural language, and limited combinations thereof. These methods have essentially achieved the goals of the machine-learning project that he had in mind – insofar as he had expressed them.

4. Regarding equations as the hidden currency of learning architecture, he might be of two minds: attracted because he was a mathematical master; repelled because the source of his mastery was the ability to deeply visualize the meaning of each equation. Neural network equations for the most part are just not that conceptually deep. In the old AI dichotomy of "scruffy" vs. "neat" research approaches, ML equations might be neat but not neat enough.

5. He would not be too worried about the large number of small, apparently unconnected hacks required to get all these successes, nor about their lack of connection to the neurobiology of living "neural" networks. For example, it was just an empirical matter to him whether computer chess programs would eventually work better by brute force search or by expert-derived heuristics. In the old AI dichotomy of "scruffy" vs. "neat" approaches, "scruffy" but functional is fine, as advocated by Minsky.

6. He might nevertheless be intrigued by conceptually deeper, less *ad hoc* architectures that connect to physics, such as manifold learning (clearly related to general relativity), and perhaps by attempts to connect artificial neural networks to real neurobiology.

7. He might be skeptical of the "causal inference by structural equations" approach to infusing AI/ML into reasoning in the softer sciences because it has no real dynamics and it does not address the main causality problem, namely, of fitting into the multiscale and reductionist perspective on science that he expounded as early as his 1964 Messenger lectures [10]. At a fundamental level at least, "causality" is better stated in quantum field theories by Feynman propagators and by the spacelike commutation and anticommutation of operators.

8. Because of his deep commitment to physics, Feynman would be quite interested in – if also a bit skeptical of – the "physics-informed machine learning" agenda, which is now being pursued with many different architectures, representations, and methods for the purpose of doing computational physics, chemistry, and

biology. If I had to pick one thing about present-day neural networks, machine learning, and AI that Richard Feynman would be most interested in, this would be it.

9. He would *certainly* have creative and potentially powerful new ideas that are not yet on anybody's list or agenda.

10.4 AI/ML for Computational Science

To a theoretically minded scientist, one of the most exciting things happening in AI/ML today (as I write in 2021) is its burgeoning adoption into computational science. The interesting neural network architectures for this purpose are those that incorporate physical or other scientific principles and constraints into their searchable function spaces. For example, many-electron quantum mechanics can be tackled using variational methods (of which Feynman was a master, e.g., [1]) with wave function bases that incorporate Slater determinants for permutation antisymmetry. One neural network generalization of this method exists under the name of "FermiNet" and has shown promise [11]. Another ML-enriched way to compute quantum potential energy functions is by training neural networks to approximate them on many density functional theory quantum electron ground state calculations [12].

Working our way up the scale hierarchy that Feynman liked to describe, a standard method for simulating multi-molecular systems in biology, biochemistry, and materials science is using molecular dynamic (MD) particle-based (non-quantum) stochastic dynamics. Methods for speeding up such simulations using machine learning now exist [13]. What is more, methods like ANI-1 or FermiNet neural networks are capable of learning potential energy functions that could appear in MD simulations, creating a truly multiscale model stack. Feynman thought "this tremendous world of interconnecting hierarchies" was important but worried that the interlevel connections were actually "a little weak. We have not thought them all through yet" [10]. Now perhaps we can think them through, aided by artificial intelligence.

A considerably further jump up the scale hierarchy leads theorists to spatial continuum models modeled as partial differential equations (PDEs), as for example, in fluid flow or elastic mechanics. The solution of these and many other PDEs can now be vastly accelerated, and indeed the PDEs can themselves be learned from data, using a variety of different neural networks and related ML methods [14]. Related work in multiscale model reduction using ML, about which I am reasonably hopeful, includes my own collaborative efforts that use a dynamical version of the Boltzmann ML algorithm on additive fundamental process operators [15], and/or graph neural networks, for that purpose. Other examples of ML for computational science are included in the review [16].

Of course, science is grounded in experiment and observation that produce data, all aspects of which are also being revolutionized by machine learning [17] and codified in deep theoretical structures that may yet be impacted as well [18]. I think the current flowering of AI/ML for multiscale, "hierarchical" computational science would have had a strong appeal and resonance for Richard Feynman and might have attracted his unique efforts.

10.5 A Mathematical Synthesis and a Return to Symbolic AI?

Could the pendulum ever swing back from numeric AI toward symbolic AI? One motivation for that possibility is the widely cited need for "explainable" versions of ML models. Another is to remember that the actual mathematical specification of most ML methods is most clearly and concisely done by (symbolic) equations. Such ML equations can be derived from statistical models and automatically differentiated and parameter optimized, although it is now possible to automatically differentiate and optimize some generic computer programs as well. So the pendulum could swing back, at least if it takes a mathematical trajectory and fully absorbs current ML methods. In that connection, there are presently two main strands of symbolic computing centered on mathematics.

Minsky's former doctoral student Joel Moses led the 1970s charge at MIT into computer algebra in the form of the Macsyma program, which was also capable in integral calculus, power series manipulation, solution of differential equations, and in many other useful applied mathematical methods. Computer algebra (CA) systems propagated modestly and comprised a continuously successful niche for symbolic computing with an "AI" flavor; they did not suffer an AI winter. At a slower burn that achieved academic but not commercial success, automatic theorem proving (ATP) systems kept improving over the intervening decades as well. For example, there was an entire *Journal of Formalized Mathematics* that automatically verified all submitted papers. More recently, a number of substantial results in mathematics such as the Kepler sphere-packing conjecture [19] have been proved using a new generation of interactive theorem verification (ITV) systems. These systems have powerful inference engines such as "hammers" that use automated theorem proving to suggest verifiable proof steps in ITV, and scalable "satisfiability modulo theories" logical solver algorithms. So, in the realm of mathematics, symbolic AI is alive and well on a very modest scale. Could symbolic AI break out from this small but critical niche (CA and/or ITV for automating mathematics) into scientific computing (which tends to be mathematical) or into general computing?

The only viable strategy for competing with ML methods at this point is to co-opt them. Fortunately, the way is clear to combining ML with both CA and ITV as shown by some recent papers that may soon be superseded by even stronger results. In the realm of computer algebra for symbolic integration and differential equation solving, there is good progress in approaching these problems by means of deep neural networks [20], although more remains to be done [21]. ML has been applied to improving the ATP inference engines in ITV [22–26] based on deep reinforcement learning, GPT-2, graph neural networks, and *k*-nearest neighbors respectively, as well as on co-training on multiple related tasks. Conversely, ITV can be used to verify ML theory (e.g., [27]) and potentially to improve the reliable generation of ML code.

Of course, better proof alone is not enough. Feynman was a mathematical innovator, but in several lectures, he expressed the physicists' sentiment that "It's possible to *know* more than you can *prove*". Another symbolic method is "symbolic regression", including an interesting experiment in rediscovering physics formulae appearing in Feynman's milestone lectures on physics [18]. My own as-yet informal proposal for deep symbolic/ML integration in computational science is outlined at the end of [28]. Together with ML in CA and ITV, these various recent works point to the possibility of powerful synergistic combinations of symbolic and numerical approaches to mathematical AI, perhaps first in applied-mathematical domains such as computational science, but eventually in general computing as well.

10.6 Conclusion

In summary, many of Richard Feynman's ideas about the future of neural network-like machine learning systems have come to pass. Since these systems work well now, he would probably have devoted time and energy to finding the next deep conceptual breakthrough they enable – particularly in the sciences.

References

1. R. P. Feynman, *Statistical Mechanics*. Reading, MA: Benjamin, 1972. https://hdl.handle.net/2027/ucl.b4227910.
2. J. J. Hopfield. Neurons with Graded Response Have Collective Computational Properties Like Those of Two-State Neurons. *Proceedings of the National Academy of Sciences of the United States of America, May* 81(10), 3088–3092, 1984. https://doi.org/10.1073/pnas.81.10.3088.
3. Carsten Peterson, Bo Söderberg. A New Method for Mapping Optimization Problems Onto Neural Networks. *International Journal of Neural Systems* 1(1), 3–22, January, 1989. https://doi.org/10.1142/S0129065789000414.
4. Anand Rangarajan, Steven Gold, Eric Mjolsness. A Novel Optimizing Neural Network Architecture with Applications. *Neural Computation* 8(5), 1041–1060, 1996.
5. J. J. Hopfield. Neural Networks and Physical Systems with Emergent Collective Computational Abilities. *Proceedings of the National Academy of Sciences of the United States of America* 79(8), 2554–2558, 1 April 1982. https://doi.org/10.1073/pnas.79.8.2554.
6. David H. Ackley, Geoffrey E. Hinton, Terrence J. Sejnowski. A Learning Algorithm for Boltzmann Machines. *Cognitive Science* 9(1), 147–169, January–March 1985.
7. Alex Krizhevsky, Ilya Sutskever, Geoffrey E. Hinton. Advances in Neural Information Processing Systems (NIPS 2012). https://papers.nips.cc/paper/4824-imagenet-classification-with-deep-convolutional-neural-networks.pdf.
8. David Silver, Aja Huang, Chris J. Maddison, Arthur Guez, Laurent Sifre, George van den Driessche, Julian Schrittwieser, Ioannis Antonoglou, Veda Panneershelvam, Marc Lanctot, Sander Dieleman, Dominik Grewe, John Nham, Nal Kalchbrenner, Ilya Sutskever, Timothy Lillicrap, Madeleine Leach, Koray Kavukcuoglu, Thore Graepel, Demis Hassabis. Mastering the Game of Go with Deep Neural Networks and Tree Search. Nature 529(7587), 484–489, 28 January 2016.
9. Alec Radford, Jeffrey Wu, Rewon Child, David Luan, Dario Amodei, Ilya Sutskever. Language Models Are Unsupervised Multitask Learners. https://cdn.openai.com/better-language-models/language_models_are_unsupervised_multitask_learners.pdf.
10. R. P. Feynman. *The Character of Physical Law*. MIT Press 1967; 2017, pp. 121–126. "a little weak …": p. 125; "… this tremendous world …": p. 126. Transcription of R. P. Feynman, Messenger Lecture #5: The Distinction of Past and Future. See Lecture 5, Section 9, Minutes 39–48 (or Start at Min. 33 for More Context). As of June 2021 a video recording is available at https://www.feynmanlectures.caltech.edu/messenger.html, although access could change.
11. David Pfau, James S. Spencer, Alexander G. D. G. Matthews, W. M. C. Foulkes. *Ab Initio* Solution of the Many-Electron Schrödinger Equation with Deep Neural Networks. *Physiological Reviews Research* 2(3), 033429–033416, September 2020. https://journals.aps.org/prresearch/abstract/10.1103/PhysRevResearch.2.033429#fulltext.

12. J. S. Smith, O. Isayev, A. E. Roitberg. ANI-1: An Extensible Neural Network Potential with DFT Accuracy at Force Field Computational Cost. *Chemical Science* 8(4), 3192–3203, 2017.

13. Weile Jia, Han Wang, Mohan Chen, Denghui Lu, Lin Lin, Roberto Car, E. Weinan, Linfeng Zhang. Pushing the Limit of Molecular Dynamics with Ab Initio Accuracy to 100 Million Atoms with Machine Learning. *Supercomputing '20*, https://dl.acm.org/doi/abs/10.5555/3433701.3433707; https://arxiv.org/abs/2005.00223.

14. M. Raissi, P. Perdikaris, G. E. Karniadakis. Physics-Informed Neural Networks: A Deep Learning Framework for Solving Forward and Inverse Problems Involving Nonlinear Partial Differential Equations. *Journal of Computational Physics* 378, 686–707, 1 February 2019. https://doi.org/10.1016/j.jcp.2018.10.045.

15. Oliver K. Ernst, Tom Bartol, Terrence Sejnowski, Eric Mjolsness. Learning Dynamic Boltzmann Distributions as Reduced Models of Spatial Chemical Kinetics. *Journal of Chemical Physics* 149, 034107, July 2018. Also arXiv 1803.01063, March 2018.

16. Giuseppe Carleo Ignacio Cirac, Kyle Cranmer, Laurent Daudet, Maria Schuld, Naftali Tishby Leslie Vogt-Maranto, Lenka Zdeborová. Machine Learning and the Physical Sciences. *Reviews of Modern Physics* 91(October–December), 2019.

17. Eric Mjolsness and Dennis DeCoste. Machine Learning for Science: State of the Art and Future Prospects. *Science* 293(5537), 2051–2055, 14 September 2001. https://doi.org/10.1126/science.293.5537.2051.

18. Silviu-Marian Udrescu, Max Tegmark. AI Feynman: A Physics-Inspired Method for Symbolic Regression. *Science Advances* 6(16), 15 April 2020. https://doi.org/10.1126/sciadv.aay2631.

19. Thomas Hales, Mark Adams, Gertrud Bauer, Tat Dat Dang, John Harrison, Hoang Le Truong, Cezary Kaliszyk, Victor Magron, Sean McLaughlin, Tat Thang Nguyen, Quang Truong Nguyen, Tobias Nipkow, Steven Obua, Joseph Pleso, Jason Rute, Alexey Solovyev, Thi Hoai An Ta, Nam Trung Tran, Thi Diep Trieu, Josef Urban, Ky Vu, Roland Zumkeller. A Formal Proof of the Kepler Conjecture. *Forum of Mathematics, Pi* 5, e2, 2017. https://doi.org/10.1017/fmp.2017.1.

20. Guillaume Lample, François Charton. *Deep Learning for Symbolic Mathematics.* https://arxiv.org/abs/1912.01412.

21. Ernest Davis. *The Use of Deep Learning for Symbolic Integration: A Review of (Lample and Charton, 2019).* https://arxiv.org/abs/1912.05752.

22. Aditya Paliwal, Sarah Loos, Markus Rabe, Kshitij Bansal, Christian Szegedy, Graph Representations for Higher-Order Logic and Theorem Proving. *Proceedings of the AAAI Conference on Artificial Intelligence.* Vol n34, No. 03, 2020. DOI: 10.1609/aaai.v34i03.5689.

23. Kshitij Bansal, Sarah M. Loos, Markus N. Rabe, Christian Szegedy, Stewart Wilcox. HOList: An Environment for Machine Learning of Higher-Order Theorem Proving. https://arxiv.org/abs/1904.03241.

24. Stanislas Polu, Ilya Sutskever. Generative Language Modeling for Automated Theorem Proving. https://arxiv.org/pdf/2009.03393.pdf.

25. Jan Jakubův, Karel Chvalovský, Miroslav Olšák, Bartosz Piotrowski, Martin Suda, Josef Urban. ENIGMA Anonymous: Symbol-Independent Inference Guiding Machine (System Description). https://arxiv.org/abs/2002.05406.

26. T. Gauthier, C. Kaliszyk, J. Urban, R. Kumar, M. Norrish. TacticToe: Learning to Prove with Tactics. *Journal of Automated Reasoning* 65(2), 257–286, 2021. https://doi.org/10.1007/s10817-020-09580-x.

27. A. Bentkamp, J. C. Blanchette, D. Klakow. A Formal Proof of the Expressiveness of Deep Learning. *Journal of Automated Reasoning* 63(2), 347–368, 2019. https://doi.org/10.1007/s10817-018-9481-5.

28. Eric Mjolsness. Prospects for Declarative Mathematical Modeling of Complex Biological Systems. *Bulletin of Mathematical Biology* 81(8), 3385–3420, August 2019. https://doi.org/10.1007/s11538-019-00628-7.

Reminiscences

Memories of Feynman at Caltech

John Preskill

Getting Acquainted

Richard Feynman and I overlapped at Caltech for four and a half years, from the start of my faculty appointment in August 1983 until Feynman's death in February 1988. Our relationship as colleagues got off to a great start.

One day soon after I arrived, I hear someone drumming on the wall while walking down the hallway, know it must be Feynman and step out of my office to greet him. Our theory group admin Helen Tuck introduces us: "Dr. Feynman, this is Dr. Preskill, our new faculty member!" Feynman replies: "What group?" Does Feynman really not know who I am? "Um ... particle theory."

Feynman: "People who say they do particle theory do many different things. What do you do?"

I ramble incoherently for a minute about the connection between particle physics and cosmology, then unwisely conclude: "And lately, I have been working, without much success, on models in which quarks and leptons are composite."

Long pause, then: "Well, your lack of success has been shared by many others." Feynman turns and disappears into his office.

So right away I knew we would be friends.

Caltech Seminar

Actually, Feynman and I had already met a few times before that, when I had come to Caltech to give seminars. Speaking at the Caltech particle theory seminar in the days of Feynman and Gell-Mann was a memorable experience. Here is how Steve Weinberg described it [1]:

> Years ago, when I was an assistant professor of physics at Berkeley [1960–1966], I used to be invited down to Caltech about once a year to

give a talk. It was usually the low point of my year. In the audience at Caltech were two leaders of modern physics, Murray Gell-Mann and Richard Feynman, who interrupted with frequent questions, ruthlessly probing to see if I really knew what I was talking about and had anything new to say. Of the two, Feynman was the more frightening. Gell-Mann was mostly interested in finding out whether there was anything in my talk that he should know about, so he was no problem if I did have anything worthwhile to say. Feynman was having fun.

Fifteen years later, when I gave my first seminar at Caltech, the experience was not quite so terrifying. By then, I discovered, one could play Feynman and Gell-Mann against one another. When Dick attacked, Murray defended me. And when Murray raised an objection, Dick would be on my side. It was an eventful seminar, but not "the low point of my year".

The World of Science

My first "encounter" with Feynman had actually occurred years earlier, when at age 9 I acquired a marvelous book called *The World of Science* by Jane Werner Watson [2]. There was a chapter about theoretical physics, which began with the story of a boy whose little red wagon has a ball in the back. The boy notices that when he pulls the wagon forward the ball rolls backward, and when he stops pulling the ball rolls forward. He asks his father why, and his father replies: "That's called inertia, but nobody knows why."

Some 20 years later, I watched the terrific interview of Feynman by Christopher Sykes, called *The Pleasure of Finding Things Out*, where Feynman tells the same story [3]. Whoa! I thought – did Feynman steal the story from the Golden Book I read as a child? Looking at the book for the first time in years, I realized what had happened. Children's author Jane Werner Watson was married to Earnest Watson, the Caltech Dean of the Faculty, and she based the book on interviews with faculty members.

What I found particularly inspiring was the discussion in that book of a discovery made just a year before the book was published – that the laws of physics governing elementary particles know the difference between left and right! That amazing fact kindled a passion for physics that eventually carried me to Caltech, where I joined Dick Feynman on the faculty 21 years later.

Talking Physics

As Caltech colleagues, Feynman and I found that we had a common interest in nonperturbative aspects of quantum chromodynamics, in particular why quarks are confined inside hadrons, and we often talked about that. Sometimes, I would impress Feynman with an idea or calculation I had learned from the literature; I would tell him the source, but he would be interested in the ideas, not the reference. Once I overheard Feynman tell Helen, as he returned to his office after our discussion: "He's like an encyclopedia. No, he's better than the encyclopedia!" That made my day. But sometimes I wondered whether Feynman knew my name, as he sometimes seemed to confuse "Preskill" and "[Michael] Peskin" – with whom, I presumed, he had discussed similar things.

Feynman and Gell-Mann had once been close, but there was evident tension between them by the 1980s. After I had gotten to know them better, I asked each one what had gone wrong. Both gave the same answer – they had gotten along well until around 1969 when Feynman was working on the parton model. Years later, Murray spoke derisively about Feynman's "put-ons," still resentful that Feynman had refused to call them "quarks." Feynman for his part recalled that Gell-Mann had scathingly ridiculed the idea that quarks would behave as nearly free particles inside hadrons. What started out as a scientific disagreement had become increasingly personal and hostile, and their relationship never fully recovered.

Feynman's Belt Trick

David Goodstein once asked Feynman to explain to him something rather esoteric in the theory of elementary particles, what we call the connection between spin and statistics [4]. Feynman promised to prepare a freshman lecture on it. But sometime later he returned and sheepishly admitted: "I couldn't do it. I couldn't reduce it to the freshman level. That means we don't really understand it."

But later he changed his mind. And Feynman, with great relish, showed me his way of explaining spin and statistics. I've been using it ever since, whenever I teach our sophomore class at Caltech, Physics 12. It goes like this.

First, he boiled down the problem to its essence. The connection between spin and statistics means this: that when two elementary particles, two electrons say, change places, that's really the same thing as having one of the particles spin around by 360 degrees. And he'd illustrate what it means to spin around by doing a pirouette. Then he would take off his belt. And he'd say:

> Look. The two ends of my belt are two electrons. What happens when I exchange them? At first you think nothing has happened to the belt. but wait, it has a twist in it! Now let's twist one end of it 360 degrees. See, the belt is the same as when it started. So exchanging the two electrons, and then twisting one of them, is the same as doing nothing! That's the connection between spin and statistics.

It's a visceral demonstration. And for the past 35 years, I've been thinking about how to interpret that belt.

In 1987, just months before he died, Feynman took a group of students to dinner at a local restaurant, where somehow the subject of spin and statistics came up.

And so, with as much relish as ever, Feynman stood up from the table and started taking off his belt. Just at that moment, the waitress came up to the table, took in what was going on, shook her finger, and scolded: "That's as far as it goes!"

The Final Blackboard

When Feynman died in 1988, the blackboard in his office was photographed, and that photo was widely disseminated. Two prominent passages appear: "What I cannot create, I do not understand" and "Know how to solve every problem that has been solved." Why those words? I may be able to shed some light on that.

In late 1986, after the Rogers Commission had finished their investigation of the Challenger disaster, Feynman was eager to dive back into physics, and particularly excited about further investigations of QCD. Feynman was interested in lattice QCD, which he recognized as a beautiful application of path integral methods (which he had invented decades earlier [5]), but he felt that computational power was then inadequate for getting accurate results and would remain inadequate for some time to come. Meanwhile, he hoped to make progress with analytic methods, or a combination of numerical and analytic methods.

In particular, he hoped that tools for solving integrable models might be helpful for treating the soft part of QCD, the physics beyond the reach of renormalization-group improved perturbation theory. He wanted some students to study integrable models with him to help him learn the subject. Well, he wanted students, and I had students, so we made an arrangement. Feynman and the students met once a week in his office, and those meetings would sometimes last all afternoon; a few times Feynman invited the students to dinner afterward.

Feynman told the students "We gotta know how to solve every problem that has been solved," and he urged them to solve the problems on their own because "What I cannot create I do not understand." To get things started he described the six-vertex model and told everyone to solve it without looking up any references [6]. That went on for weeks, without notable progress, until Feynman triumphantly unveiled his own solution. The next challenge was the eight-vertex model, but the students never solved that one, and neither did Feynman!

One of the students was Sandip Trivedi, who recalls that Feynman was becoming ill, but was "incredibly enthusiastic and extremely patient" with the students. During his final illness, he told Helen to share his notes with the students, and they were all amazed and inspired to see how meticulous and detailed the notes were, containing many intricate calculations.

We lost Richard Feynman on February 15, 1988. It was a very sad day at Caltech. Feynman was loved, admired, and held in awe by a large swath of the campus community; he was part of the soul of the place, and he's still sorely missed by those who knew him. Since then, many students have come and gone who never knew Feynman personally, but they too have been deeply influenced by his contributions, writings, ideas, and unique personality. The spirit of Richard Feynman lives on among deeply curious people everywhere.

A Poem for Feynman

I hope that readers of *Surely You're Joking, Mr. Feynman!* [7, 8] and others familiar with Feynman's life and career will appreciate this poem I wrote in honor of his 100th birthday, on May 11, 2018.

> The Feynman legend, pundits say
> Began in Queens – Far Rockaway.
> It's there a boy would stop and think
> To fix a radio on the blink.

> He grew up as a curious guy
> Who showed his sister the night sky.
> He wondered why, and wondered *why*
> He wondered why he wondered why.

New Jersey followed MIT.
The cream and lemon in his tea
Taught Mr. Feynman when to joke
And how to act like normal folk.

Cracking safes, though loads of fun,
Could not conceal from everyone,
The mind behind that grinning brow:
A new Dirac, but human now.

In New York state he spun a plate
Which led, in nineteen forty-eight
To diagrams that let us see
The processes of QED.

He left the east and made a trek
Until he landed at Caltech.
His genius brought us great acclaim.
This place would never be the same.

Dick's teaching skills were next to none
When reinventing Physics 1.
His wisdom's there for all to see
In red books numbered 1, 2, 3.

Always up and never glum
He loved to paint and play the drum.
His mind engaged with everything
For all the world is int'resting.

Dick proved that charm befits a nerd.
For papers read, and stories heard
We'll always be in Feynman's debt.
A giant we cannot forget.

Bibliography

1. Steven Weinberg. Feynman on God and His Granny; Review of the Meaning of It All, by Richard Feynman. *The Times Higher Education Supplement*, 1334:23, 1998.
2. Jane Werner Watson. *The World of Science: Scientists at Work Today in Many Challenging Fields*. Golden Press, 1958.
3. Christopher Sykes. The Pleasure of Finding Things Out. *Motion Picture*, 1981.
4. David L. Goodstein, and Richard P. Feynman. Teacher. *Physics Today*, 42(2):70, 1989.
5. Richard P. Feynman, Albert R. Hibbs, and Daniel F. Styer. *Quantum Mechanics and Path Integrals*. Courier Corporation, 2010.
6. Rodney J. Baxter. *Exactly Solved Models in Statistical Mechanics*. Elsevier, 2016.
7. Richard P. Feynman. *Surely You're Joking, Mr. Feynman!: Adventures of a Curious Character*. WW Norton & Company, 2018.
8. Richard P. Feynman, and Ralph Leighton. *What Do You Care What Other People Think?: Further Adventures of a Curious Character*. WW Norton & Company, 2001.

Physics and Computation: Learning from Feynman, Hopfield, and Sussman

Michael R. Douglas

Professor in the Simons Center for Geometry and Physics
Stony Brook University

Richard Feynman's course on computing at Caltech, and the course on neural networks taught at the same time by John Hopfield, had a profound impact on my career, indeed on my life.

I first met Richard Feynman when I visited Caltech in the spring of 1983 to decide between there and Princeton for my graduate studies. I was equally fascinated by fundamental physics and artificial intelligence, and when I learned that Feynman, Hopfield, and Carver Mead were co-teaching a course on "The Physics of Computation" I was hooked from the start. Seeing the legendary name on my appointment sheet was a bit unreal, and I felt frankly nervous when I walked into his office. But somehow Feynman quickly put me at ease, and his enthusiasm as he described the topics he planned to discuss was compelling. Soon he finished with the comments, "But you shouldn't worry about where you'll go, Caltech and the other schools you listed are all very good. You should go where your girlfriend is going!"

While I didn't have a girlfriend at the time, still this was very encouraging. And as if I needed more reasons to come to Caltech, I soon learned that Gerald Jay Sussman would be spending the next year there on sabbatical to co-teach with Feynman, study astrophysics, and build a computer. This computer, the Digital Orrery, would be specially designed to integrate the equations of celestial mechanics, and explore chaos in the solar system. Sussman was already famous for his work in AI and the combination made my decision clear.

People sometimes compare graduate school with "drinking from a fire hose", and that fall, with Feynman and Hopfield's courses, working for Gerry Sussman, occasional visits to Carver Mead's course on what would now be called "neuromorphic VLSI" and the introductory neurobiology course, not to mention quantum field theory and the required physics courses, I was busy indeed.

Looking back, it is clear to me now that more than the content of the courses, what I learned was a very different way to do science than what

I had imagined. Every student, no matter how fascinating one finds one's subject, is daunted by the sheer weight of knowledge one has to absorb. It is all too easy to fall off the track by getting so immersed in the details that one misses the big picture or by taking shortcuts that leave one without the foundations one needs to make arguments that convince others. There are many ways through this dilemma and finding them is an essential part of the formation of a scientist and helps determine his or her character.

Feynman, Hopfield, and Sussman, perhaps following the *genius loci* of Caltech, were adventurers (at least, that is how I saw them). They were willing to come into a new field relying not on their credentials and publication lists, but on their own judgment and their confidence that they had tools for thinking whose power went far beyond their origins. Such an attitude is often ascribed to physicists, indeed Feynman's course title "Potentialities and Limitations of Computing Machines" ruffled some feathers in the CS department. But Sussman shared it, coming with the rather different toolkit of a computer scientist. His belief that there were specifically computational ways of thinking that could help to solve problems in many fields was ahead of its time and influenced me deeply.

In terms of Feynman's actual course, this meant that in some way, the essential content of the computer science curriculum had to be taught in two quarters, leaving one quarter for truly novel topics. And, looking back at the syllabus, it looks hopelessly ambitious, covering hardware and logic, theory of computation, physical fabrication of devices, serial and parallel architectures, algorithms, and what was referred to as "more difficult computer applications", such as robotics, vision, speech recognition, and knowledge representation. The lectures on the physical limits of computation and quantum computing took up less than two months.

Somehow, this worked. Of course, many of the students had come in with prior CS experience and self-study, but I think it was more because the goal was not to train computer scientists but to build a deep foundation that started from basic physical laws and did not take conceptual shortcuts. It may also have something to do with the nature of computer science, as (at least for me) the corresponding attempt to present neuroscience from the ground up in Hopfield's course was less successful. More than once I heard Feynman say that "if the name of a subject contains the word 'science', you know it isn't." This is unfair on several levels (not least if we apply it to neuroscience) and I also heard him take it back. But it is true that the intellectual core of computer science is in engineering, and a computer scientist has more ability to define what is interesting and shape his or her field than a natural scientist (this is also true for mathematicians). Neuroscience has so many empirical facts, and at the time (and perhaps still) less agreement on which ones are fundamental, that presenting its foundations in a single course might simply be impossible.

As the Digital Orrery project got off the ground, I was spending more time with Gerry Sussman, and sometime that fall he invited me to join the after-class lunches with Feynman. Hearing Feynman expound on the latest scientific developments and seeing the breadth of his interests and knowledge was truly inspiring. More than in the course, this was where I got a sense of his views on artificial intelligence.

Feynman had a specific "challenge problem", which he regarded as a significant milestone for AI. It was to build a car that could drive itself down Lake Avenue in Pasadena. Self-driving cars would, of course, be the topic of the DARPA Grand Challenge in 2004 and for some years after. Despite the practical difficulty of achieving the reliability needed for widespread use, I am sure that Feynman would count this problem as solved. And as Eric Mjolsness writes in his recollections, he would surely be pleased that so much of the progress on vision and robotics has come through advances in neural networks.

On the other hand, it is not at all clear to me what Feynman would think about our progress toward realizing artificial general intelligence, the ability to perform at human level in a wide variety of domains. He was very skeptical of the 1980s' focus on symbolic methods. I learned this from his description of conversations with Doug Lenat, another famous AI researcher. Lenat had just started the CYC project and was surveying many scientists to get their insights. In a nutshell, the idea of CYC was to provide a computer with the common-sense knowledge needed to reason about everyday topics by developing a precise formal language to represent it and assembling a team of humans to type it in. One of Lenat's estimates of the size of this task was made by reflecting on the game "Twenty Questions", which suggested that the number of "commonly known" entities is about 2^{20} or one million. This was hugely optimistic and, in fact, Wikidata (a loosely similar database that is far from complete) now has over 90 million entities.

Feynman was not alone in thinking this project was crazy; most people felt that computers would have to learn their common-sense knowledge, and there is an ongoing and robust debate about how this might be done. Furthermore, Feynman felt that progress would come by discovering deep principles behind intelligence, different from but analogous to those of physics. This is another longstanding debate in AI, as many (notably Marvin Minsky) argue that intelligence results from combining a large number of problem-solving tricks and techniques.

But Feynman's skepticism seemed to go far beyond this. An encounter that stuck with me took place one weekend afternoon in spring 1984. After a long Orrery working session with Sussman, we walked over to the astro common room to take a break for tea. We were talking about whether science would ever solve the mysteries of awareness and consciousness, when Feynman walked in, if I recall with Kip Thorne. On overhearing our discussion, he immediately broke in, "What do you mean, this mysterious feeling of awareness? I don't have it, I don't even know what you mean!" Though he grinned mischievously as he said it, we were unable to get him to take it back or even to decide how seriously he meant it. But the implicit message that science could not be based on subjective feelings, no matter how widely shared, but had to be based on objective and agreed-upon observations, was very clear.

A feature of Feynman's course that greatly influenced me was his grading policy – instead of a final, most of the grade was based on individual term projects on topics we chose and which he approved. This struck me as a very good system which, while requiring more work from both students and teachers, both motivates the students and directly focuses on skills that are hard to teach in other ways, most importantly that of choosing interesting problems to work on. Feynman hired me as a grader in subsequent years and grading these projects was my main duty, finally passing on them to Feynman for his evaluation. The breadth of the students' interests was impressive and I learned a lot this way myself, and ever since my own courses follow the same policy whenever possible.

For my own final project, I asked whether a quantum computer could solve an NP hard problem, the SAT problem of finding satisfying assignments to a collection of logical propositions, in polynomial time. Of course, for classical computers, this is the famous P=?NP problem. Feynman did very clearly ask the question of whether quantum computers could solve problems faster than classical computers, though I do not recall this particular question being discussed in class. Of course, my goal was rather ambitious and I did not get very far. Although I could argue against a naive approach in which one used the many worlds picture of quantum mechanics to test all of the exponentially many possible assignments of truth values to variables, I did not have any argument against the idea that some other approach might speed up the quantum computer, and Feynman gave me low marks. Even after the breakthroughs of Shor, Grover, and others, most people today believe that quantum computers cannot solve NP problems in polynomial time, a point I will return to shortly.

As the course finished, I felt as I believe did everyone involved that we had shared a unique experience, to see one of the foremost minds of our times engage with a new field of knowledge, and ask questions that would be discussed long into the future. The Digital Orrery was working, and we were beginning the computations that would eventually demonstrate that the orbit of Pluto is chaotic. But the biggest news in fundamental physics of the summer of 1984 was the Green–Schwarz anomaly cancellation in superstring theory. Many of the Caltech theory grad students immediately began working on string theory, including myself.

As is well known, Feynman had a negative opinion of string theory. For him, as he emphasized to me more than once, the joy of physics was to understand the real world, to look at the data that experimental physicists obtained with great cleverness and skill and to evaluate it, often pointing out gaps in their analyses, and occasionally making discoveries. Feynman was not convinced that string theory would ever help to do that, and so far there are not many examples of it doing so. There are some, and one is the lower bound on viscosities of superfluids from holography of Kovtun, Son, and Starinets, which is satisfied by superfluid helium and also by the data from heavy ion collisions. String theory has had far more influence on mathematics, a story well told elsewhere. Yet, just as turning quantum computing from vision to reality is taking decades yet will someday be achieved, many physicists believe that string theory will someday be tested by confrontation with real-world data.

And although I am not sure Feynman would have liked this any better, the study of string theory, black holes, and the multiverse has led to new connections between fundamental physics and computation. My own interest in these connections was rekindled by a 2005 paper by Scott Aaronson entitled "NP-complete Problems and Physical Reality", in which he explains why not only quantum mechanics but a wide variety of its generalizations as well as other physical theories should not be expected to solve NP-complete problems. At the time, my main focus was the "landscape" approach to string theory, in which one studies the early cosmology that selected our universe among a multiverse of possibilities by combining the mathematics of string compactification with ideas used to study protein folding, fitness landscapes, and other complex systems. While the story is too long to tell here, my interest in complex systems and landscapes traces back to Hopfield's course as well, through their connections with the theory of neural networks. In 2003, I began to apply these ideas to string theory, coming up with results such as the much-quoted estimate 10^{500} for the number of vacuum configurations (possible universes).

In 2006, Frederik Denef and I showed that, much like the problem of protein folding, the traveling salesman problem, and even the problem of finding the ground state of a Hopfield model, the problem of finding vacuum configurations of string theory that agree with observation is (in a sense we explained) NP hard. In 2017, joined by Brian Greene and Claire Zukowski, we used these ideas to propose a radical reformulation of multiverse cosmology. In our picture, universes are created by a computer that is searching for solutions that satisfy the anthropic principle, which the physical laws should allow for the existence of life. Our laws of physics are selected because they are the easiest way for a computer to do this.

Many other connections between computation and physics have been proposed. Patrick Hayden and John Preskill related quantum computation and the black hole information paradox, and Lenny Susskind and collaborators argued that a black hole saturates a physical bound on computation and thus is the most powerful physically realizable computer. Susskind and Juan Maldacena have conjectured a relation between quantum entanglement and space-time wormholes, which may be a key insight toward understanding what deeper structure underlies space-time. Perhaps Feynman, as one of the pioneers of quantum gravity, would have enjoyed hearing about our speculations after all.

Much as Minsky said about intelligence, so too scientific progress over the long term comes from the combination of many approaches and many points of view. And the broader attitude I learned from Feynman's course, to do science with a spirit of adventure, served me well.

Memories of Richard Feynman

Tony Hey

Chief Data Scientist
Rutherford Appleton Laboratory
Science and Technology Facilities Council, UK

I well remember my arrival at Caltech on a sunny October morning in 1970. Fresh from Oxford where even graduate students – at that time – wore ties and shirts, I was unsure what to wear for my first meeting with Murray Gell-Mann. I gambled, wrongly, on a suit and arrived in the office of the theory group secretary, Julie Curcio, feeling more and more over-dressed and as if I had a large label dangling from my collar saying "New PhD from Oxford." I had seen Gell-Mann once before in England but was unsure if the bearded individual dressed in an open-necked shirt and sitting in Julie's office was indeed the eminent professor. A moment after I had introduced myself my doubts were dispelled by Gell-Mann putting out his hand and saying "Hi, I'm Murray." This episode illustrates only a small part of the (healthy) culture shock I experienced in California. Six years in Oxford had left me used to calling my professor "Professor Dalitz, sir." At that time, I would certainly not have dared to address Dalitz by his first name!

One of my first tasks on arrival in Pasadena was to buy a car. This was not as easy as it sounds. The used car lots in Pasadena are sprinkled down Colorado Boulevard for several miles in typical US fashion and getting to them in the days when public transport in Los Angeles was probably at its lowest ebb was not straightforward. It was only after my wife and I were stopped by the police and asked why we were walking on the streets of Pasadena that I understood the paradox that, in California, you had to have a car to buy a car. Another "chicken and egg" problem arose in connection with "ID" – a term we had not encountered before. As a matter of routine, the police demanded to see our ID and of course the only acceptable ID in deepest Pasadena at that time was a California driver's license. A British driving license without a photograph of the bearer was clearly inadequate and even our passports were looked on with suspicion. An introduction to the United States via used car salesmen is not the introduction I would recommend to my worst enemy and it is not surprising that I sought advice from the Caltech grad students. I was pointed in the direction of Steve Ellis whose advice was valued because he came from

Detroit and was believed to be worldly wise. I tracked Steve down to the seminar room where I saw he was engaged in a debate with a character who looked mildly reminiscent of the used car salesmen I had recently encountered. This was, of course, my first introduction to Dick Feynman – I did not at first recognize him from the much earlier photograph I knew from the three red books of the "Feynman Lectures". Curiously enough, even after ten years or more, I always felt more comfortable addressing him as Feynman rather than Dick.

Compared to my previous life as a graduate student in Oxford, adjusting to life at Caltech was like changing to the fast lane on a freeway. First, instead of Oxford being the center of the universe, it was evident that, to a first approximation, Europe and the UK did not exist. Second, I rapidly discovered that the ethos of the theory group of Feynman and Gell-Mann was that physics was all about attacking the outstanding fundamental problems of the day: it was not about getting the phase conventions right in a difficult but ultimately well-understood area. I remember asking George Zweig – a co-inventor of the whole quark picture of matter – for his comments on a paper of mine. This was the not-very famous "SLAC-PUB 1000", a paper that I had written with an experimenter friend at the Stanford Linear Accelerator Center (SLAC) about the analysis of three-body final states. George's uncharacteristically gentle comment to me was: "We do, after all, understand rotational invariance." In fact, the paper was both useful and correct but, on the Caltech scale of things, amounted to doodling in the margins of science. In those days I aspired to be as good a physicist as Zweig: this ambition strikes me now as similar to wanting to emulate the achievements of Jordan in the early days of quantum mechanics, rather than those of his collaborators, Heisenberg and Born.

One of the nicest things about Caltech was the sheer excitement of being around Feynman and Gell-Mann. As a post-doc from England, where we gain a rapid but narrow exposure to research, I was contemporary in age with the final year grad students and a lot of our social life was spent with them. Feynman was actively working with two of them – Finn Ravndal and Mark Kislinger, who had just been awarded his PhD – on his own version of the quark model. Perhaps because of his work with Ravndal and Kislinger, Feynman was very involved with the final-year graduate students and we all had lunch with him most days at the "Greasy" – as the Caltech self-service cafeteria was universally known. Needless to say, our table was always the center of attraction. One frequent topic for discussion was Feynman's explanation of some new experimental results obtained at SLAC on electron-proton scattering. Feynman's "parton model" – an intuitively appealing picture of the proton made up of point-like constituents – was sweeping all before it, much to Murray's annoyance. It was not surprising that I had left Oxford full of enthusiasm for working on the parton model and looking forward to hearing Feynman on the subject he had invented. Curiously, Feynman's only publication on partons was applied to proton-proton scattering. It was when he was visiting SLAC and the experimenters told him of their surprising results with electrons and protons that Feynman realized that this was a much simpler application of his parton model. There and then, Feynman gave a seminar in which he explained their results using partons. Nothing was written down by him on this, however, and it was left to Bjorken, who had been away from SLAC at the time of Feynman's visit, and Paschos, a post-doc at SLAC, to write up the analysis of the experimental results in terms of "Feynman's Parton Model".

My first encounter with Feynman on a technical level was intimidating. Two Caltech experimenters, Barry Barish and Frank Sciulli, had just had a proposal for a neutrino-proton experiment accepted. Since I liked to work with experimenters, they asked me to give an informal lunchtime seminar to their group explaining the application of the "parton model" to their experiment. Imagine my surprise when I turned up to talk to the experimental group, on finding Feynman sitting in the audience. Still, I started out and even managed to score a point off Feynman. At an early stage in the lecture, he asked how I derived a particular relation. I replied, with what now seems like foolhardy temerity: "I used Conserved Vector Current theory – you should know, you invented it!" In fact, all went well until I had nearly reached the end of the seminar. I was just outlining what further predictions could be made when Feynman said: "Stop. Draw a line. Everything above the line is the parton model – below the line are just some guesses of Bjorken and Paschos." As I rapidly became aware, the reason for Feynman's sensitivity on this point was that Murray was going around the fourth floor of Lauritsen at Caltech, growling that "Partons are stupid" and that "Anyone who wants to know what the parton model predicts needs to consult Feynman's entrails!" In fact, all the results above Feynman's line in my seminar were identical to predictions that Murray had been able to derive using much more sophisticated algebraic techniques. Feynman wanted to dissociate himself from some of the wilder "parton model" predictions of others and to stress that his simple intuitive parton approach gave identical predictions to Gell-Mann's much more fancy methods. Unfortunately for me, my lecture just happened to be a handy vehicle for him to make this point!

There were, of course, drawbacks to being in the same group as Feynman and Gell-Mann. I came to Caltech with the firm intention of pursuing research on Feynman's parton model. What I had not realized was that Caltech was the one place where one could not publish research on partons! Why was this? There was the obvious distaste of Gell-Mann for the whole approach but that would not have mattered if it had not been for the awkward fact of "Feynman's notebooks". I used to go to Feynman with some idea and proudly display my analysis on his blackboard. Each time

Feynman listened, commented, and corrected – and then proceeded to derive my "new" results several different ways, pulling in thermodynamics, rotational invariance, or what have you, and using all sorts of alternative approaches. He explained to me that once he could derive the same result by a number of different physical approaches, he felt more confident in its correctness. Although this was very educational and stimulating, it was also somewhat dispiriting and frustrating. After all, one could hardly publish a result that Feynman already knew about and had written down in his famous working "notebooks" but had not bothered to publish. So it was somewhat in desperation that I turned to Gell-Mann's algebraic approach for a more formal framework within which to work. With Jeff Mandula, then an assistant professor at Caltech, I looked at electron-proton scattering when both the electron and proton were "polarized" – with their spins all lined up in the same direction. We found a new prediction whose parton equivalent was obscure. Roughly speaking, at high energies, the spin direction of the parton is unchanged by collision with an electron. Our result concerned the probability of the parton spin changing its direction in the collision: this was related to 'spin-flip' amplitudes normally neglected in the parton model. Armed with this new result, I went to Feynman and challenged him to produce it with his parton approach. In the lectures he gave at Caltech the next term, later published as the book *Photon-Hadron Interactions*, you will find how Feynman rose to this challenge.

Life at Caltech with Feynman and Gell-Mann was never boring. Stories of their exploits abounded – many of Feynman's now preserved for posterity by his friend Ralph Leighton in *Surely You're Joking, Mr. Feynman!* There were many other stories. A friend told me of the time he was about to enter a lecture class and Gell-Mann arrived at the door to give the class. My friend was about to open the door but was stopped by Murray saying: "Wait!" There was a storm raging outside the building and at the appearance of a particularly violent flash of lightning, Gell-Mann said "Now!" and entered the class accompanied by a duly impressive peal of thunder. Another story that circulated was of Feynman giving a talk about the discovery, with Gell-Mann, of the V-A model of weak interactions. After the talk, one of the audience came up to him and said: "Excuse me, Professor Feynman, but isn't it usual in giving a talk about joint research to mention the name of your collaborator?" Feynman reportedly came back with: "Yes – but it's usual for your collaborator to have done something!" Obviously, these stories get inflated in the telling but I did ask Feynman about this one since it seemed so out of character to the Feynman I knew. He smiled and said, "Surely you don't believe I would do a thing like that!" I only knew Feynman after he had received the Nobel Prize and found happiness in his marriage to Gweneth. A somewhat more abrasive and aggressive picture of him before this time emerges from the Feynman biographies, so I am still not sure! Certainly, he enjoyed making a quick and amusing response. This feature of Feynman's was often in evidence in seminars given by visiting speakers. On one memorable occasion, the speaker started out by writing the title of his talk on the board: "Pomeron Bootstrap". Feynman shouted out: "Two absurdities" and the room dissolved into laughter. Alas for the speaker, he was deriving theoretical results supposedly valid in one energy regime but going on to apply them in another. This was just the kind of academic dishonesty that Feynman hated and on that particular occasion the speaker had a very uncomfortable time fielding brickbats thrown from the entire audience. Feynman could be restrained: on the occasion of another seminar, he leaned over to me and whispered, "If this guy wasn't a regular visitor, I would destroy him!"

It was during this time at Caltech that Feynman gave his celebrated lecture in the Beckman Auditorium on "Deciphering Mayan Hieroglyphics". Feynman's account of his honeymoon in Mexico with his second wife Mary Lou, and his efforts to decipher the Dresden Codex is contained in *Surely You're Joking, Mr. Feynman!* The lecture itself was a typical Feynman tour de force. The story perfectly illustrates Feynman's approach to tackling a new subject. Rather than look at a translation of the Codex, Feynman made believe he was the first to get hold of it. Struggling with the Mayan bars and dots in the tables, he figured out that a bar equaled five dots and found the symbol for zero. The bars and dots carried at 20 the first time but at 18 the second time, giving a cycle of 360. The number 584 was prominent in one place and was made up of periods of 236, 90, 250, and 8. Another prominent number was 2920 or 584×5 and close by there were tables of multiples of 2920 up to 13×2920. Here Feynman says he did the equivalent of looking in the back of the book. He scoured the astronomy library to find something associated with the number 584 and found out that 583.92 days is the period of Venus as it appears from the Earth. The numbers 236, 90, 250, and 8 were then connected with the different phases of Venus. There was also another table that had periods of 11,959 in the Codex, which Feynman figured out were to be used for predicting lunar eclipses. With a typical down-to-earth analogy, Feynman likened the Mayans' fascination with such "magic" numbers to our childish delight in watching the odometer of a car pass 10,000, 20,000, 30,000 miles, and so on. As Feynman says, "Murray Gell-Mann countered in the following weeks by giving a beautiful set of six lectures concerning the linguistic relations of all the languages of the world." For these lectures, Murray used to arrive clutching armfuls of books and proceed to tell his audience about the classification of languages into "Superfamilies" with a common origin. He was always fond of drawing attention to the similarities between English and German and, for example, delighted in calling George Zweig, George Twig. I still have some notes of his lectures

– with examples from the Northern, the Afro-Asiatic, the Indo-Pacific, the Niger-Kardofanian, and the Nilo-Saharian Superfamilies among others. Even though it seemed a bit strange for professional particle physicists to be attending lectures on comparative linguistics, life at Caltech was always interesting! I have always suspected that Feynman's account of his time with his father in the Catskills described in *What Do You Care What Other People Think?*, the second volume of anecdotes produced with Ralph Leighton, was partly directed at Gell-Mann's passion for languages and names. In the story, Feynman's father says, "You can know the name of that bird in all the languages of the world, but when you're finished, you'll know absolutely nothing whatever about the bird." Feynman credits his "knowing very early on the difference between knowing the name of something and knowing something" to these experiences with his father.

Other recollections of Feynman are still fresh in my memory. One time I went to get the coffee at lunch in the Greasy and returned to find that Feynman had invited my wife down to their house in Mexico for the weekend – with his family, I hasten to add. As an afterthought, he invited me too and we found ourselves strolling along the beach in Mexico, talking physics with Feynman late into the night. Feynman's advice to me on that occasion was: "You read too many novels." He had started out very narrow and focused and only later in life had his interests broadened out. Good advice perhaps, but during the years I knew Feynman I also learned how impossible he was for anyone to emulate – in his disregard for the "unimportant" things of life, like committees and administration, and in his unique ability to attack physics problems from many different angles. On another visit to Caltech many years later, sitting with him in the garden of his house in Altadena, Feynman proceeded to take off his belt and demonstrate his new understanding of the spin-statistics rule. He later wrote this up in a memorial lecture to his hero in physics, Paul Dirac, discoverer of anti-matter. This was some 20 years after the publication of *The Feynman Lectures on Physics* in which he had apologized for not being able to give an elementary explanation of this rule. As he said then: "This probably means we do not have a complete understanding of the fundamental principle involved."

What made Feynman's lectures unique? The well-known Cornell physicist David Mermin, himself noted for his thoughtful and penetrating analyses of supposedly well-understood problems in physics, was moved to say: "I would drop everything to hear him lecture on the municipal drainage system." In 1967, the *Los Angeles Times* science editor wrote:

> A lecture by Dr. Feynman is a rare treat indeed. For humor and drama, suspense and interest it often rivals Broadway stage plays. And above all, it crackles with clarity. If physics is the underlying 'melody' of science, then Dr. Feynman is its most lucid troubadour.

In the same article, the author, Irving Bengelsdorf, sums up the essence of Feynman's approach: "No matter how difficult the subject – from gravity through quantum mechanics to relativity – the words are sharp and clear. No stuffed shirt phrases, no 'snow jobs', no obfuscation." A *New York Times* article in the same year said that Feynman "uses hand gestures and intonations the way Billy Rose used beautiful women on the stage, spectacularly but with grace."

For me, it was Feynman's choice of words that made a Feynman lecture such a unique experience. The same *New York Times* article went on to say that "his lectures are couched in pithy often rough-cut phrases". There are innumerable examples to choose from. In the middle of pages of complicated mathematics Feynman deliberately lightens up the text by introducing phrases like "you can cook up two new states" or by personalizing the account by introducing imagined conversations of physicists as in "Now – said Gell-Mann and Pais – here is an interesting situation." In his invited lecture in 1971, on the occasion of the award of the Oersted medal for his services to the teaching of physics, Feynman began disarmingly by saying "I don't know anything about teaching" and then proceeded to give a fascinating account of the research problem he was working on – "What is the proton made out of? Nobody knows but that's what we're going to find out." In the talk, he likened smashing two protons together to smashing two watches together: one could look at the gearwheels and all the other bits and pieces that resulted and try to figure out what was happening. In this way, he was able to explain that smashing a simple point particle like an electron into a proton was much simpler because there was only one watch to look at. At a summer school in Erice in Italy one summer, he was asked a question about conservation laws. Feynman replied:

> If a cat were to disappear in Pasadena and at the same time appear in Erice, that would be an example of global conservation of cats. This is not the way cats are conserved. Cats or charge or baryons are conserved in a much more continuous way.

Feynman's Nobel Prize lecture should be required reading for all aspiring scientists. In it, Feynman forgoes the customary habit of removing the scaffolding that was used to construct the new theory. Instead, he tells us of all the blind alleys and wrong ideas that he had on the way to his great discoveries. The article also reveals more of Feynman's lecture technique when he says: "I shall include details of anecdotes which are of no value scientifically nor for understanding the development of the ideas. They are included only to make the lecture more entertaining." In the article, we find out how Feynman first started his attempt to answer the challenge of Dirac concerning the troublesome infinities that plagued relativistic quantum mechanics. In the last sentence of his famous book, Dirac

said: "It seems that some essentially new physical ideas are here needed."
Of his youthful idea to solve the problem Feynman says:

> That was the beginning and the idea seemed so obvious to me and so
> elegant that I fell deeply in love with it. And, like falling in love with
> a woman, it is only possible if you do not know too much about her,
> so you cannot see her faults. The faults will become apparent later, but
> after the love is strong enough to hold you to her. So, I was held to this
> theory, in spite of all difficulties, by my youthful enthusiasm.

Later in the lecture, Feynman writes: "I suddenly realized what a stupid
fellow I am; for what I had described and calculated was just ordinary
reflected light, not radiation reaction." This refreshing honesty from one
of the greatest physicists of the 20th century reminds me of another of my
heroes, Johannes Kepler – who was first to write down laws of physics
as precise, verifiable statements expressed in mathematical terms. Unlike
Copernicus and Newton, Kepler wrote down all the twists and turns in
his thought processes as he was forced to the shocking conclusion that
the orbit of Mars was not a circle but an ellipse. Kepler summed up his
struggle with the words: "Ah, what a foolish old bird I have been!"

One of the best anecdotes in the lecture concerns a physicist called
Slotnick and his encounter with "Case's theorem." This described the
moment when Feynman realized that his "diagrams" really were something new. In its full form, the story runs as follows. At a meeting of
the American Physical Society in New York, Slotnick presented a paper
comparing two different forms of the electron-neutron coupling. After a
long and complicated calculation, Slotnick concluded that the two forms
gave different results. At this point, Robert Oppenheimer rose from the
audience and remarked that Slotnick's calculation must be wrong since it
violated Case's theorem. Poor Slotnick had to admit he had never heard
of this theorem, so Oppenheimer kindly told him he could remedy his
ignorance by listening to Professor Case presenting his result the next
day. That evening, in his hotel, Feynman could not sleep so he decided to
use his new methods to repeat Slotnick's calculations. Feynman then goes
on to say:

> The next day at the meeting, I saw Slotnick and said, "Slotnick, I
> worked it out last night; I wanted to see if I got the same answers
> you do. I got a different answer for each coupling — but, I would
> like to check with you because I want to make sure of my methods."
> And he said, "What do you mean you worked it out last night, it took
> me six months!" And, when we compared the answers he looked
> at mine, and he asked, "What is that Q in there, that variable Q?"
> I said, "That's the momentum transferred by the electron, the elec-
> tron deflected by different angles." "Oh," he said, "no, I only have the

limiting value as Q approaches zero, the forward scattering." Well it was easy enough to just substitute Q equals zero in my form and I then got the same answers as he did. But it took him six months to do the case of zero momentum transfer, whereas during one evening I had done the finite and arbitrary momentum transfer. That was a thrilling moment for me, like receiving the Nobel Prize, because that convinced me, at last, I did have some kind of method and technique and understood how to do something that other people did not know how to do. That was my moment of triumph in which I realized I really had succeeded in working out something worthwhile.

What Feynman does not say in his written lecture is that he stood up at the end of Case's talk and said: "Your theorem must be wrong. I checked Slotnick's calculation last night and I agree with his results." In the days when calculations like Slotnick's could take as much as six months, this was the incident that put "Feynman's diagrams" on the map.

The other piece of required reading for students of all disciplines is Feynman's article on "Cargo Cult Science". This was originally Feynman's commencement address to new Caltech graduates in 1974 and in it, Feynman discusses science, pseudoscience, and learning how not to fool yourself. The unifying theme of the talk is Feynman's passionate belief in the necessity for "utter scientific integrity" – in not misleading funding agencies about likely applications of your research, in publishing results of experiments even if they do not support your pet theory, in giving government advice they may rather not hear, in designing unambiguous rat-running experiments, and so on. As he says, "learning how to not fool ourselves is, I'm sorry to say, something that we haven't specifically included in any particular course that I know of. We just hope you've caught on by osmosis." He concludes with one wish for the new graduates:

> the good luck to be somewhere where you are free to maintain the kind of integrity I have described, and where you do not feel forced by a need to maintain your position in the organization, or financial support, or so on, to lose your integrity.

At the risk of sounding pompous, I think the world owes a vote of thanks to Caltech for providing just such an environment for Richard Feynman. Feynman was never restricted to research in any one particular field: it is to the exercise of just this freedom that we owe these *Feynman Lectures on Computation*.

It seems appropriate to end these reminiscences with two more "Feynman stories". The first story harks back to his safecracking days at Los Alamos. At a Conference in Irvine in 1971, Feynman agreed to be on a discussion panel at the end of the conference. He was asked if he thought that physicists were getting anywhere with answering the "big questions". Feynman replied:

> You ask, are we getting anywhere. I'm reminded of a situation when I was asked the same question. I was trying to pick a safe. Somebody asked me how are you doing? Are you getting anywhere? You can't tell until you open it. But you have tried a lot of numbers that you know don't work!

The second story is the last Feynman story of all. Gweneth was by his bedside in the hospital and Feynman was in a coma. She noticed that his hand was moving as if he wanted to hold hands with her. She asked the doctor if this was possible but was told that the motion was automatic and did not mean anything. At which point, Feynman, who had been in a coma for a day and a half or so, picked up his hands, shook out his sleeves, and folded his hands behind his head. It was Feynman's way of telling the doctor that even in a coma he could hear and think – and that you should always distrust what so-called "experts" tell you!

The final word deserves to be given to James Gleick, author of a biography of Feynman. Gleick memorably summed up Feynman's philosophy toward science with the following words: "He believed in the primacy of doubt, not as a blemish upon our ability to know but as the essence of knowing."

Tony Hey
Southampton
March 1996

Afterword

Origins of the Feynman Lectures on Computation

![]

Foreword to the 1996 edition of the Lectures

Since it is now some eight years since Feynman died I feel it necessary to explain the genesis of these "Feynman Lectures on Computation". In November 1987 I received a call from Helen Tuck, Feynman's secretary of many years, saying that Feynman wanted me to write up his lecture notes on computation for publication. Sixteen years earlier, as a post-doc at Caltech, I had declined the opportunity to edit his "Parton" lectures on the grounds that it would be a distraction from my research. I had often regretted this decision so I did not take much persuading to give it a try this time around. At Caltech that first time, I was a particle physicist, but ten years later, on a sabbatical visit to Caltech in 1981, I became interested in computational physics problems – playing with variational approaches that (I later found out) were similar to techniques Feynman had used many years before. The stimulus of a Caltech colloquium on "The Future of VLSI" by Carver Mead then began my move toward parallel computing and computer science.

Feynman had an interest in computing for many years, dating back to the Manhattan project and the modeling of the plutonium implosion bomb. In "Los Alamos from Below", published in "Surely You're Joking, Mr. Feynman!", Feynman recounts how he was put in charge of the "IBM group" to calculate the energy release during implosion. Even in those days before the advent of the digital computer, Feynman and his team worked out ways to do bomb calculations in parallel. The official record at Caltech lists Feynman as joining with John Hopfield and Carver Mead in 1981 to give an interdisciplinary course entitled "The Physics of Computation". The course was given for two years and John Hopfield remembers that all three of them never managed to give the course together in the same year: one year Feynman was ill, and the second year Mead was on leave. A handout from the course of 1982/1983 reveals the flavor of the course: a basic primer on computation, computability, and information theory followed by a section entitled "Limits on Computation Arising in the Physical World and 'Fundamental' Limits on Computation". The lectures that year were given by Feynman and Hopfield with guest lectures from experts such as Marvin Minsky, John Cocke, and Charles Bennett. In the spring of 1983, through his connection with MIT and his son Carl, Feynman worked as a consultant for Danny Hillis at Thinking Machines, an ambitious, new parallel computer company.

In the fall of 1983, Feynman first gave a course on computing by himself, listed in the Caltech record as being called "Potentialities and Limitations of Computing Machines". In the years 1984/1985 and 1985/1986, the lectures were taped and it is from these tapes and Feynman's notebooks that these lecture notes have been reconstructed. In reply to Helen Tuck, I told her I was visiting Caltech in January of 1988 to talk at the "Hypercube Conference". This was a parallel computing conference that originated from the pioneering work at Caltech by Geoffrey Fox and Chuck Seitz on their "Cosmic Cube" parallel computer. I talked with Feynman in January and he was very keen that his lectures on computation should see the light of day. I agreed to take on the project and returned to Southampton with an agreement to keep in touch. Alas, Feynman died not long after this meeting and we had no chance for a more detailed dialog about the proposed content of his published lectures.

Helen Tuck had forwarded to me both a copy of the tapes and a copy of Feynman's notes for the course. It proved to be a lot of work to put his lectures in a form suitable for publication. Like the earlier course with Hopfield and Mead, there were several guest lecturers giving one or more lectures on topics ranging from the programming language "Scheme" to physics applications on the "Cosmic Cube". I also discovered that several people had attempted the task before me! However, the basic core of Feynman's contribution to the course rapidly became clear – an introductory section on computers, followed by five sections exploring the limitations of computers arising from the structure of logic gates, from mathematical logic, from the unreliability of their components, from the thermodynamics of computing, and from the physics of semiconductor technology. In a sixth section, Feynman discussed the limitations of computers due to quantum mechanics. His analysis of quantum mechanical computers was presented at a meeting in Anaheim in June of 1984 and subsequently published in the journal "Optics News" in February 1985. These sections were followed by lectures by invited speakers on a wide range of "advanced applications" of computers – robotics, AI, vision, parallel architectures, and many other topics which varied from year to year.

As advertised, Feynman's lecture course set out to explore the limitations and potentialities of computers. Although the lectures were given some ten years ago, much of the material is relatively "timeless" and represents a Feynmanesque overview of some standard topics in computer science. Taken as a whole, however, the course is unusual and genuinely interdisciplinary. Besides giving the "Feynman treatment" to subjects such as computability, Turing machines (or as Feynman says, "Mr. Turing's machines"), Shannon's theorem, and information theory, Feynman also discusses reversible computation, thermodynamics, and quantum computation. Such a wide-ranging discussion of the fundamental basis of computers is undoubtedly unique and a "sideways", Feynman-type view of the whole of computing. This does not mean to say that all aspects of computing are discussed in these lectures and there are many omissions, programming languages and operating systems, to name but two. Nevertheless, the lectures do represent a summary of our knowledge of the truly fundamental limitations of digital computers. Feynman was not a professional computer scientist and he covers a large amount of material very rapidly, emphasizing the essentials rather than exploring details. His approach to the subject is resolutely practical and this is underlined in his treatment of computability theory by his decision to approach the subject via a discussion of Turing machines. Feynman takes obvious pleasure in explaining how something apparently so simple as a Turing machine can arrive at such momentous conclusions. His philosophy of learning and discovery also comes through strongly in these lectures. Feynman constantly emphasizes the importance of working things out for yourself, trying things out and playing around before looking in the book to see how the "experts" have done things. The lectures provide a unique insight into Feynman's way of working.

I have used editorial license here and there in ways I should now explain. In some places there are footnotes labeled "RPF", which are asides that Feynman gave in the lecture that in a text are best relegated to a footnote. Other footnotes are labeled "Editors", referring to comments inserted by me and my co-editor Robin Allen. I have also changed Feynman's notation in a few places to conform to current practice, for example, in his representation of MOS transistors.

Feynman did not learn subjects in a conventional way. Typically, a colleague would tell him something that interested him and he would go off and work out the details for himself. Sometimes, by this process of working things out for himself, Feynman was able to shed new light on a subject. His analysis of quantum computation is a case in point but it also illustrates the drawback of this method for others. In the paper on quantum computation there is a footnote after the references that is typically Feynman. It says: "I would like to thank T. Toffoli for his help with the references". With his unique insight and clarity of thinking, Feynman was often able not only to make some real progress but also to clarify the basis of the whole problem. As a result, Feynman's paper on quantum computation is widely quoted to the exclusion of other lesser mortals who had made important contributions along the way. In this case, Charles Bennett is referred to frequently, since Feynman first heard about the problem from Bennett, but other pioneers such as Rolf Landauer and Paul Benioff are omitted. Since I firmly believe that Feynman had no wish to take credit from others, I have taken the liberty of correcting the historical record in a few places and refer the reader, in a footnote, to more complete histories of the subject. The plain truth was that Feynman was not interested in the history of a subject but only the actual problem to be solved!

I have exercised my editorial prerogative in one other way, namely in omitting a few lectures on topics that had become dated or superseded since the mid-1980s. However, in order to give a more accurate impression of the course, there will be a companion volume to these lectures which contains articles on "advanced topics" written by the self-same "experts" who participated in these courses at Caltech. This complementary volume will address the advances made over the past ten years and will provide a fitting memorial to Feynman's explorations of computers.

There are many acknowledgments necessary in the successful completion of a project such as this. Not least I should thank Sandy Frey and Eric Mjolsness, who both tried to bring some order to these notes before me. I am grateful to Geoffrey Fox, for trying to track down students who had taken the courses, and to Rod van Meter and Takako Matoba for sending copies of their notes. I would also like to thank Gerry Sussman, and to place on record my gratitude to the late Jan van de Sneepscheut, for their initial encouragement to me to undertake this task. Gerry had been at Caltech, on leave from MIT, when Feynman decided to go it alone, and he assisted Feynman in planning the course.

I have tried to ensure that all errors of (my) understanding have been eliminated from the final version of these lectures. In this task I have been helped by many individuals. Rolf Landauer kindly read and improved Chapter 5 on reversible computation and thermodynamics and guided me patiently through the history of the subject. Steve Furber, designer of the ARM RISC processor and now a professor at the University of Manchester, read and commented in detail on Chapter 7 on VLSI – a topic of which I have little first-hand knowledge. Several colleagues of mine at Southampton also helped me greatly with the text: Adrian Pickering and Ed Zaluska on Chapters 1 and 2; Andy Gravell on Chapter 3; Lajos Hanzo on Chapter 4; Chris Anthony on Chapter 5; and Peter Ashburn, John Hamel, Greg Parker, and Ed Zaluska on Chapter 7. David Barron, Nick Barron, and Mike Quinn, at Southampton, and Tom Knight at MIT, were kind enough to read through the entire manuscript and, thanks to their comments, many errors and obscurities have been removed. Needless to say, I take full responsibility for any remaining errors or confusions! I must also thank Bob Churchhouse of Cardiff University for information on Baconian ciphers, Bob Nesbitt of Southampton University for enlightening me about the geologist William Smith, and James Davenport of Bath University for help on references pertaining to the algorithmic solution of integrals. I am also grateful to the Optical Society of America for permission to reproduce, in slightly modified form, Feynman's classic 1985 "Optics News" paper on quantum mechanical computing as Chapter 6 of these lectures.

After Feynman died, I was greatly assisted by his wife Gweneth and a Feynman family friend, Dudley Wright, who supported me in several ways, not least by helping pay for the lecture tapes to be transcribed. I must also pay tribute to my co-editor, Robin Allen, who helped me restart the project after the long legal wrangling about ownership of the Feynman archive had been decided, and without whom this project would never have seen the light of day. Gratitude is also due to Michelle Feynman, and to Carl Feynman and his wife Paula, who have constantly supported this project through the long years of legal stalemate and who have offered me every help. A word of thanks is due to Allan Wylde, then Director of the Advanced Book Program at Addison-Wesley, who showed great faith in the project in its early stages. Latterly, Jeff Robbins and Heather Mimnaugh at Addison-Wesley Advanced Books have shown exemplary patience with the inevitable delays and my irritating persistence with seemingly unimportant details. Lastly, I must record my gratitude to Helen Tuck for her faith in me and her conviction that I would finish the job – a belief I have not always shared! I hope she likes the result.

Tony Hey
Electronics and Computer Science Department
University of Southampton
England
May 1996

Update to the 1996 Foreword

In my original Foreword to these lectures, I promised there would be "a companion volume to these lectures which contains articles on 'advanced topics' written by the self-same 'experts' who participated in these courses at Caltech. This complementary volume will address the advances made over the past ten years and will provide a fitting memorial to Feynman's explorations of computers". This companion volume to these lectures, called *Feynman and Computation: Exploring the Limits of Computers*, was published in 1999 by Westview Press in their Advanced Book Program.

This collection begins with articles from Feynman's original three collaborators for his course – John Hopfield, Carver Mead, and Gerry Sussman – and articles from many of the experts who contributed to

Feynman's course in various ways. These include Charles Bennett, Ed Fredkin, Rolf Landauer, Marvin Minsky, Norman Margolus, and Tommaso Toffoli who all had many discussions with Feynman on such topics as reversible computing and cellular automata; as well as two parallel computing gurus, Danny Hillis and Geoffrey Fox. The collection also includes reprints of Feynman's classic articles "There's Plenty of Room at the Bottom" and "Simulating Physics with Computers" as well as the transcript of his Nishina Memorial Lecture he gave in Japan in 1985 with the title "Computing Machines in the Future".

Suggested Reading

Chapters 1, 2 and 3

The Nature of Computation: An Introduction to Computer Science by Ira Pohl and Alan Shaw, Computer Science Press (1981)

Algorithmics by David Harel, Addison-Wesley Publishing Company, 2nd edition (1992)

Computer Organization and Design: The Hardware/Software Interface by John L. Hennessy and David A. Patterson, Morgan Kaufmann Publishers (1993)

Structured Computer Organization by Andrew S. Tanenbaum, Prentice-Hall, 2nd edition (1984)

Computation: Finite and Infinite Machines by Marvin L. Minsky, Prentice-Hall (1967)

Turing's World 3.0: An Introduction to Computability Theory by John Barwise and John Etchemendy, CSLI Lecture Notes 35

Introduction to Automata Theory, Languages, and Computation by John E. Hopcroft and Jeffrey D. Ullman, Addison-Wesley (1979)

'Operating Systems' by P.J. Denning and R.L. Brown, *Scientific American* (September 1984), 96

'The Problem of Integration in Finite Terms' by R.H. Risch, *Transactions of the American Mathematical Society* 139167 (1969)

'Integration of Elementary Functions' by M. Bronstein, *Journal of Symbolic Computation* 9(2) 117 (1990)

Chapter 4

Mathematical Theory of Communication by Claude E. Shannon, University of Illinois Press (1963)

Coding and Information Theory by Richard W. Hamming, Prentice-Hall (1980)

Principles and Practice of Information Theory by R.E. Blahut, Addison-Wesley (1987)

Communication Systems by A.B. Carlson, McGraw-Hill (1986)

Chapter 5

'Logical Reversibility of Computation' by Charles H. Bennett, *IBM Journal of Research and Development* 17, 525 (1973)

'Thermodynamics of Computation — A Review' by Charles H. Bennett, *International Journal of Theoretical Physics* 21, 905 (1982)

'Notes on the History of Reversible Computation' by Charles H. Bennett, *IBM Journal of Research and Development* 32(1) 16 (1988)

'Zig-Zag Path to Understanding' by Rolf Landauer, Reprint from *Proceedings of the Workshop on Physics and Computation: Physcomp* 94 IEEE Computer Society Press (1994)

Maxwell's Demon: Entropy, Information, Computing edited by Harvey S. Leff, Andrew F. Rex, and Adam Hilger (1990)

Chapter 6

'Quantum Mechanical Models of Turing Machines that Dissipate No Energy' by Paul Benioff, *Physical Review Letters* 48, 1581 (1982)

'Conservative Logic' by E. Fredkin and T. Toffoli, *International Journal of Theoretical Physics* 21, 219 (1982)

'Bicontinuous Extensions of Invertible Combinatorial Functions' by T. Toffoli *Mathematical Systems Theory* 14, 13 (1981)

'On a Simple Combinatorial Structure Sufficient for Sublying Non-Trivial Self Reproduction' by L. Priese, *Journal of Cybernetics* 6, 101 (1976)

Chapter 8

Introduction to VLSI Systems by Carver A. Mead and Lyn Conway, Addison-Wesley, (1980)

The Art of Electronics by Paul Horowitz and Winfield Hill, Cambridge University Press, 2nd edition (1989)

Physics of Semiconductor Devices by S.M. Sze, Wiley, 2nd Edition (1981)

Principles of CMOS VLSI Design: A Systems Perspective by Neil H.E. Weste and Kamran Eshraghian, Addison-Wesley, 2nd Edition (1993)

Introductory Semiconductor Device Physics by Greg Parker, Prentice-Hall (1994)

'Hot-Clock nMOS' by C.L. Seitz, A.H., Frey, S. Mattisson, S.D. Rabin, D.A. Speck and J.L.A. van de Snepscheut, Chapel Hill Conference on VLSI, 1 (1985)

'Scaling of MOS Technology to Submicrometer Feature Sizes' by Carver A. Mead, *Journal of VLSI Signal Processing* 8, 9 (1994)

'The CMOS End-point and Related Topics in Computing' by Maurice Y. Wilkes, *IEE Computing and Control Journal* 7, 101 (1996)

Index

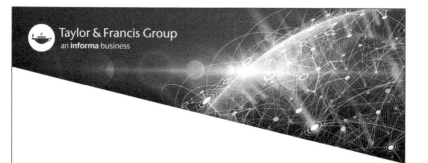